Biophysical Chemistry

P.R. Bergethon E.R. Simons

Biophysical Chemistry

Molecules to Membranes

With 108 Illustrations

Springer-Verlag
New York Berlin Heidelberg
London Paris Tokyo Hong Kong

Peter R. Bergethon
Elizabeth R. Simons
Department of Biochemistry
School of Medicine
Boston University
Boston, MA 02118
USA

574.19
B 496

Library of Congress Cataloging-in-Publication Data
Bergethon, Peter R.
Biophysical chemistry.
Includes bibliographical references (p.)
1. Biophysics. 2. Biochemistry. 3. Thermodynamics.
I. Simons, Elizabeth R. II. Title.
QH505.B397 1989 574.19 89-26095
ISBN-13:978-1-4612-7943-3

Printed on acid-free paper.

© 1990 by Springer-Verlag New York Inc.
Softcover reprint of the hardcover 1st edition 1990

All rights reserved. This work may not be translated or copied in whole or in part without the written permission of the publisher (Springer-Verlag, 175 Fifth Avneue, New York, New York 10010, USA), except for brief excerpts in connection with reviews or scholarly analysis. Use in connection with any form of information storage and retrieval, electronic adaptation, computer software, or by similar or dissimilar methodology now known or hereafter developed is forbidden. The use of general descriptive names, trade names, trademarks, etc. in this publication, even if the former are not especially identified, is not to be taken as a sign that such names, as understood by the Trade Marks and Merchandise Marks Acts, may accordingly be used freely by anyone.

Typeset by Caliber Design Planning, Inc.

ISBN-13:978-1-4612-7943-3 e-ISBN-13:978-1-4612-3270-4
DOI: 10.1007/978-1-4612-3270-4

Preface

Today there is an unprecedented explosion of interest and experimental investigation in the fields of biochemistry, physiology, and biophysical sciences. The development of the techniques of molecular biology and immunology and the availability of computer analysis have led to a phenomenally rapid pace in the acquisition of new information, especially with respect to the biology of the cell. As information relating to the control and integration of cellular processes becomes available, more interest is turning toward understanding the interdependence of cellular processes on a larger scale. The inevitable consequence of this is the growing interdisciplinary nature of modern biochemical and biophysical studies. As the focus broadens, so does the base of researchers and graduate students who are being attracted into these fields to pursue their interests in what was traditionally more the province of biology. This increased diversity of backgrounds is good, but it carries with it a cost in the heterogeneity of the preparation of modern students. Not too many years ago, a candidate for graduate studies in biochemistry or biophysics was most likely to have been a chemistry or physics major in undergraduate school and consequently had been exposed to, if had not mastered, many of the intricacies of physical processes and their treatment in quantitative terms. We have found in our teaching over the last several years that entering students are less likely to have been exposed to rigorous courses in physical chemistry and engineering physics. However, the direction of modern biochemical and biophysical studies is moving quickly toward areas where scientists will need to understand and use the disciplines of surface chemistry, electrochemistry, quantum mechanics, and mathematical modeling to understand the molecular processes in cells and at subcellular levels.

This textbook had its origins in a course entitled "Water and Its Properties" taught in the Department of Biochemistry at Boston University School of Medicine. Originally, the course was intended to instill in graduate students of biochemistry, biophysics, and physiology some understanding of the ways in which knowledge of solutions and solution thermodynamics could be used to explain and to predict the behavior of cells. This is clearly necessary in modern biochemistry if cellular events such as metabolic processes, stimulus responses, cellular differentiation, or the functional controls of cells are to be understood and the effects of perturbations, of illnesses, of drugs, etc., are to be quantitatively predicted. As the backgrounds of our students became more heterogeneous, we found that an increasing amount of prerequisite material needed to be reviewed before the complexities of cellular processes could be systematically taught. It was from this need to level the ground and still provide a challenging and meaningful graduate course leading to an appreciation of the biophysical chemistry of cellular systems that this book was born. Therein lie the reasons for the choice of material, the style of presentation, and the limitations of the coverage of areas of interest in this volume. It is worth acknowledging that we have in our minds the addition of a second volume to this text, thereby extending both the breadth and the depth of coverage of the subject.

The text is designed to be used in a single-semester course in the first or second year of graduate school but certainly could also form the basis of an advanced course for undergraduates in biophysics or chemistry. The text assumes that all students using it will be familiar with basic physics (including electrostatics), general introductory inorganic and organic chemistry, and cell biology. A background in physical chemistry is not presumed but is certainly valuable. A knowledge of calculus and algebra is assumed, but students are not expected to be particularly familiar with advanced areas such as partial differential calculus.

Because the subsequent discussions in the second and third parts depend on a thermodynamic vocabulary, the first part is a rather lengthy discussion of the basic principles of equilibrium thermodynamics. Experience has taught many teachers that students are usually left baffled by a single rigorous exposure to thermodynamics, an exposure that we do assume all readers of this book have had. We therefore have chosen to cover, in a semirigorous but conversational tone, both old ground and some ground not traditionally taught extensively in undergraduate sections on thermodynamics. The ideal use of the chapters on thermodynamics would be as a study guide to a more traditional and rigorous presentation of the subject, probably available already in a textbook in the student's library. The student or professor who already feels adequately prepared in this area might opt to simply use these chapters as a quick refresher.

The second part of the book deals with the structure and behavior of solutions of ions and molecules. The first chapter of the part begins with a qualitative look at the structure and properties of bulk water, the substance most generally considered the solvent of biological systems. It is in the second part that our dual teaching objectives will become apparent. We aim to discuss the treatment of dilute and moderately concentrated homogeneous solutions of polar and nonpolar molecules with a focus on the important role played by water as the solvent. In addition, we attempt to demonstrate the principles used in model building through the use of both thermodynamic cycles and mechanistic information and then test the resulting models against empirical evidence. Because a theory is only as good as its assumptions and only as valid as the experimental evidence supporting it, great emphasis is placed on the analysis of assumptions and the examination of the effect of modifications in the model. Such an approach leads to an analysis of ion–solvent interactions starting with Born's hypothesis and of ion–ion interactions starting with Debye and Hückel. We then go on to discuss a solvent-structure view of homogeneous solutions of organic ions, small nonpolar organic molecules, and macromolecules.

Cells and cellular organisms are not dilute, homogeneous solution systems however. The third part of the book builds on the first two as it extends the physical principles necessary to understand the nonhomogeneous and nonequilibrium nature of biological processes. To understand the formation of the cell membrane, the properties of systems containing both lipids and water are examined. This is followed by a discussion of transport phenomena in terms of both nonequilibrium thermodynamics and mechanistic models, the focus being on diffusion and conduction. We believe that one of the most important frontier interdisciplinary fields for the biological scientist is bioelectrochemistry, and so we discuss in a qualitative fashion the structure and behavior of the electrified regions at the juncture of phases. Finally, having laid the foundation for interpreting the behavior of cellular systems on the basis of principles of multiphasic, nonideal, surface-dominated physical chemistry, we consider the role of biological membranes in terms of the forces acting at the membrane and the constraints they impose on the behavior of the cellular system.

We have included a limited number of questions for problem sets and review and as well have attempted to assemble glossaries of formulas and terms so that the reader needs only to go to other texts when interested in significantly greater depth or encounters some area that needs significant remedial review. Our hope is that this text will help both teachers and students in the field to be able to appreciate, and inspire the use of, the powerful tools available through the science of physical chemistry, and will serve to help prepare biological scientists for the inevitable and exciting application of principles of materials science, quantum mechanics, and electrochemistry to biological problems.

We would like to thank our colleagues and our students for their critiques of our effort. Of all the friends and colleagues who aided our efforts, Dr. Cindy McCrone stands as the individual to be most credited for guiding this book to its final completion. Without her eternal patience with the red pencil and unerring sense of teaching clarity and style, this volume would be at best an uncut, unpolished stone. It should be noted however that any errors remaining in this book are entirely our own responsibility and in no way reflect on the readers who were kind enough to evaluate the manuscript for us.

Boston, Massachusetts

P.R. Bergethon
E.R. Simons

Contents

APPENDICES

Introduction

CHAPTER 1

Molecules, Membranes, and Modeling

Imagine yourself standing outside a country home on an early spring morning just before dawn. Take a deep breath and shiver to the taste of the sweet morning air. Listen carefully to hear the chirping of morning birds. As the sun reaches the horizon, glinting shafts of light reach your eyes. Another deep breath and you feel a peace that comes from a resonance between you and the world at your doorstep. Your eyes close and for a fleeting moment you understand the universe in its simplest most basic terms. Savor that moment, for your eyes open again and now you are drawn back to the reality—you are reading the introduction to a book on physical chemistry. If you are mildly perturbed at being returned to this apparently less appealing reality, you have just demonstrated a facility with a key and exquisitely valuable tool in the study of science, the **Gedanken experiment** (thought experiment). The use of thought trips will be of fundamental importance in the approach that this book takes toward understanding biophysical processes. That virtually any student has access to one of the most profound and sophisticated theoretical techniques available to a scientist is an important lesson to learn. This is just one of the lessons that this short experiment or mind trip has pointed out to us already. Are there other lessons?

What are the consequences of scenes imagined? For one, the reader, who is a biological organism , has interacted with the physical universe, experiencing the electric field vectors of photons (light), detecting molecular vibrations (sound), and identifying alterations in local chemical potential (smell and taste). As part of the process, the reader has converted stored energy sources to sustain the life support processes of the heart, lungs, kidneys, immunological system, etc. This support machinery has allowed the specially adapted cells of the nervous system to recall previously collected, sorted, standardized, and analyzed physical events (light, sound, smell) that were interpreted as information and to sort this information into a creative memory trace that we call imagination. It seems only rational that if the biological organism can perceive and interact with the physical forces that surround it, then at some fundamental level those physical forces and processes are the same for the organism and the physical universe. This implies that a fundamental understanding of the physics and chemistry of the universe will lead to an understanding of biological processes and behavior. It is on this assumption that the fields of biophysics and biophysical chemistry are built. The imagined observer at the door could be content with the "feeling" of universal understanding or could use the tool of Gedanken experimentation along with experimental science to attempt to understand a wide variety of questions. These might include the origins of the photons and the laws governing their propagation; the mechanism of the production of sound by the birds or perhaps the laws governing its propagation through the air and the mechanics of its perception in the ear; and the identity and molecular behavior of the molecules giving rise to the smell of the fresh air or perhaps the nature of their interaction with the specialized olfactory cells of the observer. On the other hand, the reader might choose to focus on the nature and novelty of the beast that, being able to imagine an experience such as this sunrise while actually studying a book on physical chemistry, would consider the physics and chemistry of quantum electrodynamics, wave propagation theory, and ketone, aldehyde, or amine chemistry. What is it about the human biological being, its materials, and its organization that gives it such a charmed view of the universe? How can we understand this seemingly unique biological ability and can it be understood in the same terms as the rest of the physical universe?

Understanding a biological system in terms of physical forces and materials requires some sense of what the similarities and differences are between biological systems and physical systems. As a first approximation, animate and inanimate systems share much at the basic and fundamental levels and differ mostly in the types and magnitudes of complexity. For example, at the atomic level, biological systems are made up primarily of atoms of carbon, hydrogen, nitrogen, oxygen, and a small variety of other elements. The complex variety that occurs when these simple atoms are arranged into macromolecules and multifaceted, multicompartmentalized cells and organs seems to distinguish the biological system from other systems. This complex heterogeneity is something of a matter of scale however. A galaxy, a global weather system, or a coastline certainly has enormous complexity. The complexity of biological systems seems to be one of high density; in other words, biological systems seem to concentrate complex molecules together in complex envelopes called cells that are themselves concentrated into complex systems called organs that perform complex functions in organisms that frequently live in complexly constructed communities. An astronomical magnitude of complexity can therefore be found in an ant hill or under a microscope, not to mention in a city or nation-state. Yet at every level of the complexity that derives from an enormous number of degrees of freedom in the interactions of the parts, these biological systems are subject to the same forces and laws of nature as are other systems.

It is worth explicitly mentioning that there are a number of major theoretical physical constructs used to describe the behavior of everything in the universe. These are sometimes called the great theories of physics. They include:

1. **Classical or Newtonian mechanics**, which describes the motion of large material objects, such as planets and satellites, and the trajectory of balls and missiles.
2. **Quantum mechanics**, which describes the motion and behavior of submicroscopic objects. This theory explains behavior on a scale where classical mechanics is known to be incorrect, namely, the movement of subatomic particles such as electrons and protons.
3. **Relativity**, which describes high-speed motion and is based on the fundamental concept that all aspects of nature obey the same set of invariant laws. Like the quantum theory, relativistic theory describes motion in cases where classical mechanics is known to be incorrect.
4. **Electromagnetism**, which explains the behavior of electricity, magnetism, and electromagnetic radiation.
5. **Thermodynamics**, which describes the behavior of large numbers of particles and discusses these behaviors in terms of heat, temperature, and work.

The behavior of chemical and biological systems is consistent with these theories, and they are the source of the fundamental unity that binds different systems together. It will be necessary to invoke aspects of these physical theories as an understanding of biological systems is built. Although biological systems have great complexity, the threads of these theories will be found over and over again running through the analysis. In this book, it will seem that preference is being given to the theories of thermodynamics. This is not an unjustified conclusion, but it is important that the reader does not take such a dependence on thermodynamic formulation to suggest a lack of relevance of the other theoretical constructs. For the subjects chosen and the depth and slant of their presentation, thermodynamics is the most convenient and simplest approach. Taking a thermodynamic approach is also often the best strategy for describing extremely complex systems completely. Finally, thermodynamics is one of the most valuable tools of the chemist. After mastering this material with a thermodynamic scheme, the student will invariably be ready to extend an interest to the application of the other theories to biological problems.

If biological systems are so complex, how are they ever to be accurately described, much less understood? In fact, it is more likely that an understanding of biological processes will be attained long before a detailed description will be available. Evolution is a good case in point. The form, behavior, and rules of evolutionary processes are reasonably well defined and understood, but the result of these evolutionary forces in terms of the identification, anatomy, physiology, ecology, etc., of the species produced is far from well described. In fact, even the number of species to be named is not known. So many species with so many varied habitats and behaviors means an almost unlimited number of degrees of freedom for the "system of living creatures" that must be defined before the biological system is fully described. This is a most prodigious of tasks. Interestingly, biological science has for much of its history been a predominantly descriptive science. If descriptive biology is such a Herculean task, is there any reasonable chance that biological systems can ever be understood? The history of science provides a clue. Scientific endeavor tends to go through cycles of descriptive data accumulation followed by a period of attempts to integrate the data together into a cohesive framework or theory. If the basic blocks from which all biological systems are constructed can be defined and understood, then it is reasonable to expect that this will shed light on the behavior of more complex systems. As more complex arrangements of the basic forces and materials are then examined, the fundamental principles can be carried along and augmented as necessary depending on the specific detailed complexity. Models of complex systems can be constructed from simpler systems by defining the simple models and then investigating how they can be made to interact at levels of increasing interaction. The levels of interaction provide the complexity of the system. Complexity therefore can also be considered as a simple system that links other simple systems. Model building seems to be a good candidate for synthesizing an understanding of complex systems from simple processes. This is a fundamental approach used in this book.

What is the general strategy to a model building method in science? At the outset, the following caveat must be emphasized: **a theory or model is only as good as the experimental evidence that supports it**. Models can be developed along the following lines:

1. First, the problem to be modeled must be stated as succinctly as possible; for example, what is the relationship between an ion and the molecules of a solvent into which it is dissolved?
2. Next, the model must be derived in terms of a measurable experimental parameter; for example, measure the relationship between an ion and the molecules of a solvent into which it is dissolved is determined by measuring the enthalpy of hydration of the ion.
3. Now the simplifying assumptions on which the model will be built are concisely defined; for example, as a first approximation, only electrostatic forces that have a $1/r^2$ dependency, are considered, the ion is defined as a small rigid sphere with a specific radius and specific charge, and the solvent is defined as a structureless continuum de-

scribed completely by the dielectric constant. This step is extremely important since it is to these assumptions that the modeler will return if there is disparity between the predictions of the model and the measurements of the experiments. By reconsidering and refining the assumptions, the model becomes more accurate, and relationships sought in the original question will become defined in terms of the assumptions.

4. With the assumptions and question defined, a thought or Gedanken experiment is now conducted that will provide a theoretical predictive model for the problem, usually in the form of a mathematical expression; for example, a thermodynamic cycle is constructed that will provide a mathematical expression for the enthalpy of solvation for the ion in the solvent.
5. A series of theoretical values are generated, and these values are compared with values obtained via experimentation; for example, the predicted heats of hydration for a variety of ions of certain radii are compared with the actual experimental heats of hydration for the same ions.
6. If the theoretical and experimental values do not agree within the percentage allowed for by experimental error, then reanalysis of the original assumptions is undertaken. Corrections and extensions to the assumptions are incorporated and extended into the model by conducting a further thought experiment and following through the experimentation cycle to check the validity of the modified theory. This cycle is repeated as many times as necessary until adequate accuracy is obtained.

In building biological models, the fundamental theories and techniques of physics and chemistry, as applicable to a given system, are valuable and are the logical starting point. This is the foundation on which biophysical chemistry is built. What are the biological materials to which the theories of mechanics, thermodynamics, quantum mechanics, electromagnetism, and, to a lesser degree, relativity will be applied? Biological chemistry is the chemistry of biomolecules in both solution and at surfaces. The molecules of concern can be broadly arranged into a number of categories based on their size, composition, and charge.

Small molecules of concern in biological systems include molecular water, ions, and small organic molecules including a wide variety of lipids, steroids, carbohydrates, amino acids, peptides, and nucleic acids. Larger molecules or macromolecules are constructed of polymerized chains or associated arrays of these smaller molecules. Amino acids polymerize to form proteins. Nucleic acids polymerize to form RNA and DNA. Carbohydrates polymerize to form complex sugars and starches. Lipids associate in the presence of an aqueous phase into membranes. Bulk water is, to a reasonable approximation, a hydrogen-bonded polymer of molecular H_2O molecules. The rich diversity of biological organisms results from the fact that the relatively small group of small molecules can combine in a seemingly infinite variety of distinct forms as larger molecules.

On a molar basis, water is the most prevalent chemical species found in biological systems, and this has led to the reasonable assumption that water is the predominant solvent in biological systems. Water is a unique solvent, however, by virtue of its extensive hydrogen-bonded character in bulk phase. If the description of the biological cell as a bag of water filled with a dilute aqueous solution of biomolecules were correct, the bulk properties of water would likely be very important. However, the cell is not a bag of homogeneous aqueous solution but rather a complex array of membranes, colloidal organelles, and a bewildering array of densely packed macromolecules having the character more of a hydrophilic gel than of a homogeneous solution of water. This combination means that the cell is a heterogeneous, multiphasic, surface-dominated environment rather than a homogeneous solution of many species. The true role of water in this environment cannot be easily predicted since the nature of the chemical reactions will depend to a large extent on the **nonhomogeneous** behavior of cells. The term nonhomogeneous was coined by Freeman for chemical processes that cannot occur randomly in space because at least one component is not randomly distributed in either a single or a microheterogeneous phase. In contrast to **homogeneous processes**, where components are randomly distributed in a single phase, or macroscopically **heterogeneous processes**, where components from one or more phases interact at the interface between the phases, nonhomogeneous processes may well describe the biological system most accurately. This book will focus on building first a homogeneous and then a heterogeneous chemistry. However, the expectation is that the student studying physical chemistry today will be pushing the frontiers of biophysical chemistry into the realm of nonhomogeneous chemistry tomorrow.

"The simpler and more fundamental our assumptions become, the more intricate is our mathematical tool of reasoning; the way from theory to observation becomes longer, more subtle and more complicated." Albert Einstein and Leopold Infeld in *The Evolution of Physics* (New York: Simon and Schuster, 1950).

Part 1
Review of Thermodynamics

CHAPTER 2

Thermodynamics: An Introductory Glance

2.1. Overview

Biological systems are complex because of the enormous interdependency and interaction of chemical species and physical forces. The simple molecule of water is a good example of this "systemic" complexity. Although a simple molecule, water plays an important and interesting role in biological systems. What will become clear is that this simple molecule interacts with everything around it—polar and nonpolar molecules and even surfaces, not considered to be in solution. It is important to have effective tools to predict and model the behavior of such interactions and their effect on the heterogeneous systems that make up biological cells even when the detailed nature of the interactions may not be known.

In the next several chapters, an endeavor will be made to review and develop some quantitative tools to use in the study of water and its solutions of biomolecules. Either a thermodynamic or a statistical mechanical approach can be taken to this end. Historically, the thermodynamic viewpoint preceded the mechanistic viewpoint, but both are truly interrelated, and, theoretically, thermodynamic laws should be deducible from mechanistic laws. In reality, both viewpoints are necessary since most chemical systems simply have too many degrees of freedom to be completely described in true mechanistic terms. Using a thermodynamic approach can be valuable to predict the directions of processes. On the other hand, a mechanical interpretation, even if it incorporates simplifying assumptions, can provide an understanding of the actual processes at the molecular level of these interactions.

The power of the thermodynamic approach is that it will allow the estimation and prediction of a system's behavior even if it has never been seen before or has not been understood on a molecular basis. This statement holds true best for the area of thermodynamics known as **classical thermodynamics**. In classical thermodynamics, the concern is with a system that has reached a state where the descriptive variables of the system are no longer in flux, that is, a system at equilibrium. The perceived steadiness of the state variables in this system is a macroscopic observation because on a microscopic scale the particles making up the system will be oscillating about the equilibrium value. Another area of thermodynamic study is that of **nonequilibrium thermodynamics**, and this field is concerned with describing the events that occur during the approach to equilibrium. Nonequilibrium processes, are important in studies of kinetics. Full knowledge of kinetic processes, however, requires knowledge of the mechanisms of a particular event including the identities and behavior of the intermediates through which the process passes. Such studies are intellectually quite satisfying because of their specificity but, by their very nature, do not possess the quality of predicting unknown events as does the phenomenological thermodynamic approach.

The following chapters are intended as a review and not a detailed derivation of thermodynamics. Emphasis will be on classical or equilibrium thermodynamics. The plan will be directed to developing familiarity with concepts and techniques directly applicable to the study of biochemical systems with water as the solvent. First, some of the basic concepts and vocabulary of thermodynamics will be defined, and some of the useful mathematical tools examined. Because most students who get lost are lost early in the study of thermodynamics, care will be taken in defining the basic terms and mathematical shortcuts. Next, the general laws of thermodynamics, starting with the first law, will be discussed. By starting generally and getting an intuitive feel for the thermodynamic behavior, the application of the general laws and principles of thermodynamics to biochemical studies can be easily extrapolated.

Thermodynamics is a phenomenological discipline in that it examines phenomena but not their causes. Thermodynamics is rooted in macroscopic experimental science. It is a systematic method for describing the properties and behavior of matter as observed on a macroscopic or bulk level. The observations used in thermodynamics are the properties of bulk matter, that is, pressure, volume, and temperature, rather than atomic properties. There is no attempt to explain observations in terms of mechanisms, a troubling approach for many students since most have been educated to see the universe in molecular mechanical terms. The value of thermodynamics is that the behavior of macroscopic systems can be predicted, even if little is understood about the details of how a process actually works. Obviously, this does not imply that there is no value in understanding the more microscopic behaviors that make up the macroscopic properties of bulk materials. By studying the molecular mechanisms of these microscopic behaviors, an attempt is made to explain how macroscopic properties result from microscopic properties, thereby contributing to the "understanding" of the nature of the universe. However, accounting for microscopic behavior is not a requirement in the study of a macroscopic system by thermodynamics. The converse is not true of studies of molecular mechanisms.

Thermodynamics is based on three fundamental laws from which the macroscopic behavior of matter can be understood. The focus here will be on the first and second laws. The first law of thermodynamics is concerned with **energy**, and the second law of thermodynamics is concerned with **entropy**. The fundamental macroscopic properties responsible for the properties of matter are therefore energy and entropy. As a broad perspective of the empirical laws of thermodynamics is undertaken, it should be emphasized that the laws of thermodynamic are axiomatic.

2.1.1. The First Law: "The Energy of the Universe Is Conserved"

The first law states that energy is conserved and that heat is also conserved as it passes from one body to another. The first law of thermodynamics is generalized from experience and is not derived from any other law or principle. Essentially, it is the result of centuries of experience with observing energy transformations. Experience shows that a motor may produce mechanical energy only if other energy is provided to it. The mechanical energy produced plus the frictional losses and heat lost in the transformation is always exactly equal to the energy put into the motor. The first law codifies this experience. It therefore governs the conversion and conservation of energy in the universe. Importantly, no restrictions of any sort are placed on the conversion process; the first law merely states that the total energy before and after a conversion is the same.

For example, mechanical energy can be completely converted into thermal energy if a motor turns a wheel in a body of water. That this is true can be demonstrated experimentally, and in fact this experiment was done by Joule. It turns out that all forms of energy can be completely converted into thermal energy, and this conversion can be measured as a rise in the temperature of some material, usually water. Such a conversion is entirely consistent with the first law of thermodynamics alone.

2.1.2. The Second Law: "The Entropy of the Universe Increases"

Whereas all forms of energy can be completely converted into thermal energy, the converse situation is not always true. Again, experience proves that the thermal energy of a steam boiler cannot be converted completely into mechanical work; that is, the measured energy of the thermal state will not be found to raise a weight in a gravitational field to the position (potential energy) that would be predicted by the magnitude of the thermal energy alone. This is a real life experience and not a theoretical construct. Such observations are the basis of the second law of thermodynamics.

Consider the events that occur when two objects of different temperatures are brought in contact with one another. Invariably, over time, the two objects reach the same temperature and, having done so, stay there. Just as invariably, the process by which the temperatures become equal is a result of the flow of heat from the object with the higher temperature to the object with the lower temperature. The tendency of objects of differing temperatures to behave in this fashion is the basis for thermometry and is at the root of the understanding of many thermodynamic processes. The movement of heat from hot to cold objects only, and never spontaneously from cold to hot objects, is the most classical description of the second law of thermodynamics. More formally this may be written as follows:

1. No process is possible where heat is transferred from a colder to a hotter body without causing a change in some other part of the universe, or
2. No process is possible in which the sole result is the absorption of heat (from another body) with complete conversion into work.

A rock thrown onto the ground can be shown to have the same energy if it is moving from the hand to the ground or from the ground to the hand. This is entirely consistent with the first law of thermodynamics. Yet, rocks do not leap from the ground into the hands of people spontaneously. Such things are never observed to happen, even though the first law will permit such events. There is a natural order governing events of this sort that has been observed to be consistent throughout the history of observation. This is the role that the second law of thermodynamics plays, determining the direction of natural processes. In combination with the first law, the second law allows the direction of any process to be predicted, and as a result it will allow prediction of the ultimate equilibrium.

2.2. Defining Thermodynamic Terms

So far, the thermodynamic terms such as heat, temperature, and systems have been used without precise definition. It is necessary to define the following terms:

- System, boundary, and surroundings
- Properties of a system
- State of a system
- Change in state, path, cycle, and process
- Work and heat

2.2.1. Systems, Surroundings, and Boundaries

A thermodynamic **system** is the part of the physical universe whose properties are under investigation. As for any definition, some things are included and some things are excluded in the system's definition. The system is confined to a definite place in space by the **boundary** which separates it (the included part) from the rest of the universe (the excluded part) or the **surroundings.** This boundary is very important as shall be seen repeatedly.

If the boundary prevents any interaction of the system with the surroundings, an **isolated system** is said to be present. Because it can have no effect on the surroundings, an isolated system produces no observable disturbance as viewed from the surroundings. This is a little like the old question of the sound of a tree falling in a forest when no one is present to hear it. If the experimenter is in the surroundings, any changes that might occur in an isolated system would not be measurable. A system that is not isolated may be either **open** or **closed**. A system is called **open** when material passes across the boundary and **closed** when no material passes the boundary.

The value of carefully defining a system and its boundaries cannot be over-emphasized. If the wrong system and boundaries are defined at the outset, the solutions to the problems are often embarrassingly wrong. (Would-be inventors of perpetual motion machines take note.) How a particular system is described precisely depends on the state of the system, a subject to be examined shortly.

2.2.2. Properties of a System

Because thermodynamics is an experimental science, it is important that the well-defined system with its boundaries and surroundings can be described in terms of measurable quantities, that is, macroscopic observables. The measurable attributes are called the **properties of a system**. These properties are those physical attributes that are perceived by the senses or are made perceptible by instrumentation or experimental methods of investigation. Typical examples of thermodynamic properties are temperature, pressure, concentration of chemical species, and volume. Useful **properties** would have one or both of the following characteristics:

1. Repeated measurements of the same property at equilibrium yield the same value. This is important because in kinetic studies it is the change in the value of a property as a system approaches equilibrium that is studied.
2. Different methods for measuring the same property yield the same result. This ensures that the system is being described, rather than an artifact of the experimental procedures of measurement.

Sometimes it is useful to differentiate between fundamental properties and derived properties. **Fundamental** properties are directly and easily measured, while **derived** properties are usually obtained from fundamental ones by some kind of mathematical relationship.

Properties may be either **extensive** or **intensive. Extensive** properties are additive. Their determination requires evaluation of the size of the entire system; examples are volume and mass. To describe the volume of a system, a standard is chosen such as a cubic centimeter, and each equivalent cubic centimeter in the entire system is added together to give the total volume. On the other hand, **intensive** properties are independent of the size of a system. These properties are well defined in each small region of the system. Examples are density, pressure, and temperature. Any and every region in the system, if measured, represents the same property value as any other. Intensive properties are not additive.

It is worth spending a little time with the idea of an intensive property, since, when the statement "well defined in a small region" is made, it is meant "small with respect to the total system, yet not too small." Consider the intensive property density.

$$D = \frac{m}{V} \tag{2.2-1}$$

D is the density given as a ratio of mass, m, to volume, V. The average density of a system can be determined (assuming that the density is uniform throughout the system) if a volume is chosen that is small compared to the total volume but that still contains a massive number of particles. For example, if a cubic angstrom were chosen as the volume, there would be a good chance that the volume would be filled or empty based on the fluctuations of a single molecule. In such a case, the measured density would swing drastically up and down. Defining density in such a way would not lead to a thermodynamic property. The proper sampling volume can be found by considering the volume that will give an adequate number of particles, n. It can be shown by statistical methods that the actual value of n at any given time will be a number that fluctuates around n by $n \pm \sqrt{n}$. If $n = 100$, the actual value will range from 90 to 110, while if $n = 250{,}000$, the actual volume will vary by 500 (between 249,500 and 250,500), probably an undetectable difference. If a small volume must be chosen for practical reasons, there will be cases where the volume chosen will be filled and cases where it will be empty. Then, a basic theorem of statistical thermodynamics can be used stating that averaging over (infinite) time will yield the same result as a numerical average.

2.2.3. State Functions and the State of a System

With a system and its boundaries, surroundings, and properties selected, the **state of the system** can be specified. It must be known, from experimental study or experience with similar systems, what properties must be considered in order that the state of a system can be defined with sufficient precision. The task in thermodynamics is to find the minimum number of properties necessary for a complete description of a system. The tremendous predictive power of thermodynamics is unveiled when these properties can be found and measured, and the relationship of the measured properties to other important but unmeasured terms can be known. Then, by knowing the state of a system in terms of a few easily measured parameters, the values of all other properties become known. The trick to thermodynamic problem solving is to find properties in terms of others and to find the rules for doing the transformations.

The properties of a system in equilibrium depend on its state at the time in question and not on what has happened in the past. Because of this, thermodynamic properties at equilibrium are called **functions of state.** For example, the volume of a system is denoted by V. Any infinitesimal change in volume may be denoted as dV. dV is the differential of the function of state. This means that the integral of all the infinitesimal volume changes for any process that starts in state 1, with volume V_1, and ends in state 2, with volume V_2, will lead to a total volume change that depends only on the initial and final states.

$$\text{Volume change} = \int_1^2 dV = V_2 - V_1 \tag{2.2-2}$$

The key to understanding the mathematics involved in state functions lies in the fact that the value $V_2 - V_1$ is absolutely independent of the path, route, speed, complexity, or any other factor chosen in going from state 1 to state 2. A differential of a state function can always be integrated in the traditional fashion taught in calculus courses. Such differentials are called **perfect, total, or exact differentials.** Differences in state functions, such as $V_2 - V_1$, are written as follows:

$$V_2 - V_1 = \Delta V \tag{2.2-3}$$

The delta notation (Δ) is used only for differences in the values of state functions.

2.2.4. Changes in State

If a system undergoes an alteration in state by going from a specified initial state to a specified final state, the **change in state** is completely defined when the initial and the final states are specified. When any state function changes from state 1 to state 2 and returns back to state 1 again, a cycle occurs which can be written as follows:

$$(V_2 - V_1) + (V_1 - V_2) = 0 \tag{2.24}$$

This can be written instead using the cyclical integral symbol:

$$\oint dX = 0 \tag{2.2-5}$$

Here X is a state function and $\oint$ is an integral around a closed path, one that returns to the original state. A process that returns to the original state is called a **cycle**. The cyclic integral of a change in state from an initial to a final state and back to the initial state is always equal to zero.

Notice that so far there has been absolutely no discussion of the incremental steps along the way from state 1 to state 2. As long as only state functions are being described, this is not a problem. However, the method of getting from one state to another is very important in certain situations. The **path** of the change in state is defined by giving the initial state, the sequence of the intermediate states arranged in the order traversed by the system, and the final state. A **process** is the method by which a change in state is effected. The description of a process consists in stating some or all of the following: (1) the boundary; (2) the change in state, the path followed, or the effects produced in the system during each stage of the process; and (3) the effects produced in the surroundings during each stage of the process.

2.3. Work

So far, thermodynamics has been presented as a tool that will be necessary for quantitative understanding of biochemistry. However, the value of thermodynamic thinking extends beyond its use as a scientific tool. Biological processes in general utilize the energy contained in a biochemical system efficiently to perform several kinds of work that counter the natural forces of increasing entropy. A general characteristic of biological systems therefore is the efficient use of energy to produce work and the general decrease of entropy in the system (accompanied by an increase in the entropy of the surroundings). To understand any biological process completely requires an appreciation of the role of these thermodynamic quantities.

Work in biological systems usually takes several predominant forms: mechanical, electrical, and chemical potential, and to a lesser degree, pressure–volume work. **Mechanical work** often takes the form of movement and includes such events as chromosomal movement in metaphase due to the contraction of microfibrils; ruffling of cell membranes; migration of cells; and muscular contractions that pump blood or move arms and legs. Diffusion of chemical species down a concentration gradient can be considered work done by the **chemical potential. Electrical work** is utilized to maintain concentration gradients across membranes, often countering chemical potential work; remove toxic and waste materials; provide information transfer between neural, neuromuscular, and other cells; and extract the energy from either the sun or from reduced substrates of carbon and provide chemical energy stores. **Pressure–volume work** occurs with the expansion or contraction of a gas and is important in gas-exchanging organs. Because of the historical importance of steam-driven machines, it is also the traditional concern of thermodynamics, so some attention will be given to this type of process here.

All **work** is characterized by the displacement or movement of some object by a force. All forms of work take the general form

$$\text{Work} = \text{Force} \times \text{Displacement} \tag{2.3-1}$$

By convention in thermodynamics, work done on a system is considered positive. In other words, when the surroundings do work on the system, a positive sign is given to the value. On the other hand, if the system exerts a force that causes a displacement in the surroundings, the work is said to be negative, representing work done on the surroundings.

2.3.1. Electrical Work

Electrical work is defined as the movement of charge through an electrical gradient or potential. Electrical work is defined mathematically as follows:

$$\text{Work}_{\text{(electrical)}} = \text{Electrical potential} \times \text{Charge} \tag{2.3-2a}$$

$$w_{\text{electrical}} = EQ \tag{2.3-2b}$$

where Q is the charge and E the electrical potential. This may be more practically written in terms of power since power is the measure of work expended in a unit of time:

$$\text{Power} = -EI \tag{2.3-3}$$

where E is electrical potential in volts, and I is charge in amperes or charge per unit time, $\Delta Q/\Delta t$. The sign is negative because the work is done by the system on the surroundings. To move one ampere of charge through a potential field of one volt over one second will result in a work function of one joule per second or one watt. The work done when charge is separated and transferred is of great significance throughout biochemical processes.

2.3.2. Pressure–Volume Work

Most readers of this volume will be familiar with the expression of pressure—volume work, because the classical teaching of thermodynamics and state functions is done with ideal gases and the concept of expansion work. Historically, this is quite reasonable because the work available through expansion processes is the model for the heat engine. Pressure—volume work can be important in biochemical processes as well and especially in relation to water. Water expands when cooled to freezing and work is done on the surroundings.

Consider that the system under study is enclosed in a container that has a wall that is movable, a piston. When this piston is moved up (expansion) or down (compression), work is done by or on the system with respect to the surroundings. There is a force per unit area, F/A, that is exerted on the system through the piston wall. This external force is the external pressure, P_{ext}. Now if the piston is moved a specific distance, dx, in the direction of the external force, a certain amount of work is done on the system:

$$dw = P_{ext} A\, dx \tag{2.3-4}$$

where A is the cross-sectional area of the piston. By integrating:

$$w = \int P_{ext} A\, dx \tag{2.3-5}$$

There is a volume change in the system, dV that is described by the area of the piston moving through the distance dx:

$$dV = A\, dx \tag{2.3-6}$$

Substituting gives

$$w = -\int P_{ext}\, dV \tag{2.3-7}$$

The negative sign is explained as follows. The volume change will be negative for the compression, which will give a negative value for the work done by the compression. But, by convention, work done on the system must be positive, and so a negative sign must be added to the formula. This equation will provide the work done on a system if the pressure applied and the change in volume in the system are known. In SI units, the work done by PV work is in joules, but perhaps one of the most familiar descriptions of PV power (the gasoline engine) is given in horsepower.

2.3.3. Mechanical Work

Two major forms of mechanical work must be considered for biological systems. The first is the work done by moving a weight in a gravitational field, and the second is work done when a force causes displacement of a structure acting like a spring. Consider an arm lifting a weight in the air. The action of the muscle fibers depends on work of the second kind, while the movement of the weight upward against gravity is a displacement of the first sort.

Gravitational work occurs when an object of mass m is lowered in a gravitational field (like that of the Earth) from a height h_1 to h_2. If the mass is lowered at a constant velocity, the following can be written:

$$w = mg\,(h_2 - h_1) \tag{2.3-8}$$

where g is the acceleration of gravity (9.8 m- s^{-2}). When SI units of meters and kilograms are used, the work done is in joules.

Processes in which the displacement acts like that of a spring obey **Hooke's law**. Hooke's law dictates that the force applied is directly proportional to the changed length of a spring. Consider the displacement of a spring whose length when no force is applied is x_o; when a force either compresses or expands the spring, the new length is x. Any force externally applied will be balanced by an internal force, the spring force. Hence, Hooke's law allows the following to be written:

$$\text{Spring force} = -k(x - x_o) \tag{2.3-9}$$

where k is a constant for a given spring. The external force is equal and opposite to the spring force and is given by

$$\text{External force} = k\,(x - x_o) \tag{2.3-10}$$

Because the force necessary to stretch or compress a spring varies with the length of the spring for a change in length from x_1 to x_2, the work is calculated by integrating:

$$w = \int_{x_1}^{x_2} k(x - x_o\,dx) \tag{2.3-11}$$

which gives the result

$$w = k(x_2 - x_1)\left(\frac{x_2 + x_1}{2}\right) - x_o \tag{2.3-12}$$

It is worth noting that Hooke's law in many cases is a reasonable approximation for not only the behavior of interactions of muscle fibers and locomotive proteins in microtubules, but also for the interactions of atoms with one another. It is a frequent and useful model to consider atoms as connected together by a spring.

2.3.4. A Return to the Laws

These formal terms presented above will allow for precise discussion and experimentation. Using this new knowledge, the first law can now be examined. A fuller description of heat will also be described. In the analysis that follows, the general forms of equations are derived often using a system constrained by temperature, volume, and pressure. This somewhat traditional ideal gas approach is maintained even though some of the situations considered may appear to be irrelevant to the biochemist and biophysicist. First of all, the general form of the equations is applicable to any ideal state, whether gas or liquid. It is generally a little easier to refer to the ideal state of a gas than of a liquid but both are treated identically. Since ultimately the concern in this book will be the nonideal behavior of aqueous solutions, this review may be considered as an analysis of either an ideal gas or an ideal liquid. In later chapters, a treatment of the deviations of aqueous solutions from these general equations will be undertaken, but hopefully then based on the strong foundations derived from an understanding of the ideal.

CHAPTER 3

The First Law

3.1. Understanding the First Law

In thermodynamics, a system is characterized in terms of its internal energy. Once the system is defined, the **first law** dictates that the energy of the system and its surroundings will remain constant. It is crucial to appreciate that energy is a state function and therefore its characterization depends only on the initial and final states of the system. The first law says that the total energy of the system and its surroundings remains constant, so it would be expected that some energy may be exchanged between them. This exchange occurs at the boundary of the system and will take the form of either thermal energy (heat) or work. Intuitive and mathematical comprehension of this last statement is crucial to an understanding of thermodynamics. Remember that what is being sought is a set of measurable indices that will provide information about the energy of a system. Both thermal energy and work are generally measurable, and they are both measurable in the surroundings, where the experimenter generally is. Heat and work appear at the boundary, not in the system or in the surroundings. Indeed, heat and work are either done by the system on the surroundings or by the surroundings on the system. Energy is exchanged in this way, and the effects of heat and work can be measured in the system or the surroundings, though the heat and work themselves are really only boundary characters. It is now reasonable to say that the change in internal energy, U, of a system may be related to the work, w, and heat, q, passed between the system and the surroundings. For a change in internal energy, the following equation is known as the **first law of thermodynamics**:

$$dU = đq + đw \tag{3.1-1}$$

Note that a new notation has been introduced for an infinitesimal change in heat or work. This notation will be discussed in more detail shortly. Further definition of work and heat is necessary now. Heat and work share several very important qualities. Because they are boundary entities, they are not described by state functions but instead by path functions. It is quite important that a clear distinction exists in the reader's mind between work and heat, and the energy of a system. Furthermore, by convention, both heat and work are recognized by observing changes in the surroundings and not in the system. This last point helps to highlight a very important difference between heat and work. **Work** is defined as the displacement of an object acted on by a force, thus usually converting energy into some useful form. Traditionally, this is defined as the raising (or falling) of a weight in the surroundings. Such an action leads to a rise (or fall) in the potential energy of the surroundings and hence represents an exchange of usable energy. **Heat**, on the other hand, manifests itself as energy transfer that results in a rise (or fall) in temperature of the surroundings but that is not recoverable as useful or usable energy. This statement should not be interpreted to suggest that heat is useless, a fact to which anyone near a fireplace on a cold night can attest.

Equation (3.1-1) indicates the algebraic behavior of heat and work. Since the first law dictates that the total energy of a system and its surroundings always remains constant, the transfer of energy that occurs across the boundary is either as heat or work. Since the sum of the energy in the system and the heat or work gained or lost by the system is always the same, the quantities of heat and work are treated as algebraic terms. By convention, heat that increases the temperature of the surroundings is given a negative sign (an exothermic event), while heat that raises the temperature of the system is given a positive sign (an endothermic event). Work done on the surroundings (e.g., lifting a weight in the surroundings) is called negative while work done by the surroundings on the system is said to be destroyed in the surroundings and is given a positive sign. It is worth pointing out that this convention of work done on the system being given a positive sign is the more modern convention. Older texts and papers will often use the older convention where work done on the system was given a negative sign. In these older writings, the first law is written as $\Delta U = q - w$.

3.1.1. Specialized Boundaries as Tools

A knowledge of the boundary conditions is important since any change that occurs in a system must do so through a boundary. Several forms of boundaries may be defined that help in simplifying heat—work relationships because they limit the available path functions. Such limiting conditions are useful in scientific thinking, since they frequently result in reasonable and practical approximations.

Adiabatic walls are boundaries across which no heat transfer may occur. This is the only limitation however, and anything else may be passed through an adiabatic wall between system and surroundings. Work in its many forms, whether mechanical or electrical or magnetic, may appear at an adiabatic boundary. Furthermore, material may be added or removed from the system across an adiabatic boundary that may be permeable or semipermeable. In an adiabatically bounded system, $đq = 0$ and the first law reduces to

$$dU = đw \tag{3.1-2}$$

In contrast to adiabatic boundaries, **diathermal** walls are boundaries where heat interactions are allowed to occur. It is worth restating that the concept of an **isolated system** is one in which the boundary is adiabatic, and work interactions (and therefore material transfer) are limited. For an isolated system, $q = 0$, $w = 0$, and $\Delta U = 0$.

The differential notation for work and heat has been introduced. The first law can be written in terms of infinitesimal change as $dU = đq + đw$. It is important to recognize that this differential notation does not represent an equivalence between U and w or q. It has already been stated that energy is a state property and work is a path function. This is codified in the notation dU versus $đw$. dU is a differential that when integrated gives a perfect or definite integral:

$$\int_{\text{initial}}^{\text{final}} dU = \Delta U = U_{\text{final}} - U_{\text{initial}} \tag{3.1-3}$$

a quantity that is independent of the path of integration. On the other hand, $đw$ or $đq$ are inexact differentials whose integration gives a total quantity that will depend completely on the path of integration:

$$\oint đw \neq \Delta w \neq w_{\text{final}} - w_{\text{initial}} \tag{3.1-4}$$

$$\oint đq \neq \Delta q \neq q_{\text{final}} - q_{\text{initial}} \tag{3.1-5}$$

The functions defined by an inexact integral depend not on the initial and final states but rather on the path between the states. The term given these integrals is **line integrals**. For this reason, line integrals are the appropriate mathematical expression for the path-dependent functions of both heat and work. Path functions may take any value greater than a certain minimum. The calculation of this minimum is an important aspect of thermodynamic methodology because these minima help describe reversible processes. Reversible paths are an important topic that will be covered later.

Work done on the surroundings may take many forms, all of which are equivalent because of their shared lineage in Equation (2.3-1). For example, in many cases (especially in physical chemistry textbooks) the work most discussed is the mechanical work that results from the expansion of a volume and that is described by pressure and volume changes, i.e., $-P_{\text{ext}}dV$. Many other forms of work can be considered and all are treated in the same fashion. Important representations of work in biological systems include mechanical work related to moving a weight

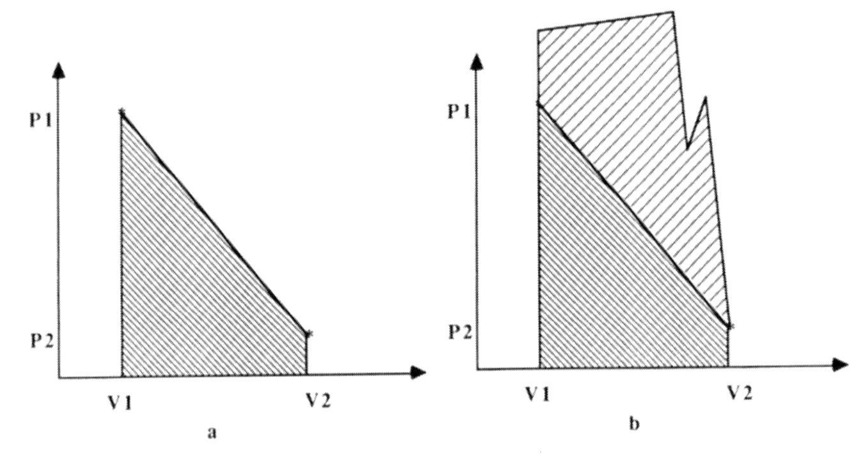

Figure 3.1-1 This figure shows that the work, represented by the area under a curve, depends on the path or curve and not on the initial and final coordinates, P_1V_1 and P_2P_2.

in a gravitational field, $mg\ dh$; electrical work resulting in charge transfer, $\Psi\ dQ$; and surface area changes, $\gamma\ dA$. A system in which all of the above forms of work could be done can be described:

$$đw = mg\ dh - \mathrm{P_{ext}}\ dV + \Psi dQ + \gamma dA + \ldots \tag{3.1-6}$$

where m is mass, g is the acceleration of gravity, h is height, Ψ is electrical potential, Q is charge, γ is the surface tension and A is the surface area. If the only concern is a single kind of work, say, pressure—volume work, then the expression can be simplified:

$$đw = -P_{ext}\ dV \tag{3.1-7}$$

Using the example of pressure—volume work, an illustration of a line integral can be made. Consider two states of a system, state 1 defined by P_1 and V_1 and state 2 defined by P_2 and V_2. The problem is to find the compression work done on the system and hence destroyed in the surroundings in going from state 1 to state 2. In Figure 3.1-1, the work done is represented as a path between state 1 and state 2. In both cases shown, the initial and final states are the same. However, the path taken to get from 1 to 2 differs between Figure 3.1-1a and 3.1-1b. Consequently, the work, which is the integral of the path and is represented by the shaded area of the graph, is dependent on the path and not on the initial or final states. It should be obvious that a specific minimum value of work must be destroyed in the surroundings for the compression to be performed, but that hardly any maximum amount of work destroyed in the surroundings exists if the path is appropriately chosen. For a cyclic function of a path, $\oint w$, the value may be zero or any other value at all, while for all state functions $\oint X = 0$.

These path properties described for work are valid as well for the behavior of heat.

3.1.2. Evaluating the Energy of a System

One fundamental goal of thermodynamics is to quantitate and describe the internal energy of a system. The first law can lead to methods for indirectly determining the energy of a system. Knowing this energy will be extremely valuable in the study of aqueous biochemistry and cellular systems, since most of the life processes are highly energy dependent. By employing boundary limitations to simplify the discussion, the relationship between work and the state of a system can be examined. Consider a system with adiabatic boundaries, so only work appears at the boundary. An infinitesimal amount of work is done on or by the surroundings, $đw$. Since $q = 0$, then

$$dU = đw \qquad (3.1\text{-}2)$$

Integrating gives

$$\Delta U = \int dU = \int đw = w \qquad (3.1\text{-}8)$$

In an adiabatic system, the only way that energy can be exchanged is through work. Importantly, since thermal interactions are not allowed, any work done on the system will directly change the related energy values; for example, if an object is subjected to an accelerating force, a new velocity will be attained without loss due to friction and heat production. In a diathermal thermodynamic system, work will not necessarily have the same effect. Here the object may be subjected to an accelerating force, but friction may prevent any change in velocity at all. The entire work destroyed may appear as a change in thermal energy. In a diathermal system, then, knowing the work expended does not necessarily lead to knowledge of the change in the system's energy.

ΔU can, however, be determined experimentally in systems that are diathermal. If an adiabatic path can be found to exist between two states, then Equation (3.1-8) becomes valid for all systems. It can generally be written for this situation

$$dU = đw_{\text{adiabatic}} \qquad (3.1\text{-}9)$$

This will provide a technique for determining ΔU. It is not possible to determine the absolute energy of a system by evaluating the path function for work. However, since it is the change in energy between states that is usually the important issue, this is adequate and very practical information.

An adiabatic path cannot always be conveniently found for a given change in state. Does this mean that knowledge about energy transformation is not available? Fortunately, the first law provides a solution to finding ΔU even for a diathermal system in which both heat and work are exchanged. The derivation follows. Consider: $\oint dX = 0$ for any state function, while $w_{\text{cyclic}} = \oint \text{đ}w \neq 0$, and $q_{\text{cyclic}} = \oint \text{đ} \neq 0$. The first law states that in a cyclic transformation, the work produced in the surroundings ($-\text{đ}w$) is equal to the heat withdrawn from the surroundings ($+\text{đ}q$). Consequently,

$$\oint (-\text{đ}w) = \oint \text{đ}q \tag{3.1-10}$$

This is equivalent to

$$\oint (\text{đ}q) + (\text{đ}w) = 0 \tag{3.1-11}$$

and:

$$\oint (\text{đ}q + \text{đ}w) = 0 \tag{3.1-12}$$

Because any cyclical function that is equal to zero must also be a state function, this equation may be defined as a state variable U, the energy of the system. Therefore,

$$dU = \text{đ}q + \text{đ}w \tag{3.1-13}$$

Integrating, using line (i.e., path) integrals where appropriate,

$$\int dU = \int \text{đ}q + \int \text{đ}w \tag{3.1-14}$$

gives

$$\Delta U = q + w \tag{3.1-15}$$

Therefore, the first law now provides a solution for finding ΔU, even if no convenient adiabatic path exists. By measuring the heat, q, and the work, w, of a path, the change in the state variable of internal energy may be found.

It is worth restating that the value of thermodynamics lies in its ability to provide information about systems and conditions that are unfamiliar and previously unseen. This power lies in the quality of state functions that constrain a real system so that by measuring just several of a system's state variables, other state variables, including the one of interest, can be determined. In the most classic case, the ideal gas equation of state

$$PV = nRT \tag{3.1-16}$$

indicates that if only P and V are known for a system, then T can be determined because of the natural constraints of the system. Likewise, if T and V are known, P can also be found for any corresponding system.

To this point, great effort has been devoted toward finding the change in the state function of the energy of a system, ΔU, because, once this variable is determined, the state of the system under study can be described in terms of other state variables. This is a very valuable property. The minimum number of state variables necessary to completely describe a particular system is given by a simple formula, the **phase rule**, which will be discussed in Section 7.1.

Consider describing a system of energy, U (found by a path method as described earlier) in terms of other state variables, for example, temperature, T, and volume, V. It follows from the discussion above that since U is a state variable as are temperature and volume, then a change in U, the energy of the system, could be described in terms of T and V. Energy is therefore a function of the temperature and volume of a system:

$$U = U(T,V) \tag{3.1-17}$$

Equation (3.1-17) is a simplification. In fact, U, or any other state function, is a function of all other constrained values in the system under study. In other words, each state variable depends to some degree on every other variable (and there is virtually an infinite number) in a given system. To be entirely accurate in describing a thermodynamic system, it would be necessary to elucidate each and every state variable:

$$U = U(T, V, P, X_i, S, G \ldots) \tag{3.1-18}$$

The symbols for mole number (X_i), entropy (S), and free energy (G) are introduced here and will be discussed subsequently. Such a choice would be a prodigious and surely frustrating task. However, in the real world such exhaustive degrees of accuracy are not necessary. Therefore, scientists are saved from the drudgery of such complete descriptions. Reasonable simplifications are almost always preferred. By using partial derivatives, the state function of interest can be written and examined in terms of the other variables:

$$dU = \left(\frac{\partial U}{\partial T}\right)_V dT + \left(\frac{\partial U}{\partial V}\right)_V dv \tag{3.1-19}$$

Expressions like the one above allow the evaluation of several variables at once. This analysis is made possible by differentiating just one variable while holding all the other variables constant. This process is repeated with each variable, until all are differentiated. The result is a differential equation that mathematically relates each of the state variables to each of the others in a simultaneous manner. One of the most important caveats that accompanies the use of partial differential statements is the need to always interpret these expressions in terms of their physical as well as mathematical meaning. This is important to ensure their effective use as well as to prevent the frustrating confusion that can be associated with the use of these equations. The point in using partial differential expressions is to find mathematically accurate expressions that can be manipulated but that are measurable in a real system.

Applying the example from above, consider the actual physical quantity referred to by these differentials. $(\partial U/\partial T)_V$ represents the change in energy of the system caused by a change in the system's temperature when the volume of the system is held constant. Obviously, such a descriptor fits exactly the quantity of heat transferred in a pure heat transfer, that is, one in which no heat appears as work (assuming the system can do only PV work, volume is constant so no work is done). Analogously the expression, $(\partial U/\partial V))_T$ is interpreted as the change in energy of the system when a change in volume occurs but at a constant temperature.

3.2. Derivation of the Heat Capacity

The thermodynamic exercise now remains to discover what property (or properties) of the system can be measured that will allow the determination of ΔU. There are two equations discussed so far that describe dU:

$$dU = đq + đw \tag{3.1-13}$$

and

$$dU = \left(\frac{\partial U}{\partial T}\right)_V dT + \left(\frac{\partial U}{\partial V}\right)_T dV \tag{3.1-19}$$

These can of course be equated, giving

$$đq + đw = \left(\frac{\partial U}{\partial T}\right)_V dT + \left(\frac{\partial U}{\partial V}\right)_T dV \tag{3.2-1}$$

For the system under discussion, only volume change work is possible, so $đw$ can be written

$$đw = -P_{\text{ext}}\, dV \tag{3.2-2}$$

Substitution gives

$$đq - P_{\text{ext}}\, dV = \left(\frac{\partial U}{\partial T}\right)_V dT + \left(\frac{\partial U}{\partial V}\right)_T dV \tag{3.2-3}$$

It is now appropriate to employ the principle of judiciously constraining the system under study. If the task is to find the change in energy of a system that is kept at constant volume, then dV becomes zero and the equations from above can be rewritten as follows:

$$dU = đq_v \tag{3.2-4}$$

and

$$dU = \left(\frac{\partial U}{\partial T}\right)_V dT \tag{3.2-5}$$

Again equating these identities gives the following result:

$$đq_v = \left(\frac{\partial U}{\partial T}\right)_V dT \tag{3.2-6}$$

This is now an expression that relates the heat drawn from the surroundings to the increase in temperature of the system at a constant volume. Because both $đq_v$ and dT are easily measured experimentally, the task of relating the equation to the real system is approaching completion. The ratio of these two experimentally accessible parameters,

$$\frac{đq_v}{dT} \tag{3.2-7}$$

is called the **heat capacity** (at constant volume) or C_v. Writing this in summary gives

$$C_v = \frac{đq_v}{dT} = \left(\frac{\partial U}{\partial T}\right)_V \tag{3.2-8}$$

Several favorable results are apparent here. First, the partial derivative discussed above, $(\partial U/\partial T)_V$, has been solved. Substituting gives

$$dU = C_v\, dT \tag{3.2-9}$$

Integrating this expression will give the following result:

$$\Delta U = \int_{T_1}^{T_2} C_v\, dT = C_v\, \Delta T \tag{3.2-10}$$

It is now possible to calculate ΔU directly from the properties of a system if the heat capacities are known. Extensive tables of heat capacities are available for many materials. C_v is always positive, and therefore whenever heat is added to a system, the temperature of the system and the system's internal energy will rise. For a system at constant volume constrained to do only PV work, temperature therefore is a direct reflection of the internal energy.

Heat capacity is expressed in units of joules per degree mole in the SI system. Much literature still uses the units of calories per degree mole however. The heat capacity relates on a molar basis just how much heat it takes to raise the temperature of a system one degree Celsius or Kelvin. Heat capacities often vary with temperature and depend upon the volume and pressure. In the example above, the derivation of heat capacity was for a system held at constant volume. In a system that is held at constant pressure, the heat capacity is expressed as C_p. C_v is just about equal to C_p for liquids and solids. For a gas, heat capacities under these differing conditions vary by a factor of R, so that $C_p \equiv C_v + R$.

3.3. A System Constrained by Pressure: Defining Enthalpy

Just as C_v was derived by considering a system at constant volume, the constraints on a system can be and should be guided by the practical aspects of a specific system under study. Defining systems at constant pressure is practical to most biochemists and life scientists because most of the real systems studied in the organism or in vitro are constant-pressure systems themselves. An analysis of the first law under this constant-pressure constraint will lead to a new state function, **enthalpy, *H*.** Remember that the goal is to find an experimentally measurable value that can inform on the internal energy state of the system under study, given the constraints chosen.

Starting again with the first law:

$$dU = đq + đw \tag{3.1-1}$$

and examining a system that does only pressure–volume work, so $đw = -P\,dV$, now:

$$dU = đq - P\,dV \tag{3.3-1}$$

Pressure is a constant in this system and so integration can be immediately carried out:

$$\int_1^2 dU = \int_1^2 đq_p - \int_1^2 P\,dV \tag{3.3-2}$$

which gives

$$U_2 - U_1 = q_p - P(V_2 - V_1) \tag{3.3-3}$$

By algebra and by substitution of the identities $P_1 = P_2 = P$ at constant pressure:

$$(U_2 + P_2V_2) - (U_1 + P_1V_1) = q_p \tag{3.3-4}$$

Now both P and V are already defined as state functions, and a product of state functions is itself dependent only on the state of the system. Consequently, the expression

$$U + PV \tag{3.3-5}$$

is a state function. This new state function is given the symbol $\boldsymbol{H}$ and the name, **enthalpy:**

$$H = U + PV \tag{3.3-6}$$

Substituting H for $U + PV$ in Equation (3.3-4) above gives the following:

$$(H_2 - H_1) = q_p = \Delta H \tag{3.3-7}$$

Having this result, that the heat withdrawn from the surroundings is equal to the increased enthalpy of the system is especially valuable if the energy change, ΔU, is now considered with a real simplification of most aqueous phase and biological processes, $\Delta V = 0$. Consider that

$$\Delta U + P\Delta V = q_p \tag{3.3-8}$$

and with $\Delta V = 0$, a reasonable approximation can be written that

$$\Delta U = q_p \tag{3.3-9}$$

and since

$$q_p = \Delta H \tag{3.3-10}$$

then

$$\Delta U = \Delta H = q_p \tag{3.3-11}$$

Enthalpy is a valuable state function because it provides a method for determining a realistically constrained biological or aqueous phase system's energy simply by measuring the heat exchanged with the surroundings. Finally, by writing enthalpy in terms of temperature and pressure, $H = H(T, P)$, in a fashion similar to that done earlier for C_v, it can be shown that the heat capacity at constant pressure, C_p, is related to enthalpy as follows:

$$\Delta H = C_p \, \Delta T \tag{3.3-12}$$

The actual derivation of this is left to the reader as an exercise.

Enthalpy is a useful state function to the biochemist for a variety of practical reasons. First, most chemical and biological processes of interest occur at constant pressure and with no change in volume, and hence the internal energy of a system is easily and quite accurately approximated by measuring the enthalpy. Secondly, given these constraints, the experimental measurement of q_p is both easy and reasonably precise. Finally, enthalpy is easily predicted from a state property, C_p, thus allowing inference of changes in enthalpy for systems never seen before. As indicated in the introduction to this section, this is precisely the valuable kind of tool that thermodynamics can provide the practical scientist.

A great deal of information based on enthalpy is available, including tables listing heats of formation, fusion, solution, dilution, reaction, etc. These enthalpic values are valuable precisely because they indicate the change in energy state of the system under study. Most readers of this volume will already be familiar with the use of enthalpy to determine heats of reaction and formation of compounds as well as heats associated with phase changes and solvation. Most texts in introductory chemistry and physical chemistry provide excellent coverage of the practical use of this state function including the use of Hess's law, standard states, and calculation of heats of formation, phase change, etc.

Enthalpy will play an important role because of its ability to provide a good picture of the energy changes associated with a process. But enthalpy and the first law alone will not adequately predict the spontaneous direction of a reaction. However, when combined with the state function entropy, a function that represents the second law, a new state function, the **Gibbs free energy**, is the result. The Gibbs free energy indicates the spontaneous tendency of a system to move in one direction or another. Enthalpy will reappear therefore after a short examination of the second law of thermodynamics.

CHAPTER 4

The Second Law

4.1. Understanding the Second Law of Thermodynamics

It is common experience that watches run down, bouncing balls stop bouncing, and even the most perfectly designed and crafted engine will eventually cease operating. This universal tendency toward stasis is the essence of the second law of thermodynamics. There are multiple possible statements of the second law, some useful and some needlessly confusing. Some of the more useful paraphrases are included in the following list; of these, the first two expressions are familiar from the discussion in the first chapter and are the more traditional representations of the **second law**:

1) No process is possible where heat is transferred from a colder to a hotter body without causing a change in some other part of the universe.
2) No process is possible in which the sole result is the absorption of heat (from another body) with complete conversion into work.
3) A system will always move in the direction that maximizes choice.
4) Systems tend toward greater disorder.
5) The macroscopic properties of an isolated system eventually assume constant values.

The last definition (#5) is particularly useful because it leads naturally toward a definition of **equilibrium**. When the state values of an isolated system no longer change with time, the system may be said to be at equilibrium. Movement of a system toward equilibrium is considered the "natural" direction.

A state function that could indicate whether a system is moving as predicted toward a set of unchanging state variables would be of great value. Such a state function exists and is the state function of the second law, **entropy**, and it is given the symbol S. Traditionally, the derivation of S is accomplished by analysis of a Carnot cycle. For a detailed analysis of the Carnot cycle, the reader is referred to any number of traditional thermodynamics texts. The approach that leads to the Carnot cycle does have value, especially given the exercises of the previous chapter, and will be introduced as a brief thought problem that will lead to equations that describe entropy.

Following this discussion, a restatement of the second law based on a statistical thermodynamic approach will lead to the same mathematical description as a Carnot cycle derivation. In the statistical approach, it is considered that a system will tend to maximize choice. Sometimes, this statement of the second law is written as "systems tend toward increased disorder" or "systems tend toward increased randomness." These paraphrases of the second law are useful, but care should be taken with their use because the terms disorder and randomness often lead to confusion. Thinking of the natural direction of a system in terms of its ultimate uniformity or choice is a practical and the ultimately intuitive treatment of entropy.

4.2. A Thought Problem: Designing a Perfect Heat Engine

The Carnot cycle requires the use of a **reversible** cyclic path for its derivation. A definition of this concept is important. The reader is reminded that the use of reversibility is not limited to a Carnot cycle and that the concepts developed in the following section are completely general.

4.2.1. Reversible Versus Irreversible Path

The ideas of reversible and irreversible paths will be developed here for a system in which pressure–volume work is being done. Again, this choice is simply one of convenience and is practical for the heat engine under consideration. A **reversible process** is one in which the steps of the path occur in a series of infinitesimal steps. The path is such that direction may be reversed by an infinitesimal change in the external conditions prevailing at the time of each step.

The system under study is an isothermal expansion in which a piston encloses a system of a gas with temperature T_1, pressure P_1, and volume V_1. On top of the piston is a weight of mass M. The mass is part of the surroundings, and if it is moved upward, work will be done by the system on the surroundings. Conversely, if the mass pushes the piston head down, work will be done on the system. The container containing this system is specially designed for this example and has a series of stops that may be inserted through the side of the container (Figure 4.2-1). The power stroke, that is, the raising of the weight in the surroundings, is complete when the piston head reaches the top set of stops. At this point, the state of the system will be defined as V_f, P_f, T_1. When the system is in the initial state, the volume occupied by the gas is smaller than in any

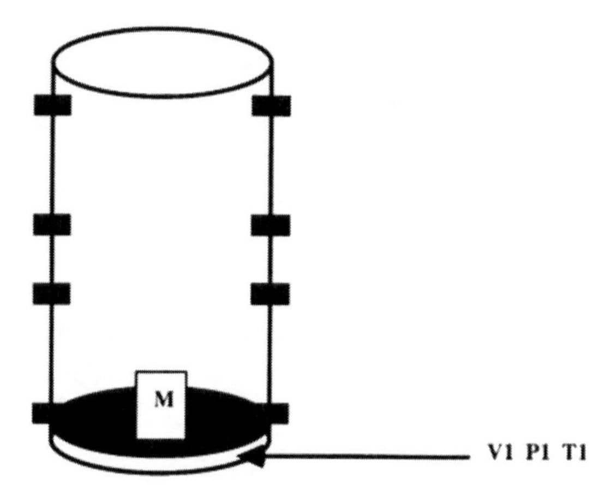

Figure 4.2-1 Special container for reversible isothermal expansions and compressions.

other state and the pressure of the gas is higher than in any other state (each state represented by a set of stops). As long as the internal pressure of the gas is greater than the external pressure, as represented by the weight on the piston, the gas will expand and the piston will be driven upward. Since the internal pressure will be greater closer to the initial state due to the higher compression of the gas, the external pressure on the piston can be set higher at the early stages of the expansion. By continually decreasing the weight on the piston as the gas inside expands, the power stroke of the piston can be sustained until the final state is achieved.

Consider the following experiments with this device. Initially, two sets of stops are inserted (see Figure 4.2-2). The first set maintains the position

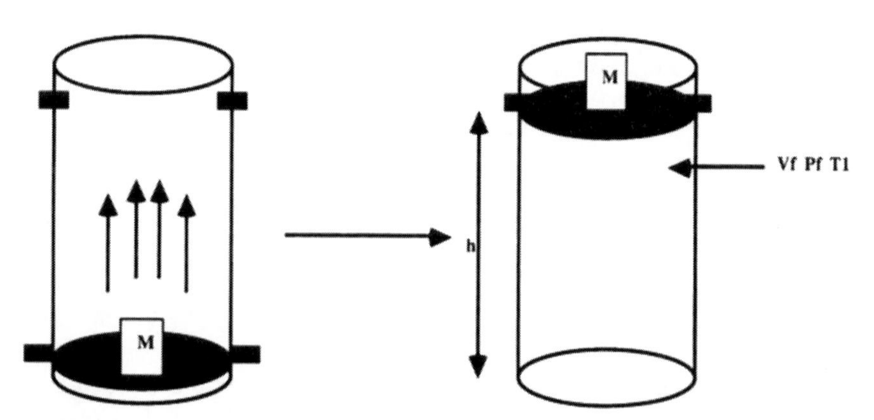

Figure 4.2-2 Single-step expansion where $w = -P_{ext}\,\Delta V$.

of the piston head at the intial state of V_1, P_1, T_1. The second set of stops restrains the system to the state described by V_f, P_f, T_1. Since this two-stop system will allow motion from the initial to the final step, the maximum weight that can be placed on the piston head must exert a pressure just slightly less than the final pressure, P_f. Note that this weight is significantly smaller than the weight (or P_{ext}) that could be moved against by the initial pressure, P_1. Now, the lower set of stops is removed and the piston moves upward halting at the second set of stops. The new state of the system is now V_f, P_f, T_1. The expansion to the new state has resulted in the raising of the mass, M, in a gravitational field. The work done is Mgh, and since pressure is simply force over an area, and the external pressure (including the mass, M) is the force opposing the movement of the piston, the following can be written:

$$\frac{Mg}{A} = P_{ext} \tag{4.2-1}$$

$$w = Mgh = P_{ext}\,Ah \tag{4.2-2}$$

Ah is the volume change $V_2 - V_1$, and so this is a restatement of the problem derived earlier in the section on pressure–volume work, and the resulting equation holds:

$$w = -P_{ext}\,\Delta V \tag{4.2-3}$$

This is the correct mathematical statement for all real expansions as long as the external or opposing pressure, in this example provided by the mass on the piston, remains the same throughout the expansion. The work function for the two-stop system having now been described, the experiment is repeated but this time with a third set of stops in place (Figure 4.2-3). Now, a weight is chosen that will allow expansion to the state V_2, P_2, T_1. Because this intermediate state has a pressure $P_2 > P_f$, the weight chosen for the initial expansion is larger than the weight chosen for the one-step expansion. Therefore, for the distance traveled, more work is done in the first part of the expansion than occurred in the one-step case. Now, the piston is allowed to move to the position described by V_2, P_2, T_1. The weight is now changed to allow the full expansion to the final state. Because the final internal pressure will be lower at V_f, P_f, T_1, the weight must be reduced before the two-step expansion is completed. After the weight is changed, the piston is allowed to move to the final position with state functions V_f, P_f, T_1.

As can be seen in Figure 4.2-4, the work done on the surroundings that results from the two-step expansion is more than that resulting from the one-step expansion, even though the final state is the same in both cases.

Now, if an infinite series of infinitesimal changes in the movement of the piston are made to occur and are accompanied by the continuous readjustment of the weight on the piston (and hence P_{ext}), the maximum

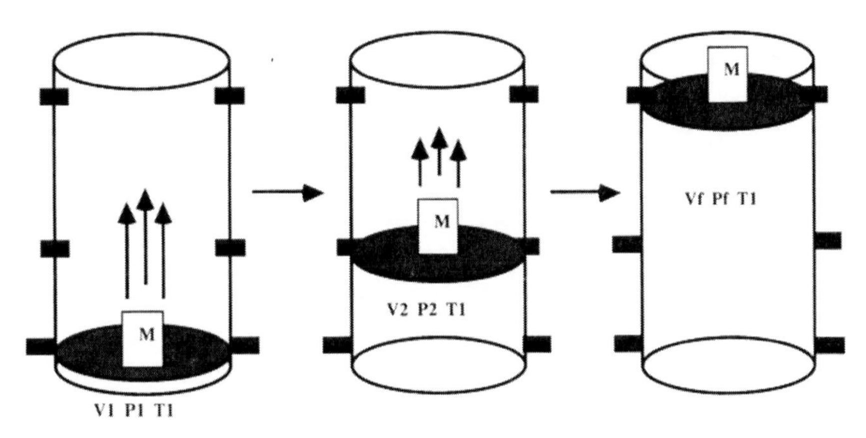

Figure 4.2-3 Two-step expansion where weights are changed at each step of the expansion.

work available from the expansion will result. Two points are obvious. First, such an expansion would encompass an infinite number of steps and would as a result take an infinite amount of time to complete. Secondly, the internal equilibrium of such a process will be disturbed only infinitesimally, and in the limit no disturbance at all will take place. It is reasonable to consider therefore that this system does not depart from

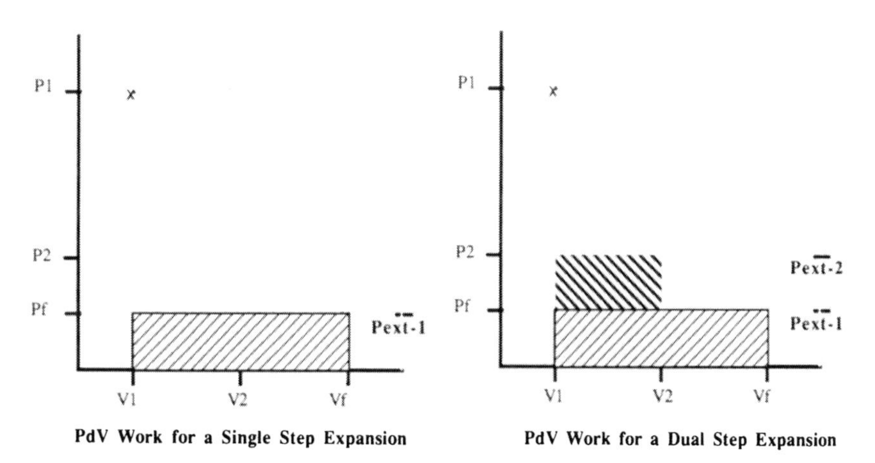

Figure 4.2-4 Work function graph showing the work produced for a one- and a two-step expansion. In the one-step expansion, the work is found equal to $-P_{ext-1}$ ΔV and, since P_{ext-1} is just an infinitesimal less than P_f, the maximum amount of work available is limited. In the two-step process, a higher weight, represented by P_{ext-2} is used for the initial expansion step, then the weight on the piston is reduced to P_{ext-1} and the expansion is completed. More work is extracted in the two-step case because of the change in weight along the path.

equilibrium, except infinitesimally, throughout the entire path. Consequently, the state variable describing a system that is undergoing a reversible process can be treated as if the system were at equilibrium.

In summary, for a reversible process, the path is such that it will take an infinite time to complete, but the maximum work will result only from this reversible path. Furthermore, reversible processes are essentially at equilibrium except for infinitesimal perturbations and may be treated as if they are equilibrium conditions. Because of these properties, in a reversible process the external forces, such as P_{ext} in the example, become uniquely related to the internal forces, that is,

$$P_{ext} = P_{int} \pm \text{infinitesimal} \tag{4.2-4}$$

Thus:

$$P_{ext} = P_{int} = P \tag{4.2-5}$$

Work for a process like this can be found in the fashion

$$đw = -P\,dV \tag{4.2-6}$$

which by integration gives

$$\int_1^2 đw = -\int_{V_1}^{V_2} P\,dV \tag{4.2-7}$$

By using the equation of state $PV = nRT$ and the identity $P_1V_1 = P_2V_2$ and substituting, the following equation is derived:

$$w = -nRT \ln \frac{P_1}{P_2} \tag{4.2-8}$$

This is the maximum work that is available from an isothermal expansion and therefore from a reversible path.

Where does the energy come from for this process? Recalling that this is an isothermal expansion, $\Delta T = 0$ for the entire process. Previously, it was shown that

$$\Delta U = C_v \Delta T \tag{4.2-9}$$

Using this equation,

$$\Delta U = 0 \tag{4.2-10}$$

By the first law of thermodynamics,

$$\Delta U = q + w \tag{4.2-11}$$

so

$$0 = q + w \tag{4.2-12}$$

and

$$w = -q \tag{4.2-13}$$

for an isothermal reversible expansion. Therefore, in a reversible isothermal process, there is a perfect exchange of heat from the surroundings for work done on the surroundings.

A few moments of reflection on this discussion should make it obvious why the second law of thermodynamics can be stated with such clarity and certainty . Since only a reversible process can lead to the 100% efficient transformation between heat and work, any process that is not reversible will lead to a less than 100% interconversion. Under no circumstances can even a reversible process be more than 100% efficient; in other words, work can never be produced for free. The reason that all processes generate less work than the energy they consume is also explained by the nature of the perfect transforming process, the reversible path. Any path that is reversible must have an infinite series of infinitesimal steps. Obviously, such a process would never be completed in the real world, and hence any real process can only approach reversibility. At best, the entropy of the universe remains constant by having only reversible processes at work, but in reality no reversible process can be completed in a finite time frame, and hence any real process leads to an increase in entropy.

Knowing that real, and therefore **irreversible, processes** inevitably lead to an increase in entropy is all the more valuable if a state function can be derived that will indicate in what direction and how far a system is from equilibrium. Recall that a state function is one whose cyclic integral is equal to zero:

$$\oint X = 0 \tag{4.2-14}$$

Specifically, an entropy state function is required such that

$$\oint S = 0 \tag{4.2-15}$$

The change in entropy should be zero at equilibrium and should be positive or negative otherwise to indicate the necessary direction to reach equilibrium.

4.2.2. A Carnot Cycle

The Carnot cycle is a series of expansions and compressions for a reversible machine. First, an isothermal expansion and then an adiabatic expansion are performed. These are followed by an isothermal compression and an adiabatic compression. The heat transferred in the Carnot cycle occurs between two heat reservoirs of different temperatures. A heat engine performs work by the transfer of heat between two reservoirs of different temperature. If a negative amount of work is done (i.e., work is done by the system on the surroundings), heat is drawn from the first reservoir and a certain amount of heat is placed in the second reservoir; this heat has a negative value. The efficiency of a heat engine is equivalent to the work performed divided by the heat extracted from the first reservoir. The work done by the machine will be equal to the heat drawn from the first reservoir minus the heat wasted, which is placed into the second reservoir. Therefore, the greater the heat drawn from the first reservoir compared to the heat placed in the second reservoir, the greater is the efficiency of the engine. It can be shown that the efficiency of the machine will depend only on the temperature of the two reservoirs. Consequently, for the efficiency of the interconversion of heat to work to be 100%, i.e.:

$$\text{Efficiency} = 1 - \frac{T_2}{T_1} \tag{4.2-16}$$

the temperature difference between the reservoirs must be infinite.

Working through the four exchanges in this cyclic process leads to the following result which will be given here without derivation:

$$\oint \frac{đq}{T} = 0 \tag{4.2-17}$$

Since what is being sought is an expression for this cycle where the cyclic integral is 0, this expression can be rewritten as

$$dS = \frac{đq_{\text{rev}}}{T} \tag{4.2-18}$$

This is the expression of entropy for a reversible process, that is, one that has a 100% efficiency. For any other irreversible process,

$$dS > \frac{đq_{\text{rev}}}{T} \tag{4.2-19}$$

The general case can be written after integrating:

$$\Delta S = \frac{q_{\text{rev}}}{T} \tag{4.2-20}$$

Recalling the result for the work obtainable from a reversible isothermal expansion:

$$w = -q = -nRT \ln \frac{V_2}{V_1} \tag{4.2-21}$$

rearranging gives

$$\frac{q_{\text{rev}}}{T} = nR \ln \frac{V_2}{V_1} \tag{4.2-22}$$

which is the same as writing

$$\Delta S = nR \ln \frac{V_2}{V_1} \tag{4.2-23}$$

for an ideal gas expansion. Consider the practical chemical and biochemical case of a system at constant pressure, and recall that heat exchanged was equal to the enthalpy change of a system. Thus,

$$q_p = \Delta H \tag{4.2-24}$$

This enables equations of the following sort to be written:

$$\frac{\Delta H_{\text{fusion}}}{T_{\text{melt}}} = \Delta S_{\text{fusion}} \tag{4.2-25}$$

which is the entropy of a phase change, where T is the melting point. Similarly, other equations such as

$$\frac{\Delta H_{\text{vaporization}}}{T_{\text{boil}}} = \Delta S_{\text{vaporization}} \tag{4.2-26}$$

can be written.

These general formulas will be of great value as shall be seen in the coming chapters. It is instructive and intellectually satisfying that the same mathematical expressions can be derived from a statistical approach. As pointed out in the introduction, it would be expected that the mechanistic approach of statistical mechanics would be able to provide the same mathematical formulation of thermodynamic laws as does the empirically derived classical treatment.

4.3. Statistical Derivation of Entropy

A statistical approach to deriving the state function of entropy, S, will lend mathematical form to the paraphrase that "a system will always move in the direction that maximizes choice." It will now be shown that the state function S can be defined in terms of a statistical variable, W:

$$S = k \ln W \tag{4.3-1}$$

4.3.1. Limits of the Second Law

The second law of thermodynamics is only applicable to macroscopic systems. It is, in fact, not able to describe accurately too small a system. Following the example of Poincare, consider the following system.

There are two connected containers that contain a number of particles. The probability that any single particle will be in either of the two containers at any instant is 1, while the probability that a single given particle is in a specific one of the containers is 1/2. The probability that all the particles in the system, N, will be in one of the containers is $(1/2)^N$. For a single particle, the chance that it would be in container A is 1/2 at any instant or 50% of the time. For two particles, the chance that both particles are found in A at one time is 1/4 or 25% of the time. For 10 particles, if container A is examined 1024 times, all 10 particles will be found in container A at the same time just once. However, if the number of particles in the system is increased to 100, the chance of finding all 100 particles in container A at one instant is less than 1 in 10^{30}. Even if the system is looked at again and again, each look taking a microsecond, it would take just over 4×10^{14} centuries to find all 100 particles in container A simultaneously just once. For a fluctuation away from equilibrium, that is, such as finding all the particles in container A, the larger the number N, the less likely it is that a statistically possible fluctuation will occur that will invalidate the second law. For a molar quantity, large fluctuations simply do not occur in the real temporal framework of the universe. From this discussion, entropy should be recognized as a macroscopic property and must always be considered with those limitations.

4.3.2. Statistical Distributions

In the derivation of a statistical approach to entropy, it is necessary to be able to express the possible distributions of a set of particles among states of a defined energy level. The number of particles is given by N_1, the energy level may be denoted as E_1, and the degeneracy of this energy level is defined as g_1. **Degeneracy** is defined as the number of distinguishable states of the same energy level. The question with regard to distributions then is, How many possible ways can N_1 particles be distributed into g_1 boxes at energy level E_1? Depending on whether or not certain restrictions apply to the particles, that is, whether there are characteristics that distinguish the particles comprising N_1 or whether there are rules that specify how many particles may be placed in each box, certain mathematical expressions that will describe the distributions may be written.

In the case where the particles are distinguishable but no restriction applies to the number of particles found in each box, the distribution is defined by **Boltzmann statistics.** Boltzmann statistics take the form of

$$t_1 = g_1^{N_1} \tag{4.3-2}$$

where t_1 is the number of ways that the N_1 particles can be arranged in g_1 boxes. In describing a complete system, there are N_1 particles distributed in g_1 boxes, N_2 particles distributed in g_2 boxes, N_3 particles distributed in g_3 boxes, and so on, where the different subscripts are used to designate different energy levels. The population of particles in this system will be given by N, where

$$N = \sum_i N_1 + N_2 + N_3 + \ldots + N_i \tag{4.3-3}$$

The total number of ways, T, that the entire population, N, can be distributed will be given by

$$T_{\text{Boltzmann}} = N! \prod_i \frac{g_i^{N_i}}{N_i!} \tag{4.3-4}$$

This equation obviously does not descend directly from the previous notation. The conditions for Boltzmann statistics are based on the distinguishable nature of each particle. Because of this condition, many arrangements can be made by forming subgroups from the population N. These subgroups must be taken into account in overall summation. In general, the number of ways to group N objects into subgroups n_1, n_2, n_3 . . . is given by

$$\frac{N!}{n_1!\ n_2!\ n_3!\ n_i!} \tag{4.3-5}$$

Boltzmann statistics are based on assumptions that are not consistent with the generally regarded dogmas of quantum mechanics. Other statistical distributions are possible. When the particles comprising N are considered to be indistinguishable but no limit is placed on the number of particles contained in each box, **Bose–Einstein statistics** result. **Fermi–Dirac statistics** result when the particles are indistinguishable and there is a limit of one particle per box. Both of these conditions are actually more accurate in the world of real particles than Boltzmann statistics. Fermi–Dirac statistics apply to particles that obey the Pauli exclusion principle, such as electrons and positrons, while Bose–Einstein statistics apply to particles that have **integral spin**, such as photons, and are not subject to the Pauli exclusion principle. Boltzmann distributions in fact do not relate to any real particle. It can be shown however, and it is left as an exercise for the reader to do so, that for a sufficiently dilute system, where $N_i \ll g_i$, the T for Fermi–Dirac and Bose–Einstein statistics are equal to $T_{\text{Boltzmann}}/N!$. This results in the following expression:

$$T_{\text{special Boltzmann}} = \frac{T_{\text{Boltzmann}}}{N!} = \prod_i \frac{g_i^{N_i}}{N_i!} \tag{4.3-6}$$

The advantage of this is that the mathematics of this modified expression are easier to work with than those of the Fermi–Dirac and Bose–Einstein statistics. The reader is referred to Appendix I for references to standard texts of statistical thermodynamics for a more complete discussion of these topics.

4.3.3. The Boltzmann Distribution

It has already been stated that for a macroscopic system described by thermodynamics, the state of equilibrium is reached when the state functions no longer are changing with time. Furthermore, it has been pointed out that the equilibrium position is the most probable arrangement of the particles that make up the system. For a large system, as would be appropriately described by thermodynamics, there is a very large number of degeneracies. Therefore, finding the equilibrium position is the same as finding the most probable distribution of the particles in this system, which is precisely what the Boltzmann equation provides.

The system for which a most probable distribution of states is sought is generally constrained in several ways. First of all, it is an isolated system, and so its energy, U, is constant. Furthermore, in an isolated system at equilibrium, the total number of particles, N, may also be considered constant. The mathematics for solving this problem are not suitably derived here and so the result alone will be given:

$$\frac{N_i}{N} = \frac{g_i\, e^{-U_i/kT}}{Z} \tag{4.3-7}$$

This is the **Boltzmann distribution,** where k is Boltzmann's constant and Z is the partition function, a function that sums over all the system's energy levels. Z is defined as follows:

$$Z = \sum_i g_i\, e^{U_i/kT} \tag{4.3-8}$$

Often, what is of most interest is the ratio of the number of particles or molecules in two different energy levels, and this is given by

$$\frac{N_i}{N_j} = \frac{g_i}{g_j}\, e^{-(U_i-U_j)/kT} \tag{4.3-9}$$

This indicates that lower energy states will generally have more molecules than higher ones, assuming that a rapid increase in degeneracy with increased energy does not occur. With this as background, the following problem is considered.

4.3.4. A Statistical Mechanical Problem in Entropy

The distribution of a group of molecules among a set of energy states in an isolated system can be undertaken as a simple example of the most probable arrangement. By isolating the system, the energy of the system will remain unchanged through the analysis. The system considered contains three molecules, each of which may occupy a set of specific quantitized levels, 0, 1, 2, and 3. All start at the zero energy level, 0. The system is given three quanta of energy. The problem now is to find the possible arrangements for these three molecules in the system at constant energy. As can be seen from Figure 4.3-1, there are 10 possible arrangements. Each of the 10 arrangements is called a **microstate**. The groupings a, b, and c are called **configurations** while the whole group of microstates

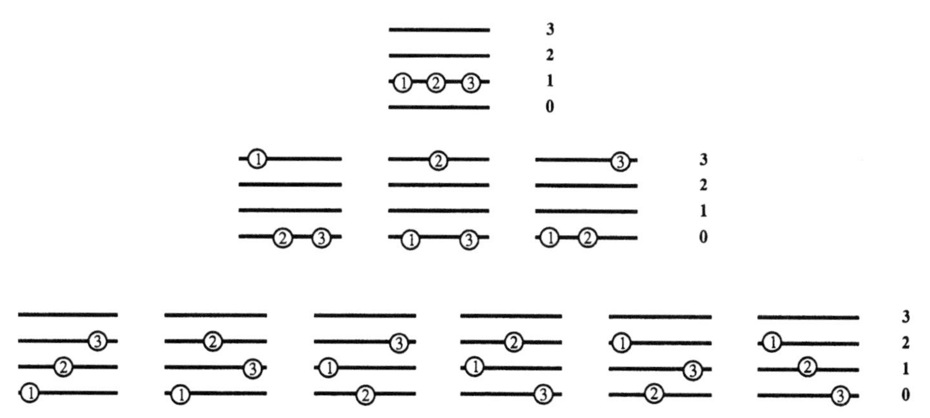

Figure 4.3-1 Microstates and configurations of an ensemble of three molecules with total energy 3.

is called an **ensemble**. If the total number of particles is N, how many ways can they be distributed among these energy levels 0 through 3 so that the total energy is constant? Particle #1 has N choices, particle #2 has $(n-1)$ choices, and so on. The total number of choices available can be given by a variable, W. If a group of particles, n, as subset of the total N, occupies the same energy level, the following expression can be written:

$$W = \frac{N!}{n!} \qquad (4.3\text{-}10)$$

This value for W will give the number of possible microstates contained in a particular configuration. The larger the value of W, the more likely it is that a particular configuration will be found. Therefore, W can be used to indicate the most probable configuration for the particles in the ensemble.

Specifically, consider the example of three particles. Prior to the addition of the energy, each particle resides in the same energy level, 0. There is only one configuration that makes this initial state possible, and

$$W = \frac{3!}{3!} = 1 \qquad (4.3\text{-}11)$$

Alternatively, following the addition of energy, the number of microstates that can be contained in a configuration with two molecules in the same energy level, $n = 2$, corresponds to configuration b in the figure. This is given by

$$W = \frac{3!}{2!} = 3 \qquad (4.3\text{-}12)$$

Table 4.3-1. Values of W for different values of N.

N	n	W	N	n	W	N	n	W	N	n	W	N	n	W
2	1	2	3	1	6	4	1	24	5	1	120	10	1	3,628,000
2	2	1	3	2	3	4	2	12	5	2	60	10	2	1,814,400
			3	3	1	4	3	4	5	3	20	10	3	604,800
						4	4	1	5	4	5	10	4	151,200
									5	5	1	10	5	30,240
												10	6	5,040
												10	7	720
												10	8	90
												10	9	10
												10	10	1

Similarly, if a single molecule is found in a given energy level, $n = 1$, then configuration c is the result:

$$W = \frac{3!}{1!} = 6 \tag{4.3-13}$$

Finally, if all three molecules are found in a single energy level (obviously not 0 though), this probability is given by configuration a:

$$W = \frac{3!}{3!} = 1 \tag{4.3-14}$$

In this example therefore, the most probable distribution will be the one which corresponds to c.

The results for a similar problem with 2, 3, 4, 5, or 10 particles are given in Table 4.3-1. This table shows that the configurations containing the greatest number of microstates and therefore the most probable configurations are those that have the broadest distribution of molecules in the various energy levels, that is, those that contain $n = 1$ molecule per level (if possible).

Clearly, none of these systems containing just several molecules are good thermodynamic systems for the reasons discussed earlier. However, if N is made large enough, for example, the number of molecules in a mole (Avogadro's number, 6×10^{23}), a good thermodynamic system would be under study. Calculating W from numbers this size is quite difficult; hence, a mathematical approximation called the **Stirling approximation** is used:

$$\ln N! = N \ln N - N \tag{4.3-15}$$

W is of interest because the maximum value of W, represents both the most probable distribution of particles and the distribution at equilibrium. When W has attained a maximal value

$$dW = 0 \tag{4.3-16}$$

This statement implies what is already obvious from the earlier example, namely, that for a thermodynamic system at equilibrium, an infinitesimal change in any direction will not result in any significant or detectable change in the distribution of the particles in microstates.

If W is maximal at equilibrium, then it follows that for a system moving in a spontaneous direction toward equilibrium, the change in W from state 1 to state 2 (the thermodynamic states now under discussion) will be given by

$$W_2 - W_1 \leq 0 \tag{4.3-17}$$

In this statement, W obviously has the form of a state function. It is an extensive state function and is one that increases in a spontaneous process. Interestingly, both of these are qualities for the state function of entropy, S.

The state functions S and W can be related:

$$S = k \ln W \tag{4.3-18}$$

and

$$\Delta S = k \ln \frac{W_2}{W_1} \tag{4.3-19}$$

where k is Boltzmann's constant.

Returning again to the example of the three molecules presented in Figure 4.3-1, initially the entropy of the system was 0, because $W = 1$. Such a result is consistent with the third law of thermodynamics, which states that when all molecules in a system are in their lowest energy states, the entropy of the system is zero. Equilibrium will be found when entropy is maximal, which in this case is

$$S = k \ln (6) \tag{4.3-20}$$

The physical interpretation of Equation (4.3-20) is that as a result of the addition of the energy, there are more choices available to describe the arrangement of the system, and in fact less is now known about the system than before.

It can be shown that the statistical expression of entropy, S_{se},

$$S_{se} = k \ln W \tag{4.3-21}$$

is equal to the traditionally defined entropy,

$$S = \frac{q_{rev}}{T} \tag{4.3-22}$$

and if the expansion of an ideal gas is considered, it can be shown that

$$\Delta S = k \ln \frac{W_2}{W_1} = Nk \ln \frac{V_2}{V_1} \tag{4.3-23}$$

4.4. The Third Law and Entropy

The **third law** states that at absolute zero, the entropy of all pure, perfect crystals is 0:

$$S_{\text{pure-perfect xtal}}\ (0\ \text{K}) = 0 \tag{4.4-1}$$

This makes some sense if it is considered that at zero degrees it is possible to imagine that a crystal would be perfectly arranged so that perfect order reigned everywhere, and hence the location of every atom could be known. Such a system would be like that of the three particles when they were all in energy level 0. Generally, only a crystal could have this structure, not a liquid or a gas. Also, if the system were not pure, the entropy would not be at a minimum because by separating the components of the mixture, the entropy would fall. (The entropy of mixing will be taken up in Section 6.3.)

The temperature dependence of entropy is easily found:

$$\Delta S = S_2 - S_1 = \int_{T_1}^{T_2} \frac{q_{\text{rev}}}{T}\, dT \tag{4.4-2}$$

At constant pressure, $q = H$ and $\Delta H/T = S$, while $\Delta H = C_p \Delta T$, so that:

$$\Delta S = \int_{T_1}^{T_2} C_p \frac{dT}{T} \tag{4.4-3}$$

which is equal to

$$\Delta S = C_p \ln \frac{T_2}{T_1} \tag{4.4-5}$$

Since C_p is always positive, the entropy always increases as the temperature is increased. This is a result that makes sense when considered in terms of the three-particle example, where the addition of energy led to a new equilibrium with a larger W. The tendency of systems to move in the direction of greatest choice as characterized by entropy is important throughout the biological universe.

At this juncture, with a good understanding of the first and second laws, attention will be focused on what is perhaps the root equation and concept for the study of biological physics, the concept of free energy and the total differential equation that represents it.

CHAPTER 5

Free Energy

5.1. The Gibbs Free Energy

While the analysis so far has treated the first and second laws separately, obviously in a real system both energy conservation and entropy play a role. Furthermore, the role of entropy in determining the spontaneous direction of a system was derived for an isolated system that exchanged neither energy nor material with its surroundings. There are few biological systems of interest that fulfill the requirements of this isolation. What is necessary then for dealing with biological systems is a state variable that will indicate the direction and equilibrium position of a system that undergoes radical changes in energy (usually measured as enthalpy) and entropy together. Such a state function has been derived and is the **Gibbs free energy, *G*.**

The Gibbs free energy is applicable to systems that are constrained to be at constant temperature and pressure and therefore is ideally suited to biological systems. G is an extensive variable that relates enthalpy, entropy, and temperature in such fashion that the spontaneity of a system may be determined:

$$G = H - TS \qquad (5.1\text{-}1)$$

If ΔG for a change in a system is negative, then that change can occur spontaneously. Conversely, a change in state with a positive ΔG is associated with a nonspontaneous process. It follows therefore that when ΔG has a value of 0, the system is at equilibrium. Like enthalpy and energy, G has units of joules in the SI system. It is worth emphasizing that ΔG depends only on the initial and final states of the system under study and that it does not provide information about the rate or kinetics of the change.

The point of having a variable such as G is to determine the "natural" direction or spontaneous direction of a system undergoing a change toward equilibrium. Differentiating Equation (5.1-1) gives

$$dG = dH - T\,dS - S\,dT \qquad (5.1\text{-}2)$$

The restrictions of constant temperature ($dT = 0$) and pressure (which gives $dH = đq$) are now applied:

$$dG = đq - T\,dS \qquad (5.1\text{-}3)$$

Entropy can be defined as $dS = đq_{rev}/T$, and this gives the interesting result

$$dG = đq - đq_{rev} \qquad (5.1\text{-}4)$$

Now if

$$đq = đq_{rev} \qquad (5.1\text{-}5)$$

then $dG = 0$ (at constant temperature and pressure).

In the last chapter, a reversible process was defined as one that is at equilibrium, so this solution fulfills the criterion that equilibrium is defined when ΔG is 0. For the other case, in an irreversible process:

$$đq < đq_{rev} \text{ so } dG < 0 \qquad (5.1\text{-}6)$$

(See Appendix 5-1 for the derivation of $đq < đq_{rev}$.)

This result also makes sense since any real or natural process has already been demonstrated to be an irreversible one, and a negative value for ΔG is expected for such a case.

5.2. A Moment of Retrospection Before Pushing On

The relationships derived for the Gibbs free energy are general and applicable for any thermodynamic system appropriately constrained by temperature and pressure. The analysis has advanced enough at this point so that it is worth stating explicitly that all of the derivations so far have been for "ideal" systems. Very shortly, the real and "nonideal" world will necessarily be confronted and equations will need to be derived for this. It will turn out that the forms of the simple equations derived so far can be kept even though the forces operating in nonideal systems are more complex than those in the ideal systems.

It is worth pointing out that thermodynamic systems have many functions that can be considered to define the state. In the system considered so far, the state has been adequately defined by the pressure (P), temperature (T), and volume (V), which have allowed the determination of the fundamental properties of energy (U) and entropy (S) as defined by the laws of thermodynamics. By taking into account various constraints, two useful, derived state variables have been defined, namely, enthalpy (H) and Gibbs free energy (G). Other derived variables exist, but will not be considered here, such as A, the Helmholtz free energy. For the most part, these state functions will be adequate for the remaining discussion, though some further permutation is to be expected.

The task lying ahead is to examine the properties of the Gibbs free energy and then to relate this function to various conditions and constraints in systems of biological interest. Remember that the free energy of a system indicates the tendency of the system to move spontaneously in a particular direction. It in no way provides information on the rate of the process or its mechanism. Furthermore, while the free energy provides a great deal of information about the predicted equilibrium state of a system, it is reasonable to make the broad generalization that in biological systems equilibrium is never reached except at death.

5.3. The Properties of the Gibbs Free Energy

In the formulation of the Gibbs free energy, both temperature and pressure were constrained to be constant. While such a simplification is useful in many cases, there are also many situations where it is valuable to know what the free energy of a system will be when changes of either temperature or pressure occur. A very practical reason to know such a relationship lies in the simple fact that standard tables of free energy (as well as enthalpy and entropy) are written in the standard state. Clearly, if a scientist is interested in a system at 310 K and has tables that provide free energies in a standard state at 298 K, then to be able to use the tabulated values, the temperature dependence of G needs to be considered and calculated. Beyond this practical (and time-saving) reasoning, there are many questions in biological systems that need to consider altered pressures and temperature. For example, what effect does the increased temperature of a fever have on the spontaneity and equilibrium of the steps in a metabolic process? How does an increase in temperature affect the free energy of a conformational change in a protein or a lipid bilayer? How about the effect on metabolic processes when the temperature drops as in a hibernating bear or in a cold water immersion (children for example have been "drowned" for greater parts of an hour in icy water, appeared clinically dead, and been brought back to life with apparently no serious side effects)? The role of pressure in a biological system comes into play when considering gas exchange in the lung and tissues, both events that do not occur at standard or constant pressures. The **oncotic** pressure of the body fluids is crucial for the normal retention of intravascular water in the high hydrodynamic pressure environment of the bloodstream; loss of protein in protein-wasting nephropathies can lead to drastic pathologies. What, then, are the properties of G, related to the variables of state so far considered?

The problem can be approached by starting with the definition of G:

$$G = H - TS \qquad (5.3\text{-}1)$$

This is then differentiated, which gives the following, the fundamental equation:

$$dG = dH - T\,dS - S\,dT \tag{5.3-2}$$

A number of substitutions can be made. First, because $dH = dU + P\,dV + V\,dP$, the following is written:

$$dG = dU + P\,dV + V\,dP - T\,dS - S\,dT \tag{5.3-3}$$

By the definition of the first law, dU can be written $đq + đw$ so

$$dG = đq + đw + P\,dV + V\,dP - T\,dS - S\,dT \tag{5.3-4}$$

For a reversible process, it can be shown that $đ = T\,dS$, and it can be assumed that $đw$ is the PV work or $-P\,dV$, if only pressure–volume work is performed. Substituting then gives

$$dG = T\,dS - P\,dV + P\,dV + V\,dP - T\,dS - S\,dT \tag{5.3-5}$$

Combining the terms gives the following result:

$$dG = V\,dP - S\,dT \tag{5.3-6}$$

This formulation now provides the dependence of G on temperature and pressure.

Equation (5.3-6) is appropriate when only pressure–volume work is being considered. As previously stated, much of the work done in or by a cell or organism takes forms other than PV work. Therefore, a new term, w_{other}, can be added and defined as needed in various systems:

$$dG = V\,dP - S\,dT + dw_{\text{other}} \tag{5.3-7}$$

Again, considering Equation (5.3-6), an equivalent equation can be written using partial derivatives:

$$dG = \left(\frac{\partial G}{\partial P}\right)_T dP + \left(\frac{\partial G}{\partial T}\right)_P dT \tag{5.3-8}$$

This equation also relates the change in free energy to changes in pressure at constant temperature and to changes in temperature at constant pressure. An important result of having both Equations (5.3-7) and (5.3-8) is that the following identities are revealed:

$$\left(\frac{\partial G}{\partial P}\right)_T = V \tag{5.3-9}$$

$$\left(\frac{\partial G}{\partial T}\right)_P = -S \tag{5.3-10}$$

Remember that the trick to making effective use of partial differentials is to identify their physical meaning. In the case of Equation (5.3-9), volume is always positive, so the free energy of a system at constant temperature always increases as the pressure increases. On the other hand, Equation (5.3-10) indicates that the free energy of a system decreases as temperature increases in a system at constant pressure. This result occurs because entropy is always positive. Equation (5.3-10) provides the answer to the problem of finding the temperature dependence of free energy. A number of further manipulations of Equation (5.3-10) are possible, the use of each depending on whether or not it is useful for a particular problem. One of the most useful expressions is the **Gibbs–Helmholtz equation**, which is used in situations where enthalpy or entropy varies with temperature:

$$\left(\frac{\partial(G/T)}{\partial T}\right)_P = -\frac{H}{T^2} \tag{5.3-11}$$

For the effect on G of a change from T_1 to T_2, the following version of Equation (5.3-11) is of value:

$$\frac{\Delta G_{(T_2)}}{T_2} - \frac{\Delta G_{(T_1)}}{T_1} = -\int_{T_1}^{T_2} \frac{\Delta H_{(T)}}{T^2} dT \tag{5.3-12}$$

In many biological situations, the temperature dependence of entropy or enthalpy, and hence of the free energy, is minimal. In such cases, usually when the temperature difference between states is not very great, the difference can be ignored, and the following approximation can be used:

$$\frac{\Delta G_{(T)}}{T} - \frac{\Delta G^{\circ}}{298} = \left(\frac{1}{T} - \frac{1}{298}\right)\Delta H^{\circ} \tag{5.3-13}$$

The pressure dependence of G depends greatly on whether the system contains only liquids and solids or gases. This can be shown easily if it is considered that ΔG for any pure material is simply the integral of Equation (5.3-6) from the standard state (1 atm) at constant temperature (state 1 in Equation 5.3-12). If $dT = 0$, then Equation (5.3-6) becomes

$$dG = V\,dP \tag{5.3-14}$$

Integrating gives:

$$\int_{P_1}^{P_2} dG = \int_{P_1}^{P_2} V d\,P \tag{5.3-15}$$

Since state 1 is defined here as G°, the following can be written:

$$G - G^{\circ} = \int_{P_1}^{P_2} V\,dP \tag{5.3-16}$$

Again, using the identity for an ideal gas that $V = nRT/P$:

$$G - G^{\circ} = \int_{P_1}^{P_2} \frac{nRT}{P}\, dP \tag{5.3-17}$$

which evaluates to

$$G - G^{\circ} = nRT \ln \frac{P_2}{P_1} \tag{5.3-18}$$

and finally

$$G = G^{\circ} + nRT \ln \frac{P_2}{P_1} \tag{5.3-19}$$

This is the result for any gas that can be treated as an ideal gas.

For solids or liquids however, the volume change with pressure under all but the most extreme circumstances is negligible, and volume can be removed from the integral in Equation (5.3-16). This provides the following:

$$G = G^{\circ} + V(P - 1) \tag{5.3-20}$$

However, on a molar basis, the volume of condensed phases is generally very small and so the expression $V(P - 1)$ is quite small and the following equation can be written:

$$G = G^{\circ} \tag{5.3-21}$$

This says that, for condensed phases, G depends on temperature alone and not significantly on pressure.

5.4. Introduction of μ, the Free Energy per Mole

In Equation (5.3-17), the use of the ideal gas law led to the introduction of the mole quantity. This is an important consequence for questions that involve chemical processes, as most questions in biological science do. The relationship of the free energy on a per mole basis is so important that a special symbol, μ, is used to denote the relationship:

$$\mu_i = \left(\frac{\partial G}{\partial n_i}\right)_{T,P,n_{j \neq i}} \tag{5.4-1}$$

Here the term n_i represents the concentration of the component under consideration and n_j represents all other components in the system. Substituting μ into Equation (5.3-19), results in an expression describing the **molar free energy**

$$\mu = \mu^{\circ} + RT \ln \frac{P_2}{P_1} \tag{5.4-2}$$

Since P_1 has already been defined as the standard state where $P_1 = 1$ atm, this equation becomes

$$\mu = \mu^{o} + RT \ln P \qquad (5.4\text{-}3)$$

5.5. Transforming the General Ideal Equation to a General Real Equation

So far, the point has been made and emphasized repeatedly that the equations being derived are general for any ideal substance. The point of fact is that few materials behave in an ideal manner, and especially, few systems that involve water can be described in this ideal manner. It would seem that the efforts of the last chapters have been an exercise in futility. Fortunately, this is not the case. For the case of the ideal gas described by Equations (5.4-3) and (5.3-19), the free energy is related to its pressure. To continue to use the same general mathematics, which are simple and elegant, all that needs to be done is to invent a new state function that is related to the free energy of a real gas or any real substance. In the case of the gas, the new state function is the **fugacity**, ***f***, and substituting it into Equation (5.4-3) gives the following:

$$\mu = \mu^{o} + RT \ln f \qquad (5.5\text{-}1)$$

It is obviously important that this new function, f, can be related to some real measurable property, and it turns out that in the case of gases it can be. For the curious reader, a further discussion about fugacity is available in standard physical chemistry texts.

In the next sections, it will be shown that the free energy in an ideal solution is a function of concentration and concentration can be treated just as pressure was for the ideal gas. Consequently, when the move is made from an ideal to a real solution, inventing a new state function will allow the same mathematics to be used for ideal gases and solutions and real gases and solutions. The new function in the case of the solution will be called the **activity** and given the symbol ***a***.

The fortunate aspect of the ability to define a new state function as necessary is the powerful result that the relationships developed to this point are general and may be used in specific cases as long as the appropriate conditions and constraints are met and defined. It is now time to expand consideration toward aspects of systems that are relevant to biological study. The first expansion will be to consider the effects of a change in the components of a system as a change in state occurs.

Appendix 5.1. Derivation of the Statement, $q_{rev} > q_{irrev}$

In Section 4.2, it was demonstrated that for a process producing work in the surroundings, the maximal work can be obtained from a reversible process only, allowing the following to be written:

$$w_{rev} > w_{irrev} \quad \text{(A5.1-1)}$$

Consider a change in state carried out first reversibly, and then subsequently irreversibly. By the first law:

$$\Delta U = q_{rev} + w_{rev} \quad \text{and} \quad \Delta U = q_{irrev} + w_{irrev} \quad \text{(A5.1-2)}$$

However, since work is done on the surroundings, w_{rev} and w_{irrev} are given a negative sign, hence

$$\Delta U = q_{rev} + (-w_{rev}) \quad \text{and} \quad \Delta U = q_{irrev} + (-w_{irrev}) \quad \text{(A5.1-3)}$$

In both cases, the change in state is identical, so these equations can be equated:

$$q_{rev} - w_{rev} = q_{irrev} - w_{irrev} \quad \text{(A5.1-4)}$$

which is the same as writing

$$q_{rev} - q_{irrev} = w_{rev} - w_{irrev} \quad \text{(A5.1-5)}$$

But

$$w_{rev} > w_{irrev} \quad \text{(A5.1-1)}$$

so

$$q_{rev} - q_{irrev} > 0 \quad \text{(A5.1-6)}$$

This result leads to

$$q_{rev} > q_{irrev} \quad \text{(A5.1-7)}$$

which is the solution used in Equation (5.1-6).

CHAPTER 6

Multiple-Component Systems

6.1. New Systems, More Components

So far, the systems that have been considered have been implicitly defined as having a single component. In the study of biological systems, this is not a particularly useful simplification. Not only are most of the systems of concern comprised of a number of different materials but frequently the amounts of these materials change as a system changes state because of chemical reactions that take place. Therefore, it now becomes necessary to consider explicitly the various components that make up a system and what their relationship will be to the free energy of the system.

For a pure substance, the free energy has been defined as

$$dG = V\,dP - S\,dT \tag{5.3-6}$$

Now, consider a system that is comprised of a number of components present in different quantities, i.e., $n_1, n_2, n_3, \ldots, n_i$. It is already apparent from the general case described in Equation (5.4-1) that there is a dependence of G on the mole quantity. Therefore, for this expanded system of several components, what is of interest is G in terms of $G(T,P,n_1,n_2,n_3,\ldots n_i)$. The total differential is now

$$dG = \left(\frac{\partial G}{\partial T}\right)_{P,\,n_i} dT + \left(\frac{\partial G}{\partial P}\right)_{T,\,n_i} dP + \left(\frac{\partial G}{\partial n_1}\right)_{P,n_j \neq n_i} dn_1 + \left(\frac{\partial G}{\partial n_2}\right)_{P,\,n_j \neq n_i} dn_2 + \ldots \tag{6.1-1}$$

where n_i represents all components, and n_j represents all components except the one under consideration. This equation is simply an expanded and therefore more general case but simplifies if appropriately constrained; i.e., if $dn_1 = 0$ and $dn_2 = 0$, etc., then the following familiar equation results:

$$dG = \left(\frac{\partial G}{\partial T}\right)_{P,\,n_i} dT + \left(\frac{\partial G}{\partial P}\right)_{T,\,n_i} dP \tag{6.1-2}$$

This equation looks similar to Equation (5.3-8) except that it applies to the case where the system contains different components provided the number of moles of each, n_i, is kept constant.

Equation (5.4-1) defined the molar free energy, μ. This is an expression that is similar to the new terms for the total differential equation in Equation (6.1-1), so that, as in the identities established for $-S$ and V, the following will be defined:

$$\left(\frac{\partial G}{\partial n_i}\right)_{P,T,\ n_j} = \mu_i \tag{6.1-3}$$

Now, rewriting the total differential of $G(T,P,n_1,n_2,n_3,\ldots,n_i)$

$$dG = V\,dP - S\,dT + \mu_1\,dn_1 + \mu_2\,dn_2 + \ldots \tag{6.1-4}$$

is the same as writing

$$dG = V\,dP - S\,dT + \sum_i \mu_i dn_i \tag{6.1-5}$$

What has been accomplished here is significant. A relationship between the free energy of a system and the components that make it up has been described and given mathematical form. This quantity, μ, is called the **chemical potential,** and the property μ_i defines the chemical potential of substance i.

6.2. Chemical Potential and Chemical Systems

Chemical potential defines the property of matter that leads to the flow of a material from a region of high potential to a region of low potential just as gravity causes the movement of a mass from a region of high potential to one of lower potential. Looking at the partial derivative for μ:

$$\left(\frac{\partial G}{\partial n_i}\right)_{P,T,n_j} = \mu_i \tag{6.1-3}$$

makes it clear that, as the molar quantity of an added material to a system is increased, the free energy is also increased. Following the addition, the material will seek to be distributed equally throughout the mixture so that no part of the volume that contains the mixture will be at a higher chemical potential than any other part. This is a very important quality of the equilibrium condition when dealing with chemical potential. **At equilibrium, μ_i must have the same value throughout the system.** From the discussion earlier in Section 2.2, it is found that intensive variables are uniform throughout a system. Therefore, it is to be expected that μ_i would have the same value everywhere in the system at equilibrium.

Consider now the final equation in the previous chapter:

$$\mu = \mu^{\circ} + RT \ln f \tag{5.5-1}$$

which was derived from the ideal statement

$$\mu = \mu^{\circ} + RT \ln P \tag{5.4-3}$$

Because of the uniform properties of μ at equilibrium, it follows that the pressure and its related quantity fugacity are also uniform throughout the system at equilibrium. If a system of various (gaseous) components is considered, the chemical potential of each component will be given by

$$\mu_i = \mu_i^{\circ}(T) + RT \ln P_i \tag{6.2-1}$$

where $\mu_i^{\circ}(T)$ is the chemical potential for a pure gas at temperature T and one atmosphere of pressure. P_i is equivalent to writing the mole fraction of substance i at pressure P, i.e., $P_i = X_i P$. By substituting this identity for P_i and algebraically manipulating Equation (6.1-1), a generalized relation can be found:

$$\mu_i = \mu_i^{\circ}(T,P) + RT \ln X_i \tag{6.2-2}$$

This equation provides the result that a pure substance ($\mu_{i(T,P)}^{\circ}$) in the same form (gas, liquid, or solid) as a mixture will flow spontaneously into the mixture. This provides the argument that different gases, liquids, and solids will diffuse one into the other.

6.2.1. Characteristics of μ

Several qualities of the chemical potential are worth highlighting. Consider a system in which both temperature and pressure are kept constant. Under these conditions, the free energy of the system will be determined solely by its composition. The following can be written:

$$G = \sum_i n_i \mu_i \tag{6.2-3}$$

This characterizes the **additive** nature of the chemical potential. In the appropriate system at specified temperature and pressure, the free energy of the system can be computed if the molar quantities of the components are known.

Another important relationship of the chemical potential is derived from Equation (6.2-3). If this equation is differentiated, the following result is obtained:

$$dG = \sum_i n_i \, d\mu_i + \sum_i \mu_i \, dn_i \tag{6.2-4}$$

However, by Equation (6.1-5), the following identity can be written:

$$\sum_i n_i \, d\mu_i + \sum_i \mu_i \, dn_i = V\,dP - S\,dT + \sum_i \mu_i \, dn_i \tag{6.2-5}$$

Solving the algebra yields

$$\sum_i n_i \, d\mu_i = V\,dP - S\,dT \tag{6.2-6}$$

This is the **Gibbs–Duhem** equation, a special case of which may be written when pressure and temperature are constrained to be constant. Then, Equation (6.2-6) reduces to

$$\sum_i n_i \, d\mu_i = 0 \tag{6.2-7}$$

at constant T, P. This result makes it clear that there is a dependent relationship between the components making up a system. If a system is made up of several constituents and the chemical potential of one of them changes, then the others must also change coincidentally in some fashion, since the sum of the chemical potentials must remain zero.

6.2.2. An Immediate Biological Relevance of the Chemical Potential

Because the chemical potential seeks to be uniform throughout a system, an important result is found when equilibrium is prevented. Consider the effect of a barrier, such as a permeable membrane, on a chemical species whose μ on the left side is not equal to that on the right side. In this situation, there will be a force measured by ΔG that exists across the membrane. This force will appear as a driving force and will have the ability to perform work in the system. This is an extremely important consequence, as shall be seen in later chapters, of the state function, μ.

6.3. The Entropy and Enthalpy and Free Energy of Mixing

The discussion so far has really been about the free energy change when one substance is mixed with another. Free energy however is the result of a relationship between the entropy and enthalpy of a system. It will be of value to examine the constituent parts of the **free energy of mixing,** the entropy and enthalpy of mixing, while considering the free energy of mixing. In this derivation, the model system will be a container that contains two pure substances that, in the initial state, are each separately found at constant T under pressure P. In the final state, the substances will be at the same T and the same P as found initially, but mixed together.

For the pure substances, $G_1 = n_1\mu_1^\circ$ and $G_2 = n_2\mu_2^\circ$. Therefore, the free energy of the initial state will be found by

$$G_{\text{initial}} = G_1 + G_2 = n_1\mu_1^\circ + n_2\mu_2^\circ = \sum_i n_i\mu_i^\circ \tag{6.3-1}$$

In the final state, Equation (6.2-3) will provide the value of G by adding together each part of $n_i\mu_i$, giving

$$G_{\text{final}} = \sum_i n_i\mu_i \tag{6.3-2}$$

ΔG_{mix} is equal simply to the initial value of G subtracted from the final value of G:

$$\Delta G_{mix} = \sum_i n_i (\mu_i - \mu_i^{\circ}) \tag{6.3-3}$$

By Equation (6.2-2), the term $\mu_i - \mu_i^{\circ}$ can be evaluated and substituted:

$$\Delta G_{mix} = RT\,(n_1 \ln X_1 + n_2 \ln X_2) = \sum_i RTn_i \ln X_i \tag{6.3-4}$$

Sometimes, this is more conveniently written by factoring out n_i by the relationship $n_i = x_i N$, where N is the total number of moles in the mixture and X_i is the mole fraction of component i. This gives:

$$\Delta G_{mix} = NRT \sum_i X_i \ln X_i \tag{6.3-5}$$

Solving this equation for a two-component mixture would show that the maximal free energy decrease occurs when equal amounts of two components are added to make the final mixture. In the case of four components, the maximal decrease will be for the case when the components each represent one fourth of the final mixture, and so forth.

The **entropy of mixing** is easily found if ΔG is differentiated with respect to temperature:

$$\left(\frac{\partial \Delta G_{mix}}{\partial T}\right)_{P,\,n_i} = -\Delta S \tag{6.3-6}$$

Thus, differentiating Equation (6.3-5) with respect to temperature:

$$\left(\frac{\partial \Delta G_{mix}}{\partial T}\right)_{P,\,n_i} = NR \sum_i X_i \ln X_i \tag{6.3-7}$$

will give

$$\Delta S = -NR \sum_i X_i \ln X_i \tag{6.3-8}$$

Since $X_i < 1$, this solution dictates that the entropy of mixing is always positive and the free energy of mixing is always negative. The argument that entropy always increases with mixing makes sense in the statistical sense since there is far more choice in the mixed system, where many arrangements of different molecules look identical, compared to the initial state, where there is only one possible arrangement of the different molecules. Like the solution of Equation (6.3-5), the entropy of mixing is greater for a binary mixture if the components are each equal to half of the final mixture, and so on.

Finally, the **heat of mixing** can be found most easily simply by solving the fundamental equation

$$\Delta G_{\text{mix}} = \Delta H_{\text{mix}} - T\Delta S_{\text{mix}} \tag{6.3-9}$$

Using the work already completed provides the result that

$$NRT \sum_i X_i \ln X_i = \Delta H_{\text{mix}} - T(-NR \sum_i X_i \ln X_i) \tag{6.3-10}$$

which reduces to

$$NRT \sum_i X_i \ln X_i = \Delta H_{\text{mix}} + NRT \sum_i X_i \ln X_i \tag{6.3-11}$$

and finally

$$\Delta H_{\text{mix}} = 0 \tag{6.3-12}$$

Therefore, for an ideal mixture, there is no heat associated with its formation. This holds true for any components of the ideal mixture whether the components are ions, atoms, or macromolecules. But what about a nonideal mixture?

To answer this question, the following is considered. First, rewrite the fundamental equation for the free energy of mixing using the results to this point:

$$-\Delta G = T\Delta S \tag{6.3-13}$$

The spontaneous process of mixing is wholly driven by entropy according to Equation (6.3-13). If Equation (6.3-8) is solved for its maximum (using a binary mixture, let $x = \frac{1}{2}$), a value of approximately 1.377 eu is the result. For a biological system at 310 K, this gives a value for ΔG of -102.12 J. This is not a large number, and in nonideal mixtures if the heat of mixing is more than a little positive, the two solutions will remain separate in immiscible layers.

6.4. Free Energy When Components Change Concentration

To this point, the discussion has focused on the question of mixing pure substances together. What happens when the substances in the initial state react to form different components in the final state. In other words, how does free energy relate to chemical reactions? Consider a simple chemical reaction:

$$a\text{A} + b\text{B} \rightarrow c\text{C} + d\text{D} \tag{6.4-1}$$

Because $\Delta G = G_{\text{final}} - G_{\text{initial}}$ or $\Delta G = G_{\text{products}} - G_{\text{reactants}}$, the following may be written:

$$\Delta G = (cG_{\text{C}} + dG_{\text{D}}) - (aG_{\text{A}} + bG_{\text{B}}) \tag{6.4-2}$$

This is the familiar statement of **Hess's law** that has been mentioned. It will be useful to be able to write ΔG in terms of a change in product–reactant behavior as was done earlier:

$$G = G^{\circ} + nRT \ln \frac{P_2}{P_1} \tag{5.3-19}$$

A look at Equation (5.3-19), however, finds the equation in the form for an ideal gas. This is not a problem since in fact this equation is the general form for any ideal substance, whether gas, liquid, or solid. Since the majority of the cases of interest for the remainder of this book will deal with solutions, it is a worthwhile side trip to derive a general form of Equation (5.3-19) now so that it may be used henceforth.

6.4.1. A Side Trip: Derivation of a General Term, the Activity

In the discussion of the properties of the chemical potential for a given species, μ_i, it was noted that at equilibrium, μ_i would be equal throughout a mixture. In the case of an ideal gas, this dictated that the partial pressure exerted by a component of the mixture would therefore be equal everywhere at equilibrium and that it would represent the free energy directly. In the case of a nonideal or real gas, the pressure could be correctly substituted with the term fugacity and the same relationship between the intensive property μ and the fugacity would exist as did for pressure and μ. The general form of the equation will hold true for any system, ideal or real, if a term is introduced that will have the same properties with respect to μ as does pressure or fugacity. This new quantity is called the **activity, *a***. If activity is substituted into Equation (5.4-3), which, it should be recalled, was derived from Equation (5.3-19) directly,

$$\mu_i = \mu_i^{\circ} + RT \ln P_i \tag{5.4-3}$$

the following new equation is the result:

$$\mu_i = \mu_i^{\circ} + RT \ln a_i \tag{6.4-3}$$

This is clearly the molar free energy for a single component, i, in a system and, because of the identity of μ, may be rewritten as follows:

$$\overline{G}_i = \overline{G}_i^{\circ} + RT \ln a_i \tag{6.4-4}$$

where $\overline{G}_i$ is the free energy per mole of component i in a mixture, $\overline{G}_i^{\circ}$ standard free energy of a mole of component i, and a_i is the activity of component i in the mixture.

6.4.2. Activity of the Standard State

The activity of a substance in its standard state is easily found through the use of Equation (6.4-4). This equation shows that the activity of a component is related to its standard state in the following manner:

$$\ln a_i = \frac{\overline{G}_i - \overline{G}_i^{\circ}}{RT} \tag{6.4-5}$$

Since ln 1 is zero, it follows that the activity of a material in its standard state is always equal to unity. This presents the question, What are the standard states for substances? For an ideal gas, the standard state is defined as a gas with a partial pressure equal to one. This is consistent with Equation (5.3-19). For a real gas, the fugacity is given by a coefficient of activity, γ, multiplied times the partial pressure of the gas:

$$f = \gamma_i P_i \tag{6.4-6}$$

At pressures approaching zero, γ becomes equal to unity, and it is for this reason that in low-pressure systems a real gas may be treated as an ideal gas, so that pressure directly indicates the fugacity. For a pure solid or liquid, the activity is unity at one atmosphere of pressure. Because the free energy of a solid has been shown to be essentially independent of pressure (Equations 5.3-20 and 5.3-21) it is generally correct to write that at any pressure the activity of a pure solid or liquid is unity.

The standard state of a solution is a little more difficult to define than that of a pure substance. Since in Part II a great deal of attention will be focused on solutions, let it be sufficient at this point to indicate that the activity of a solution depends on an activity coefficient and the concentration of the component in the solution. One of the problems in defining a standard state arises from the inclusion of the concentration. There are quite a few methods of measuring concentration in solutions, and hence confusion can be easily generated. As an example of the general form for the activity of a component, *i*, of a solution, the specific formula for the mole fraction and the solvent standard state will be written:

$$a_i = \gamma X_i \tag{6.4-7}$$

Here γ is the activity coefficient for this standard state, and X_i is the mole fraction of component *i*. Mole fractions are a convenient measure of concentration and represent the number of moles of component *i* in a mixture divided by the total number of moles found in the mixture. As the value of X_i approaches unity, the number of moles of component *i* becomes closer and closer to the total number of moles that comprise the entire system. The activity coefficient is such that when the mole fraction X_i is equal to unity, the activity of component *i* in the solution is also unity. This is a particularly valuable quality, since when a solution is a single pure component—in other words, the initial state of the solvent itself—it will have an activity of one. This result is in agreement with the result for the standard state of a pure liquid already given.

6.4.3. Returning to the Problem at Hand

Now, the problem of defining the free energy for Equation (6.4-2) can be addressed. Using Equation (6.4-4), which defines G in terms of activity, the following can be written:

$$\Delta G = (c\overline{G}^{\circ}_{C} + d\overline{G}^{\circ}_{D}) - (a\overline{G}^{\circ}_{A} + b\overline{G}^{\circ}_{B}) + (cRT \ln a_{C} + dRT \ln a_{D}) - (a\,RT \ln a_{A} + b\,RT \ln a_{B}) \tag{6.4-8}$$

The terms can be collected and Equation (6.4-8) can be written in the more convenient form

$$\Delta G = \Delta G^{\circ} + RT \ln \frac{(a_{C})^{c}\,(a_{D})^{d}}{(a_{A})^{a}\,(a_{B})^{b}} \tag{6.4-9}$$

This equation relates the free energy change found when reactants and products with any arbitrary value of activities are mixed together or are formed by a chemical reaction. The change in G is taken relative to the standard state and so the symbol G° is used. The ratio of activities in Equation (6.4-8) is called Q:

$$Q = \frac{(a_{C})^{c}\,(a_{D})^{d})}{(a_{A})^{a}\,(a_{B})^{b}} \tag{6.4-10}$$

Q is the ratio of the activities of the components each raised to the power of its coefficient in the chemical reaction. Changing mole numbers arithmetically will change Q geometrically; that is, doubling the mole number will square Q. When the product activities are large relative to the reactant activities, Q will be large and its contribution to the free energy of the reaction will be positive, thus making the reaction less likely to occur spontaneously. Conversely, a small Q will result in a negative contribution to the free energy and thus will make the reaction more likely to occur. Q is not necessarily representative of a system at equilibrium, but such a case can be considered.

At equilibrium, the ratios of products and reactants are given the special designation K, the equilibrium constant:

$$K = \frac{(a_{C})^{c}\,(a_{D})^{d}}{(a_{A})^{a}\,(a_{B})^{b})} \tag{6.4-11}$$

The equilibrium constant K is a special case of Q, and in this case, $\Delta G = 0$. Using the equilibrium constant in place of Q:

$$0 = \Delta G^{\circ} + RT \ln K \tag{6.4-12}$$

which is equal to

$$\Delta G^{\circ} = -RT \ln K \tag{6.4-13}$$

Several other algebraic manipulations provide convenient formulas:

$$K = e^{-\Delta G^{\circ}/RT} \tag{6.4-14}$$

and

$$K = 10^{-\Delta G^{\circ}/2.303RT} \tag{6.4-15}$$

All of these equations are extremely valuable because they allow calculation of the standard free energy change for a reaction simply by measuring the amounts of products and reactants in a system. This is precisely the kind of information that is valuable in biochemical studies.

It is implicit in this discussion that the free energy of a system depends on the concentration of the components that make up the system. Changes in the concentration of various products and/or reactants can lead to the spontaneous progression of a reaction forward or backward. When the activity of various components is changed, the system responds by attempting to move toward a new equilibrium position. Such a response is the same as that predicted by the Le Chatelier principle.

At this point, two clear methods exist to predict the movement of a system in one direction or another depending on either the "natural" tendencies or on work being done on or by the system. The experimenter may either use the equilibrium constant to predict the free energy change or use the equation

$$\Delta G = \Delta H - T\Delta S \tag{6.4-16}$$

and solve for ΔG. Enthalpy may be obtained from either a table and Hess's law or from careful thermometry and calorimetry using $\Delta H = q$. ΔS may also be found using Hess's law or by careful thermal measurements of heat capacities on pure substances. A third and very valuable method of finding ΔG is also available and is an electrochemical method. In many cases, this method is preferred, and its derivation and use will be considered following a discussion of the temperature dependence of the equilibrium constant.

A very useful relationship allows the determination of entropy and enthalpy changes for a reaction by measuring the temperature dependence of the equilibrium constant. Consider the equilibrium constant for each of two temperatures, T_1 and T_2:

$$\ln K_{(T_1)} = \frac{-\Delta G^{\circ}_{(T_1)}}{RT_1} \tag{6.4-17}$$

and

$$\ln K_{(T_2)} = \frac{-\Delta G^{\circ}_{(T_2)}}{RT_2} \tag{6.4-18}$$

Their relationship can be found by subtracting one from the other:

$$\ln K_{(T_2)} - \ln K_{(T_1)} = \frac{-\Delta G^{\circ}_{(T_2)}}{RT_2} - \left[\frac{-\Delta G^{\circ}_{(T_1)}}{RT_1}\right] \tag{6.4-19}$$

This equation can be rearranged so that for ln K:

$$\ln K_{(T_2)} = \ln K_{(T_1)} - \left[\frac{-\Delta G^{\circ}_{(T_2)}}{RT_2} - \frac{\Delta G^{\circ}_{(T_1)}}{RT_1}\right] \tag{6.4-20}$$

Using $\Delta G = \Delta H - T\Delta S$ and rewriting Equation (6.4-20) in terms of entropy and enthalpy:

$$\ln K_{(T_2)} = \ln K_{(T_1)} - \left[\frac{-\Delta H^{\circ}_{(T_2)}}{RT_2} - \frac{\Delta H^{\circ}_{(T_1)}}{RT_1} - \frac{-\Delta S^{\circ}_{(T_2)}}{R} - \frac{\Delta S^{\circ}_{(T_1)}}{R}\right] \quad (6.4\text{-}21)$$

Assuming that ΔH and ΔS do not change significantly over the temperature range under consideration, Equation (6.4-21) can be simplified:

$$\ln K_{(T_2)} = \ln K_{(T_1)} - \left[\frac{\Delta H^{\circ}_{(T_1)}}{R}\left(\frac{1}{T_2} - \frac{1}{T_2}\right)\right] \quad (6.4\text{-}22)$$

$$\ln K_{(T_2)} = \ln K_{(T_1)} - \left[\frac{\Delta H^{\circ}_{(T_1)}}{RT_1T_2}\left(T_2 - T_1\right)\right] \quad (6.4\text{-}23)$$

Equation (6.4-23) indicates the shift in equilibrium with a variation in temperature. If a reaction is exothermic ($\Delta H < 0$), then an increase in temperature will favor the reactants; but if the reaction is endothermic ($\Delta H > 0$), an increase in temperature will favor product formation.

An exact result can be found for cases when the approximations leading to Equation (6.4-22) are not valid by using the Gibbs–Helmholtz equation (Equation 5.3-11):

$$\left(\frac{\partial(G/T)}{\partial T}\right)_P = -\frac{H}{T^2} \quad (5.3\text{-}11)$$

which gives the van't Hoff equation:

$$\frac{d \ln K}{dT} = \frac{-\Delta H^{\circ}_{(T)}}{RT^2} \quad (6.4\text{-}24)$$

Graphing ln K versus $1/T$ gives a slope of $-\Delta H^{\circ}/R$ Therefore, by measuring equilibrium concentrations over a range of temperatures and using the van't Hoff equation, the enthalpy of a reaction may be found without resorting to calorimetry.

6.5. The Thermodynamics of Galvanic Cells

A system consisting of an electrolyte and two electrodes, with the two electrodes connected together through an external electric circuit, is called a **galvanic cell**. When a chemical process occurs in the cell that causes electricity to flow, the flow of electricity can be sensed by a voltmeter placed in the external circuit of the cell. The potential that is established depends directly on the free energy change of the process. If appropriate electrodes are chosen that are sensitive to a given process under study, a galvanic cell can be used to determine ΔG at equilibrium. A galvanic cell can be used in the study of, for example, oxidation–reduction reactions, concentration differences of a species separated by a membrane, formation of complex ions, acid–base dissociations, and solubilities of salts. For a galvanic cell to measure ΔG, the cell must be acting reversibly; that is, the application of an infinitesimal potential across the cell will move the reaction equally in one direction or another. Stated another way, if the resistance of the external measuring circuit is ∞, no current will flow in the external circuit. The voltage measured between the two electrodes then represents the free energy change exactly. All of the energy resulting from such a system is theoretically available to do work, and, as described earlier (cf. Section 2.3), electrical work is given by the product of charge and potential:

$$w_{\text{electrical}} = -QE = -QE \tag{6.5-1}$$

This equation can be rewritten as

$$w = -nF\epsilon \tag{6.5-2}$$

where ϵ is equal to E in volts, and Q is written as the product of the electrical equivalents per mole, n, and the faraday, F. The faraday, F, is the charge associated with one mole of electrons or 96,487 C mol^{-1}. The sign is negative because the reversible work is done by the system on the surroundings and reduces the free energy of the system. Therefore, for a system at constant temperature and pressure:

$$\Delta G = -nF\epsilon \tag{6.5-3}$$

If the reaction in the cell is given by

$$a\text{A} + b\text{B} \rightarrow c\text{C} + d\text{D} \tag{6.4-1}$$

then the free energy change for the reaction is given by Equation (6.4-9):

$$\Delta G = \Delta G^{\circ} + RT \ln \frac{(a_{\text{C}})^{c}\,(a_{\text{D}})^{d}}{(a_{\text{A}})^{a}\,(a_{\text{B}})^{b}} \tag{6.4-8}$$

If Equations (6.4-9) and (6.5-3) are equated, then the following equation results:

$$-n\text{F}\epsilon = \Delta G^{\circ} + RT \ln \frac{(a_{\text{C}})^{c}\,(a_{\text{D}})^{d}}{(a_{\text{A}})^{a}\,(a_{\text{B}})^{b})} \tag{6.5-4}$$

Dividing by $-nF$ gives the potential for any reaction occurring that is not in the standard state:

$$\epsilon = \frac{-\Delta G^{\circ}}{nF} - \frac{RT}{nF} \ln \frac{(a_{\text{C}})^{c}\,(a_{\text{D}})^{d}}{(a_{\text{A}})^{a}\,(a_{\text{B}})^{b}} \tag{6.5-5}$$

A standard emf potential, ϵ°, for a cell in which the components are in a defined standard state can be written:

$$\epsilon^{\circ} = \frac{-\Delta G^{\circ}}{nF} \tag{6.5-6}$$

The standard state is defined when the activity of each of the components is equal to unity. The biochemist's standard state for ϵ^o is at constant pressure of 1 atm, 298 K, and pH = 7. Equation (6.5-6), is the same as the first term in Equation (6.5-5), and substitution gives

$$\epsilon = \epsilon^o - \frac{RT}{nF} \ln \frac{(a_C)^c (a_D)^d}{(a_A)^a (a_B)^b} \tag{6.5-7}$$

This is the **Nernst equation** and relates the activity of the components in the system comprising the galvanic cell to the standard state emf, ϵ^o. The Nernst equation will be the starting point for the discussion of membrane potentials in Chapter 22. ϵ^o is significant since at equilibrium ϵ^o can be written in terms of K, i.e.:

$$\epsilon^o = -\frac{RT}{nF} \ln K \tag{6.5-8}$$

which relates the standard cell emf to the equilibrium constant of the cell reaction.

Knowing the temperature dependence of the emf allows calculation of ΔH^o and ΔS^o. If Equation (6.5-3) is written for the standard state:

$$\Delta G^o = -nF\epsilon^o \tag{6.5-9}$$

and this equation is differentiated with respect to temperature, the following is the result:

$$\left(\frac{\partial \Delta G^o}{\partial T}\right)_P = -nF\left(\frac{\partial \epsilon^o}{\partial T}\right)_P \tag{6.5-10}$$

Because $(\partial \Delta G^o/\partial T)_P$ is equal to $-\Delta S^o$, the following can be written:

$$\Delta S^o = nF\left(\frac{\partial \epsilon^o}{\partial T}\right)_P \tag{6.5-11}$$

Using the relation for enthalpy, $\Delta H^o = \Delta G^o + T\Delta S^o$, the enthalpy, ΔH^o, can be found by

$$\Delta H^o = -nF\epsilon^o + nFT\left(\frac{\partial \epsilon^o}{\partial T}\right)_P \tag{6.5-12}$$

The activity of a component can also be measured directly with an electrode system in terms of the mean ionic activity, $a_\pm$, and mean ionic molality, $m_\pm$. The activities of components are explicitly described by the Nernst equation, and if ϵ is measured at a particular concentration of an electrolyte, m, then the mean ionic activity of the electrolyte can be determined if ϵ^o is also known. The concept of the mean ionic activity will be explained in Section 12.2.

CHAPTER 7

Phase Equilibria

7.1. Principles of Phase Equilibria

Up to this point, the systems under consideration have been homogeneous solutions. For the biological scientist, these solutions represent an extremely simplified set of conditions, those of a system that, while containing a number of constituents, contains them in a single homogeneous **phase**. (A phase is considered to be a state in which there is both chemical and physical uniformity.) This simplification has been valuable because through it general concepts have been developed and introduced. However, in a biological system, such a presumption is not realistic. The nature of biology is that of heterogeneous systems. Many of the processes in cells and living organisms involve the transfer of chemical species from one phase to another. For example, the movement of ions across a membrane such as the cell membrane or an intracellular organelle is often treated as transport between two phases, one inside and the other outside. The nature of the equilibria that can exist between phases will be the focus of this section. When different phases come in contact with each other, an interface between them occurs. This interface is a surface, and the properties of a surface are different from those of either of the phases responsible for creating it. The thermodynamic treatment of the properties of surfaces will be the subsequent focus of this section.

The second law of thermodynamics indicates that the chemical potential of each of the components is equal everywhere in any system in equilibrium. If two or more phases are in equilibrium, then the chemical potentials of each species in each phase of the system are equal. This can be written:

$$\mu_i \text{ (phase 1)} = \mu_i \text{ (phase 2)} = \ldots \qquad (7.1\text{-}1)$$

An equivalent relationship can be written for each component or species of the system. Since the chemical potential depends on the activity of each species in each phase, Equation (7.1-1) can be written as

$$\mu_i = \mu_i^{\circ} + RT \ln a_{i\,(\text{phase 1})} = \mu_i^{\circ} + RT \ln a_{i\,(\text{phase 2})} \qquad (7.1\text{-}2)$$

If the standard states of each component are equivalent, this equation expresses the important result that the activity of the species will be the same in each phase. Since the activity is defined as

$$a_i = \gamma_i [i] \tag{7.1-3}$$

where $[i]$ is the concentration of species i, this means that if the activity coefficients, γ, for the component are the same in each phase, the concentrations in each phase must also be equal. This is an important principle in biochemical research for it forms the basis of equilibrium dialysis studies (cf. Appendix 7.1). Although this principle of phase equilibrium has great practical significance in biological studies, it must be used with the greatest caution in approaching cellular problems. The activity coefficient, γ, is a formal term that represents the deviation from ideality of a particular component and is determined largely by the environment of the molecule and its interaction with that environment.

In a cellular system, the activity coefficient may reasonably be considered to always deviate from unity, and hence the activity of virtually all species will vary considerably from the ideal. Many phases in cellular systems are separated by membranes, and the resulting compartmentalization leads to environments that can differ radically. It is therefore generally difficult to presume that the activity coefficients in each phase are identical. In Part II, some of the forces that are reflected in the activity coefficient will be discussed, primarily from the viewpoint of the solute. It is crucial to recognize that a common assumption about cellular chemistry is that, due to the small size and high molar concentration of liquid water, its activity can generally be taken as unity and its concentration as unchanging. The actual activities of water in the multiple compartmentalized microenvironments of the cell are presently unknown. However, it is probably reasonable to assume that the activity coefficients of all components, including the aqueous solvent, deviate from unity.

7.1.1. Thermodynamics of Transfer Between Phases

The function and survival of a cell, and of the organisms comprised by cells are dependent to a great extent on the uptake and elimination of various molecules into and out of the intracellular environment. Inside most cells (with the exception of the red blood cell), the environment is dominated by membrane-defined subcompartments, thus generating multiple intraorganelle phases that are related to the cytoplasmic phase. It is therefore often convenient to consider the cell (or a portion thereof) as a two-phase system with a membrane separating the phases. At constant temperature and pressure, the free energy change associated with the transfer of components between the phases depends on the chemical potential of each component:

$$\Delta G = \mu_{(\text{phase 2})} - \mu_{(\text{phase 1})} \tag{7.1-4}$$

or

$$\Delta G = RT \ln \frac{a_{(\text{phase 2})}}{a_{(\text{phase 1})}} \tag{7.1-5}$$

In the case where activity coefficients are equal in the different phases, then Equation (7.1-5) may be written in terms of concentration alone. However, as discussed above, it is generally wiser to use Equation (7.1-5) and explicitly express the activity coefficient. Equation (7.1-5) is adequate for determining the free energy of transfer between phases as long as the molecule is uncharged. If a molecule or ion has an electric charge and is in the presence of an electric field (as is almost always the case in cells), then a further term must be added to the free energy expression:

$$\Delta G = RT \ln \frac{a_{(\text{phase 1})}}{a_{(\text{phase 2})}} + nF\epsilon \tag{7.1-6}$$

where n is the valence number of the ion, F is the faraday and ϵ is the potential field in volts. The potential field is expressed with reference to phase 1, and therefore the sign of ϵ is positive if phase 2 is at a higher potential than phase 1.

7.1.2. The Phase Rule

The systems described so far in this chapter have been more complex than those considered earlier, notably because they have been said to consist of several phases. The reflective reader will be thinking at this point, "Just what is a phase and how is it defined anyhow? And how does one know just how many phases are present in a system?" In the analysis necessary to answer these questions, another very important question will also be answered: What are the constraints that apply to the thermodynamics of a system comprised of multiple phases? In other words, how many degrees of freedom exist that can change the variables of state in a system that is composed of a number of components and a number of phases at equilibrium? The solution to this problem was found in the 19th century by J. Willard Gibbs, the father of chemical thermodynamics, and stands as one of the most elegant results in the whole of modern science. The **phase rule** states that the maximum number of phases that can exist at equilibrium in the absence of external fields is equal to the number of components plus two. For a one-component system such as water, if external forces such as gravitational and electric fields are ignored, three-phases may exist at equilibrium: ice, water, and steam. However, nature dictates that this three-phase equilibrium may only be achieved at the expense of the freedom to change the variables of state. If a single component such as water exists in a three-phase equilibrium, the state variables are defined; that is, the temperature and pressure of the system are constrained and may not be varied, as indicated by the result of zero obtained from the following equation:

$$F = 3 - P \tag{7.1-7}$$

where F is the number of degrees of freedom, i.e., temperature and pressure, that can be varied, and P represents the number of phases that exist in equilibrium. As will be described later in this chapter, this point is called the **triple point**. Water can exist in numerous forms of crystalline ice at various temperatures and pressures, but the phase rule says that only a total of three of these phases, including water and steam, can exist together at equilibrium.

The more general case is considered for a system of more than one component. Again ignoring external fields, the phase rule is written in its classical form:

$$F = C - P + 2 \tag{7.1-8}$$

where C represents the number of components of the system. Typically, the variables considered are the temperature, the pressure, and the chemical potential of the components. For example, if a two-component system of NaCl and water is considered, Equation (7.1-8) can be used to determine the maximum number of possible phases. The maximum number of phases occurs when the number of degrees of freedom to manipulate the system variables is minimum, i.e., $F = 0$:

$$\begin{aligned} 0 &= 2 - P + 2 \\ P &= 4 \end{aligned} \tag{7.1-9}$$

These phases might include solid NaCl, liquid electrolyte solution, ice, and water vapor. If, on the other hand, only three phases existed in equilibirum, i.e., solid NaCl, liquid electrolyte, and water vapor, a single degree of freedom would be available. Consequently, this three-phase system could exist in equilibrium at a range of temperatures or pressures. For each external force (gravitational, magnetic, electrical) imposed on the system, the constant 2 in Equation (7.1-8) is increased by one.

This discussion has answered the reflective reader's questions except for how a phase is defined or identified. This in fact is usually simple but be warned that with a little imagination it is possible to be tricked and get into thermodynamic quicksand. Gibbs himself wrote " a phase . . . is a state of matter that is uniform throughout not only in chemical composition but also in physical state." A homogeneous solution of NaCl is a single phase since it is a uniform mixture (on a reasonable time-averaged scale) of Na^+ and Cl^- ions and water molecules. Any reasonable or thermodynamic sample of the solution will accurately represent the whole solution. What are the gray areas for the application of the phase rule? Unfortunately for the biological worker, a microscopic examination of a **dispersion**, a material that is uniform on a macroscopic scale but is made up of distinct components embedded in a matrix, is such a gray area for the application of the phase rule. Another gray area is consideration of what constitutes a phase in a gravitational or electric field. If a biological membrane is considered, the description of a dispersion fits it well, and in addition the presence of strong electric fields is a well-proven fact. Care must be taken therefore when applying the phase rule in cellular biological systems.

7.2. Pure Substances and Colligative Properties

A pure substance such as water will have equilibrium relationships with its various phases, solid, liquid, and vapor. The equilibria of these phase phenomena depend on the state variables. Traditionally, the variables that are related to the phase changes of a pure substance are temperature and pressure. These relationships are important in biological systems. The approach to phase transitions of a pure substance is useful in understanding the changes in conformation of macromolecules such as proteins or DNA. The equilibrium between the phases of a pure substance is also important because it forms the basis for defining the behavior of ideal solutions, the activity of solvents, osmotic pressure, and colligative properties of solutions. It must be noted that while phase transitions are most formally described in terms of temperature and pressure variations, the phase transition will be affected if other forces are present, for example, electric or magnetic fields.

The equilibrium between phases of a pure substance is given by the **Clapeyron equation**:

$$\frac{dP}{dT} = \frac{\Delta S}{\Delta V} \tag{7.2-1}$$

This equation is derived from the condition that

$$\mu(T, P)_{\mathrm{a}} = \mu(T, P)_{\mathrm{b}} \tag{7.2-2}$$

where a and b represent two phases in equilibrium. Writing this in terms of the fundamental equation ($\mu = -S\,dT + V\,dP$) gives

$$(-S\,dT + V\,dP)_{\mathrm{a}} = (-S\,dT + V\,dP)_{\mathrm{b}} \tag{7.2-3}$$

which is

$$(S_{\mathrm{b}} - S_{\mathrm{a}})dT = (V_{\mathrm{b}} - V_{\mathrm{a}})dP \tag{7.2-4}$$

The expressions in parentheses are ΔS and ΔV respectively, and therefore Equation (7.2-4) can be written

$$\Delta S \, dT = \Delta V \, dP \tag{7.2-5}$$

Rearrangement gives Equation (7.2-1), the Clapeyron equation. The Clapeyron equation describes the relationship between the equilibrium temperature and pressure quantitatively. Phase diagrams of the equilibrium pressure against temperature can be constructed using this equation. A phase diagram for water, for example, could be constructed by applying the Clapeyron equation to the phase transitions between solid–liquid, liquid–gas, and gas–solid states.

Consider first, the biochemically relevant case of water at moderate pressure. The terms ΔS and ΔV are given by ΔS_{fusion} and ΔV_{fusion}. At the equilibrium temperature, ΔS_{fusion} is reversible and is equal to

$$\Delta S_{\text{fusion}} = \frac{\Delta H_{\text{fusion}}}{T} \tag{7.2-6}$$

Heat is always absorbed as the state changes from solid to liquid or gas ($\Delta H_{\text{fusion}} > 0$). ΔS_{fusion} is therefore positive for all substances. It is worth noting that for all substances undergoing phase transformations from solid to gas (sublimation) and liquid to gas (vaporization) as well as solid to liquid, ΔH is greater than 0 and therefore ΔS for any of these phase transitions is always positive. The term ΔV_{fusion} will be positive or negative depending on whether the density of the solid or the liquid phase is greater. For most compounds, the solid phase is slightly more dense than the liquid phase so ΔV_{fusion} is positive, and hence $\Delta S_{\text{fusion}}/\Delta V_{\text{fusion}}$, the slope of the line for the solid–liquid phase transition, is positive. In the case of water, however, the density of ice is less than that of liquid water and the slope of the line describing the phase transition between ice and water is negative (see Figure 7.2-1). The variation in density between liquid and solid is usually small, and consequently, the slope of the line described for this phase transition is generally quite steep.

Similar lines may be drawn on the graph of Figure 7.2-1 for the other phase transitions. In the case of liquid–gas phase transitions, the value of $\Delta H_{\text{vaporization}}$ is always greater than 0 since the density of a gas is less than that of a liquid in every case. The change in volume is generally quite significant, and so the slope of the line will be small compared to that of the solid–liquid transition. Finally, a line may be drawn on the phase diagram representing the equilibrium relationship between solid and gas phases. This line also will have a positive slope. As the phase diagram for water (Figure 7.2-1) shows, there are a series of equilibrium points described by the line between two phases, where ice and water vapor, ice and liquid water, and liquid water and water vapor are in equilibrium. Since the points that comprise the lines represent points of equilibrium between two phases, a single phase is represented by a point

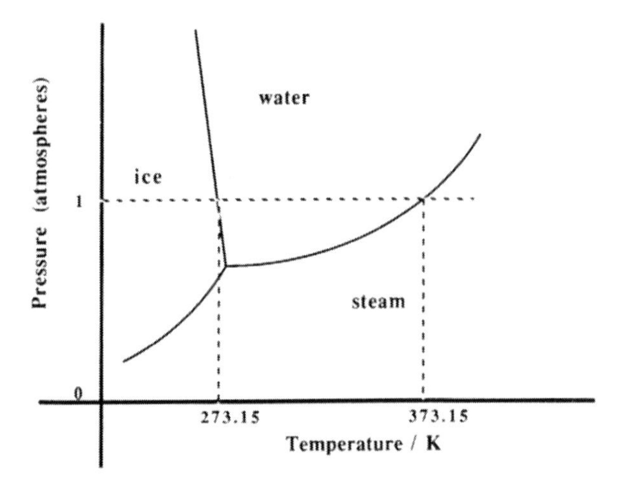

Figure 7.2-1 Phase diagram of water. A phase diagram represents the variety of states that a system may have, constrained by the variables on the axes. In this figure, the state of a system of water is defined for each x, y corrdinate representing a particular value of temperature and pressure. The line on the graph define the special case where an equilibrium exists between the various phases of the water. For example, there are numerous states defined by a particular temperature or pressure where water exists as either a gas or liquid. These states have x, y co-ordinates either above or below the line on the diagram. The systems defined at the points which fall on the line separating the gas and liquid states are the equilibrium states where both gas and liquid may coexist.

on the graph that does not fall on the line. At the intersection of the three lines, a unique state is described where all three phases of a substance may coexist. This is the **triple point.** Only one triple point can exist for a specific set of three phases. The phase diagram of water in Figure 7.2-1 explains the thermodynamics involved in the sport of ice skating. It can be seen that the added pressure on the ice from the skate alters the local phase equilibria and the ice melts, providing a lubricating film of water. However, if the ambient temperature is too low, even the added pressure is not sufficient to generate the lubricating phase of water, and the skater will stick to the ice surface and fall.

The Clapeyron equation can also be used to predict the effect of a change in one system variable such as pressure on another, such as the phase transition temperature. This has practical value in predicting changes in the boiling point ($T_{\text{vaporization}}$) at various pressures and is applied in pressure cookers and autoclaves. By integrating the Clapeyron equation, expressions can be derived for solid–liquid and condensed phase–vapor phase equilibria. These are as follows for cases of solid-to-liquid transition:

$$\Delta P = \frac{\Delta H_{\text{fusion}}}{\Delta V_{\text{fusion}}} \frac{\Delta T}{T_{\text{melt}}} \tag{7.2-7}$$

Another useful expression is the **Clausius–Clapeyron equation.** It is useful for cases of condensed phase–gas equilibria:

$$\ln \frac{P_2}{P_1} = \frac{-\Delta H_{\text{vaporization}}}{R}\left(\frac{1}{T_2} - \frac{1}{T_1}\right) \tag{7.2-8}$$

It is important to recognize that the equation assumes that $\Delta H_{\text{vaporization}}$ is independent of temperature. Another derivation of this equation is given in Appendix 7.2 in which the relationship of the activity of the phases is discussed more explicitly. The Clausius - Clapeyron equation quantitates the increase in the activity of the substance in the condensed phase with respect to the pressure in the system. For example, the activity of oxygen in the body tissues is increased under hyperbaric conditions which change the equilibria of oxygen binding in numerous enzyme systems and in oxygen binding proteins such as hemoglobin. The concept behind hyperbaric oxygen treatment for diseases such as severe anaerobic infections and sickle cell anemia is based on the presumed beneficial increase in oxygen activity.

As mentioned earlier, a very important aspect of phase equilibria in biological science is the use of the colligative properties (vapor pressure, osmotic pressure, boiling point, and freezing point) of a solution. The relationship of the colligative properties of a solution to the solute concentration is used to define an ideal solution and also to explicitly measure the activity of components in a solution. Such measurements can be used in biological applications for purposes that include the determination of the molecular weight of a macromolecule and of the osmolarity of a patient's serum.

7.2.1. Colligative Properties and the Ideal Solution

The colligative properties of a solution vary with the amount of solute. The boiling point is increased and the freezing point is depressed as the amount of solute added into solution is increased. An ideal solution is defined as one in which the colligative properties vary linearly with solute concentration at all concentrations. If sufficiently dilute, all solutions show this linear variation and can be considered to act ideally. If a non-volatile solute, such as sodium chloride or glycine, is added to a volatile solvent, such as water, the vapor pressure of the solvent is decreased. The linear relationship of this decrease to the mole fraction of the solvent in solution was formulated by Raoult and is **Raoult's law**:

$$P_{\text{A}} = X_{\text{A}} P^{\circ}_{\text{A}} \tag{7.2-9}$$

where P_{A} is the vapor pressure of the solvent, X_{A} is the mole fraction of the solvent, and P°_{A} is the vapor pressure of pure solvent. If a solution obeys Raoult's law, it is defined as **ideal.**

All of the colligative properties are interrelated, and consequently information about the osmotic pressure, vapor pressure, freezing point, and

Table 7.2-1. Values of k_B for Henry's law.

Compound	k_B (in atmospheres @ 298 K)
N_2	86.0×10^3
CO_2	1.6×10^3
O_2	43.0×10^3
CH_4	41.0×10^3

boiling point can be determined from any of the other properties. The relationship of the colligative properties to solute concentration is summarized in the following equations. The freezing point depression for dilute solutions can be expressed as

$$T_o - T_f = k_f m \quad (7.2\text{-}10)$$

where k_f is

$$k_f = \frac{M_A R T_0^2}{1000 \Delta H_{fusion}} \quad (7.2\text{-}11)$$

where m is the molality of the solute, and M_A is the molecular weight of the solvent. For water, $k_f = 1.86$. Equation (7.2-10) is derived from a simplification of the **virial equation**. The virial equation is described in the Glossary and will not be discussed in this treatment.

The boiling point elevation is given by

$$T_b - T_o = k_b m \quad (7.2\text{-}12)$$

with k_b given by:

$$k_b = \frac{M_A R T_0^2}{1000 \Delta H_{vaporization}} \quad (7.2\text{-}13)$$

For water, $k_b = 0.51$. For a volatile solute, the vapor pressure of the solute fraction is given by **Henry's law**:

$$P_B = k_B X_B \quad (7.2\text{-}14)$$

where k_B is the **Henry's law constant** (representative values are listed in Table 7.2-1).

The osmotic pressure plays an important role in the behavior of cells and in their physiology. Osmotic pressure is defined as the external pressure that must be exerted on a solution to maintain the activity of the solvent at the same value as that of pure solvent (at constant temperature). Consider a system that is comprised of a solution of glucose in water separated by a semipermeable membrane (permeable to water but not to glucose) from a quantity of pure solvent (water in this case) (Figure 7.2-2). As already described, the addition of solute to pure solvent lowers the activity of the solvent as reflected in alterations in its colligative

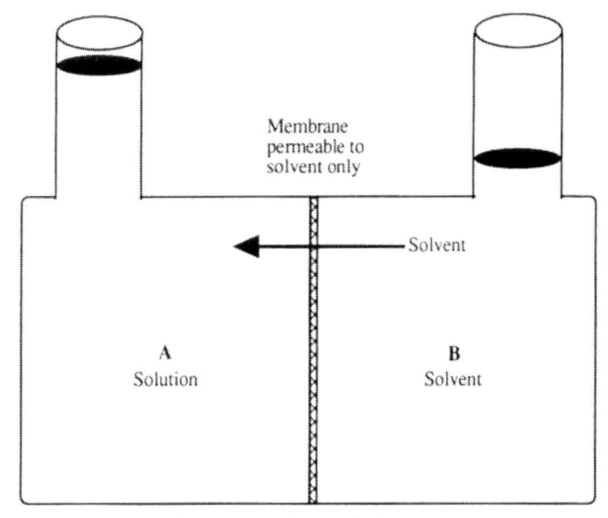

Osmometer without added pressure

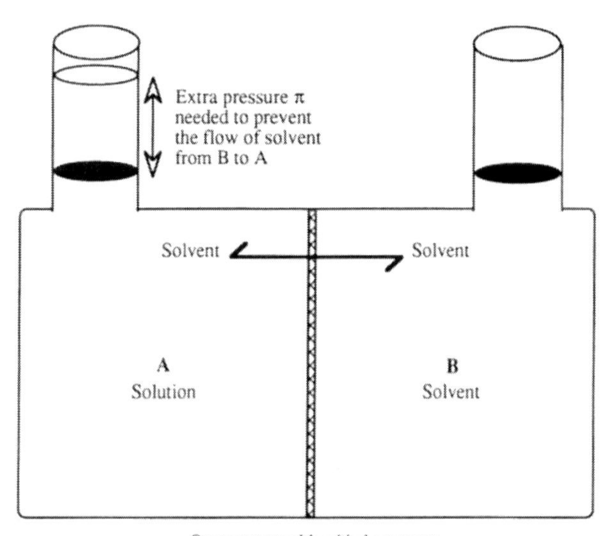

Osmometer with added pressure

Figure 7.2-2 Diagram of an instrument that measures osmotic pressure, called an osmometer. An osmometer consists of two chambers separated by a membrane permeable to the solvent, but not the solute. A solute added to compartment A lowers the activity of the solvent in compartment A compared to that in compartment B. Solvent will flow from B to A (osmosis) unless the activity of the solvent in A can be restored to its previously higher level. Measuring the amount of pressure, π, that must be applied to compartment A to raise the activity of the solvent in A to the same level as in B gives the osmotic pressure.

properties. Therefore, the activity of the water component in the glucose solution is lower than that of the water on the pure solvent side of the apparatus. At equilibrium, the chemical potential of each of the components must be equal, and since only water is free to move across the membrane, water must move from the pure solvent compartment into the solution compartment. The movement of the solvent across the membrane is called **osmosis.** In order to prevent this flow of solvent, the activity of the water on the solution side must be raised so that the equivalence of the chemical potential is restored. The activity of the water in the solution phase can be raised by exerting pressure on the solution side. This is the osmotic pressure, π.

For solutions that are dilute, the **van't Hoff equation for osmotic pressure** may be used to find π:

$$\pi = \frac{nRT}{V} \quad \text{or} \quad \pi = cRT \tag{7.2-15}$$

where c is the molar concentration, equal to n/V. The van't Hoff equation is derived in Appendix 7.3. Readers are frequently admonished to observe the similarity between the van't Hoff relation and the ideal gas equation, and a pleasing symmetry in thermodynamics awaits those who do. If the solute molecules and the gas molecules are considered analogously, and the solvent and the vacuum of the container in which these molecules are free to move are equated, it is easy to see why both systems can be treated equivalently. Activity and fugacity in these ideal systems are equivalent to concentration and pressure, respectively. When there are interactions between the solute molecules or between the solute and solvent, the activities of the solvent and the solute change. A key point about osmotic pressure must be emphasized: the osmotic pressure must not be considered as a pressure exerted by the solute; rather it is the external pressure required to equalize the chemical potential of the solvent, *because it is the solvent to which the membrane is permeable and that will move to equalize the chemical potential.* The process by which the solvent moves through the membrane is not important in finding the osmotic pressure, since the equilibrium result is the same regardless of the mechanism.

All of the colligative properties can be used in the laboratory to find molecular weights and activities of solutions. Osmotic pressure measurements are well suited to measuring the activity of solvents and of molecular weights of macromolecules because of the extremely large magnitude of the osmotic pressure. For example, the osmotic pressure associated with a 1-molar solution is 24.4 atm, which corresponds to a mercury column of 18.54 m. Molecular weights can be found using osmotic pressure measurements with the following formula:

$$\text{MW}_{(x)} = \frac{RTW_{(x)(\text{in grams})}}{\pi} \tag{7.2-16}$$

where again the virial equation is simplified.

7.2.2. Measurements of the Activity Coefficient Using Colligative Properties

All of the colligative properties are related to one another because of their shared common parentage in Equation (7.1-1): the chemical potential of a component in different phases is always equal at equilibrium. The activity of the solvent can be calculated from the colligative properties of a solvent. The most useful of these are vapor pressure and osmotic pressure measurements since they can be done at any temperature. The activity of a solvent is directly available from osmotic pressure measurements:

$$\ln a = \frac{-\pi \overline{V_A}}{RT} \tag{7.2-17}$$

This formulation resulted as part of the derivation of Equation (7.2-15) and is shown in Appendix 7.3. Derivation of a formula to find activity from vapor pressure measurements is left as an exercise for the reader.

7.3. Surface Phenomena

When different phases come in contact with each other, an interface between them occurs. This interface is a surface, and the properties of a surface are different from the properties of either of the phases responsible for creating it. The region that includes the surface and the immediately adjacent anisotropic boundary is called the **interphase**. Some of the general thermodynamic properties of surfaces will be the focus of this section, and the associated interphases will be discussed in Chapter 19.

Why are surfaces any different from the bulk phase from which they are derived? In one sense, they differ because they define where the phase begins or ends. Surfaces are boundaries, and boundaries are anisotropic. They are different from the bulk phase simply because they have an inside that is bulk phase and an outside that is not. The molecules making up a surface are different from those of the bulk phase because they are not surrounded by bulk phase molecules. Consider a bulk phase of water molecules. These water molecules may be visualized as being arranged in a tetrahedral array (cf. Chapter 9). The cohesive energy from the association of the molecules leads to an energy of association per mole, U. Each water molecule therefore has an energy associated with it:

$$\frac{U}{N_A} = \chi \tag{7.3-1}$$

where N_A is Avogadro's number. In the tetrahedral array, each molecule is bound to four others and the bond energy is $\chi/4$. Consider that the water molecules at the phase boundary or surface will have three rather

Table 7.3-1. Surface tensions of liquids (in air).

Compound	γ (dyn cm^{-1})
H_2O @ 298 K	71.97
H_2O @ 373 K	58.85
Hg	487.0
Benzene	28.9
Acetone	23.7
n-Hexane	18.4

than four neighbors. The bonding energy of the surface molecules will be $3(\chi/4)$ rather than $4(\chi/4)$. The bonding energy of the surface molecule is therefore less than the bonding energy associated with a bulk phase molecule, and the energy of the surface molecule is therefore higher than that of molecules in the bulk phase. It follows that work must be done to move a molecule from the interior bulk phase of the water to the surface. Because the presence of a surface requires the expenditure of work and a higher energy, all materials attempt to minimize the surface area. The force in the surface that attempts to keep the surface area at a minimum is called the **surface tension.** As a surface is stretched, the free energy of the surface increases, and the increase in free energy (usually given in erg cm^{-2}) is related to the change in area through the surface tension (designated by the symbol γ and usually given in dyn cm^{-1}). The more highly associated a substance is, the higher the surface tension will be. The surface tensions for various materials are compared with that of water in Table 7.3-1.

What happens to the surface tension of a pure solvent such as water (though it must be emphasized that some of the most important surfaces in biological systems are not comprised of water) when a solute is added? The answer depends on the solute. If the substance decreases the free energy of the surface, then the solute will concentrate at the surface, and the surface tension will decrease. If, on the other hand, the solute were to raise the surface tension of the solution, it would be energetically unfavorable for it to concentrate at the surface, and it will instead shun the surface, preferring the bulk phase. Therefore, substances that are capable of lowering the surface tension will do so quite dramatically, because it is energetically favorable to do so. Material that would be able to raise the surface tension will at best only slightly raise the surface tension, because it is energetically unfavorable for it to be found at the surface. Generally then, the addition of solute results in either a lowering or a slight increase in surface tension.

The concentration of solute molecules at a surface is different than in the bulk. The surface excess concentration, denoted by the symbol Γ, represents the quantity of solute adsorbed to the surface. The surface excess can be quantified through the **Gibbs adsorption isotherm**:

$$\Gamma_i = \frac{-1}{RT}\left(\frac{\partial \gamma}{\partial \ln a_i}\right) \tag{7.3-2}$$

Γ_i has the units of mol m^{-2}. The surface tension is γ and the activity of the solute, i, in solution is a_i. This expression indicates that a solute that decreases the surface tension will concentrate at the surface, since the sign of the excess concentration is opposite that of the change in surface tension.

Some of the most important surface-active compounds in biological systems are the amphiphilic long-chain fatty acids. By virtue of their hydrophilic—COOH and hydrophobic hydrocarbon groups, these molecules will preferentially adsorb to the water surface with the hydrophobic tail sticking out of the water phase. They act to lower the surface tension of the aqueous phase significantly and are important, for example, in the lungs where they increase the surface area available for gas exchange. These amphiphilic fatty acids or detergents will form surface films of monomolecular dimension if allowed to spread on the surface of water. The change in surface tension that results from the formation of these monolayers can be easily measured through the use of a Langmuir film balance or similar devices. Multilayers of these detergent molecules can be formed by the repeated dipping of a glass slide through a monolayer. As the slide is repeatedly passed through the monolayer, layer after single layer of molecules is deposited, one upon the other, polar to polar and nonpolar to nonpolar portions of the molecule orienting toward each other. The resultant film is called a Langmuir–Blodgett film, and its thickness can be directly measured by optical interference methods. Since the number of dips is related to the number of monolayers, the length of the detergent molecule can be directly calculated. Biological membranes are generally considered as being comprised of a bilayer of amphiphilic phospholipid molecules. These phospholipids are composed of a glycerol backbone, nonpolar hydrocarbon chains, and a polar head group, and thus resemble a Langmuir–Blodgett film generated by two passes through a monolayer. In the biological membrane, there are a variety of different lipid components, as well as proteins and polysaccharides, making up the membrane and giving it a wide range of properties. The formation and properties of the biological membrane will be taken up in some detail in Chapter 15.

The change in surface tension is reflective of a change in surface free energy, and consequently many other properties of the surface are changed as well. Extremely important biological effects are seen in the varied transport of small molecules such as water and ions and in the fluidity of the membrane, which affects the lateral diffusion of large molecules such as proteins that act as receptors. Membranes can have compositions that result in physical properties similar to those found in liquids or in solids and can undergo phase transitions. Since change in surface tension is a direct measure of free energy change, the thermodynamics of the phase changes can be determined through the use of measurements in devices such as the Langmuir film balance. In Part III, a much more detailed picture of the role of surfaces and membranes will be developed with regard to their role in transport, information transduction, mechanical work functions, and modulation of cellular function. In addition, the properties and forces acting in the interphase region, a very important area that is associated with surfaces, will be examined. The properties of the interphase are fundamental to the accurate description of biochemistry and biophysics of cellular behavior, because the volume of the cell represented by the interphase region is enormous. This is the result of the simple fact that cells have an enormous surface area attributable to the large number of cellular membranes.

The discussion is now ready to move on from this slightly formal review of the fundamentals of thermodynamics to their use in attempting a deeper understanding of the aqueous solutions and systems that make up the biological world. Before going on, however, a final look at the complete differential equation that will provide the backbone of the study of these systems is in order. The differential that has been derived thus far has considered the effect on G of temperature, pressure, volume, components, surface area, and other forms of work (a generality that now needs definition). Mathematically,

$$
\begin{aligned}
dG = \left(\frac{\partial G}{\partial T}\right)dT + \left(\frac{\partial G}{\partial P}\right)dP + \left(\frac{\partial G}{\partial V}\right)dV \\
+ \left(\frac{\partial G}{\partial A}\right)dA + \sum_i \left(\frac{\partial G}{\partial n_i}\right)dn_i + \left(\frac{\partial G}{\partial Q}\right)dQ
\end{aligned} \tag{7.3-3}
$$

Each of these terms plays a role in biological aqueous systems. The first term, $(\partial G/\partial T)$, is equivalent to $-S$, the entropy change. The second term, $(\partial G/\partial P)$, represents the osmotic pressure in a biological system. The third term, $(\partial G/\partial V)$, represents PV work and has a varying degree of importance depending on the system studied. The fourth term, $(\partial G/\partial A)$, represents the contribution made by the surface tension in the system. The fifth term, $(\partial G/\partial n_i)$, is the change associated with a change in chemical composition and represents the chemical potential. The final term, $(\partial G/\partial Q)$, defines the reversible work component in transferring charge (Q) in the system.

The great value of the total differential equation lies in the simplifications that occur when aspects of the system under study are constrained. For example, in many biological systems, it is reasonable to assume no change in temperature, pressure, and volume or surface area. Consequently, the corresponding four terms will become zero, leaving only chemical potential and charge-related phenomena as the thermodynamic factors of concern, i.e.:

$$dG = \sum_i \left(\frac{\partial G}{\partial n_i}\right) dn_i + \left(\frac{\partial G}{\partial Q}\right) dQ \tag{7.3-4}$$

Depending on the system under study, either or both of these terms may also be eliminated. It is time then to begin an examination of the aqueous biological world and discover the role of each of these physical forces in the function of life processes.

Appendix 7.1. Equilibrium Dialysis and Scatchard Plots

In an equilibrium dialysis experiment, the binding of a small molecule, i, to a macromolecule, M, is studied by establishing two phases through the use of a dialysis membrane. The macromolecular species is placed inside the dialysis bag and it is suspended in a large volume (usually > 100 times the volume in the dialysis bag). There are therefore two phases, phase a which is *inside* the bag and phase b which is *outside* the bag. The concentration of the small molecule is the measured quantity, C_i. At equilibrium, the chemical potential of i is the same in both phases:

$$\mu_{i(\mathrm{a})} = \mu_{i(\mathrm{b})} \tag{A7.1-1}$$

The activity of i in each phase can also be equated:

$$a_{i(\mathrm{a})} = a_{i(\mathrm{b})} \tag{A7.1-2}$$

Since the activity is equal to the product of the activity coefficient, γ, and the concentration of i, c_i, the following relation can be written:

$$\gamma_i c_{i(\mathrm{a})} = \gamma_i c_{i(\mathrm{b})} \tag{A7.1-3}$$

The solution in phase b is made dilute enough so that the activity coefficient γ_i^{b} can be considered equal to unity; therefore, Equation (A7.1-3) can be simplified and rearranged:

$$\gamma_{i(\mathrm{a})} = \frac{c_{i(\mathrm{b})}}{c_{i(\mathrm{a})}} \tag{A7.1-4}$$

The activity coefficient of i inside the bag is generally less than unity, indicating that there is a greater concentration of i in phase a than in phase b. The interpretation of this thermodynamic measurement is that the difference in the amount of i is due to the binding of i to the macromolecule. If the activity coefficient of i that is free inside the bag is assumed to be unity, then the concentration of $i_{\text{free-inside}}$ is equal to the concentration of i_{outside}:

$$c_{i\text{-free(a)}} = c_{i\text{(b)}} \tag{A7.1-5}$$

but

$$c_{\text{inside(a)}} = c_{\text{bound(a)}} + c_{\text{free(a)}} \tag{A7.1-6}$$

Rearranging Equations (A7.1-5) and (A7.1-6) gives the result

$$c_{i\text{-bound(a)}} = c_{i\text{-inside(a)}} - c_{i\text{(b)}} \tag{A7.1-7}$$

By measuring the total concentration of i inside and outside and subtracting, the difference gives the amount of i bound. The amount bound, v, per macromolecule, M, can then be expressed:

$$v = \frac{c_{i\text{-bound(a)}}}{c_M} \tag{A7.1-8}$$

where v is equal to the average number of molecules of i bound to each macromolecule.

As will be seen later (cf. Section 14.2), v will depend on the binding constant at equilibrium, the number of binding sites on a macromolecule, and the actual concentrations of M and i. If the total number of sites available for binding is given by N, then a term, Θ, can be defined that describes the fraction of sites bound:

$$\Theta = \frac{v}{N} \tag{A7.1-9}$$

This analysis will assume that the binding sites are independent and identical to one another, and so the relationship derived for a single binding site is the same for all binding sites. A case where this condition of identical and independent binding is not true will be considered in Section 14.2. Ignoring activity coefficients, at equilibrium, a binding constant K can be written for the reaction

$$M + i \rightleftharpoons M\text{—}i \tag{A7.1-10}$$

giving

$$K = \frac{[M\text{—}i]}{[M][i]} \tag{A7.1-11}$$

Equation (A7.1-9) could be rewritten in terms of the concentrations of M and i:

$$\Theta = \frac{[M\text{—}i]}{[M] + [M\text{—}i]} \tag{A7.1-12}$$

Now, Equations (A7.1-11) and (A7.1-12) can be combined to give the

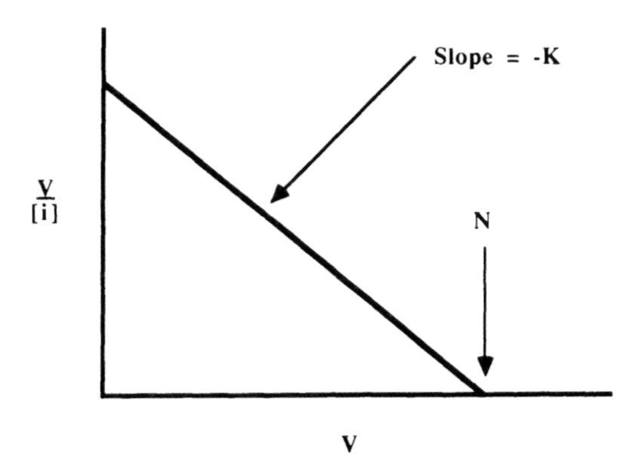

Figure A7.1-1 Idealized Scatchard plot analysis.

fraction of sites bound in terms of the equilibrium binding constant, K, for all N when binding sites are equivalent:

$$\Theta = \frac{K[i]}{1 + K[i]} \tag{A7.1-13}$$

The binding constant of $[i]$ is given by

$$\frac{\Theta}{1-\Theta} = K[i] \tag{A7.1-14}$$

Since the measured quantities in an equilibrium binding experiment are the concentrations of i inside and outside the dialysis bag, and N and K are generally the coefficients being sought, Equation (A7.1-14) is rewritten in terms of v and N:

$$\frac{v}{N - v} = K[i] \tag{A7.1-15}$$

The **Scatchard equation** is obtained by rearranging into a $y = mx + b$ form:

$$\frac{v}{[i]} = K(N - v) \tag{A7.1-16}$$

Plotting $v/[i]$ on the y axis against v on the x axis (numbers easily available from experiment) will give a straight line, at least at low values of v (Figure A7.1-1). The x intercept of this plot is N, the number of binding sites per molecule, and the slope of the line is $-K$. A nonlinear plot of Equation (A7.1-16) is evidence that the binding sites on the macromolecule under study are not identical, equivalent, or independent; that is, they exhibit cooperativity (cf. Chapter 14.2).

Appendix 7.2. Derivation of the Clausius–Clapeyron Equation

The relationship between pressure and the boiling point of a pure solvent can be simply derived by considering the equilibrium constant for the process

$$X_{(\text{liquid})} \longrightarrow X_{(\text{gas})} \quad \text{(A7.2-1)}$$

The activity of the pure liquid is unity, and if the gas is dilute enough to be considered an ideal gas, its activity is simply its pressure. K may be written in this case as

$$K = \frac{a_{(\text{gas})}}{a_{(\text{liquid})}} = \frac{P}{1} \quad \text{(A7.2-2)}$$

The temperature dependence of the equilibrium constant can be given by (cf. Section 6.4)

$$\ln \frac{K_2}{K_1} = \frac{-\Delta H}{R}\left(\frac{1}{T_2} - \frac{1}{T_1}\right) \quad \text{(A7.2-3)}$$

Combining these last two equations gives the Clausius–Clapeyron equation:

$$\ln \frac{P_2}{P_1} = \frac{-\Delta H_{\text{vaporization}}}{R}\left(\frac{1}{T_2} - \frac{1}{T_1}\right) \quad \text{(A7.2-4)}$$

Appendix 7.3. Derivation of the van't Hoff Equation for Osmotic Pressure

At equilibrium, there will be no net flow of solvent through the membrane. If one phase is the solvent in standard state, μ°, and the other phase, μ, is the solvent, A, with the added solute, b, and the added external pressure is $+\pi$, the following can be written:

$$\mu^{\circ}{}_{(T,\,P)} = \mu_{(T,P+\pi)} \quad \text{(A7.3-1)}$$

This can be rewritten

$$\mu^{\circ}{}_{(T,\,P)} = \mu^{\circ}{}_{(T,P+\pi)} + RT \ln a_{A} \quad \text{(A7.3-2)}$$

The solute that is added to make the solution has decreased the activity of the solvent by the amount $RT \ln a_A$:

$$\mu^{\circ}{}_{(T,\,P)} - \mu^{\circ}{}_{(T,P+\pi)} = RT \ln a_{A} = \Delta\mu \quad \text{(A7.3-3)}$$

Therefore, the problem now is to determine how much external pressure must be added to the solution side to raise the activity of the solvent by the amount $RT \ln a_A$. At constant temperature, the fundamental equation reduces to:

$$\Delta G = \Delta\mu = \int_{P_1}^{P_2} \overline{V}_A \, dP = \overline{V}_A(P_2 - P_1) \qquad \text{(A7.3-4)}$$

$\overline{V}$ is the molar volume of the solvent. Since $P_2 - P_1$ is equal to $-\pi$, Equations (A7.3-3) and (A7.3-4) can be combined to give

$$\Delta\mu = RT \ln a_A = \overline{V}_A \, \pi \qquad \text{(A7.3-5)}$$

This allows direct evaluation of the activity of the mole fraction of the solvent from the osmotic pressure:

$$\ln a_A = \frac{-\overline{V}_A \pi}{RT} \qquad \text{(A7.3-6)}$$

For a dilute solution, $\ln a_A$ can be approximated by writing first the activity in terms of the mole fraction of A:

$$\ln a_A = \ln X_A \qquad \text{(A7.3-7)}$$

but $\ln X_A$ can be written in terms of the mole fraction of the solute, X_b:

$$\ln X_A = \ln (1 - X_b) \qquad \text{(A7.3-8)}$$

For a dilute solution, $X_b \ll 1$, and therefore the logarithm can be expanded and all but the first term ignored:

$$\ln (1 - X_b) \approx -X_b \approx \frac{-n_b}{n_A} + n_b \approx \frac{-n_b}{n_A} \qquad \text{(A7.3-9)}$$

since $n_b \ll n_A$. Substituting this result into Equation (A7.3-6) gives the following result:

$$\pi = \frac{n_b RT}{n_A \overline{V}_A} \qquad \text{(A7.3-10)}$$

The volume of the solute is very small and can be ignored, so the total volume, V, can reasonably be considered equal to the solvent molar volume, $n_A \overline{V}_A$. This gives the result sought:

$$\pi = \frac{n_b RT}{V} \quad \text{or} \quad \pi = cRT \qquad \text{(A7.3-11)}$$

the van't Hoff equation for osmotic pressure.

CHAPTER 8

Engineering the Cell: A Modeling Approach to Biological Problem Solving

It would be useful to stop at this point and look back at what has already been accomplished and to cast a glance ahead toward what is yet to come. The initial premise of the value inherent in a (reasonably) complete understanding of the physics and chemistry that occurs in a biological system leads to a qualitative look at what is generally the most prevalent chemical species in biological systems, water. The structure of bulk water will be described as deriving from the extensive hydrogen bonding interactions between the individual molecules of water. The fact that water, as a liquid, is so highly structured will turn out to be an extremely important property in systems that are based in an aqueous solvent. In fact, when water is not the predominant species in a system, it is still an extremely important chemical species with important properties, as will be seen in the forthcoming chapters.

Because the behavior of a biological system will depend to a great extent on the properties of the aqueous solvent system, a semiquantitative review of the principles of thermodynamics was presented in order that further discussion of the nature of the physical biochemistry of living systems can be made on quantitative grounds. This discussion has emphasized that many of the properties used to measure the overall energy of a system are constrained in biological systems and that therefore the mathematics involved become greatly simplified. Importantly, it has been shown that completely general and ideal mathematical descriptions of a system can be altered to describe the real and often "nonideal" system without the loss of the general solution already derived. This allows generalized solutions to be applied to specific instances of interest, thus allowing integration of special cases with more general knowledge.

In the coming chapters, it will be discovered that biological systems are multiphasic, heterogeneous systems. The value of a good grounding in the application of the ideal thermodynamic equations will become apparent as these more complex and biologically relevant systems are considered. The background already acquired will help in the description of ionic interactions in an aqueous solvent, the interactions of macromolecules such as proteins and complex carbohydrates in water, and the development of multiphase systems separated by membranes. The development of multiphase systems will lead to a consideration of separation not only of chemical species but of charge and hence electrical forces acting on the systems. Further, since the presence of membranes in the real biological environment leads to charge separations, consideration of the role of these electrical forces on the processes of the organism must be considered. The presence of membranes further creates an environment often dominated by surfaces and boundary conditions. In biological systems, these surface environments are themselves nonhomogeneous and lead to complex microenvironments. These microenvironments must be appreciated because of the tremendous deviation from the relatively "idealized" behavior in real solutions. For the remainder of the book, the purpose then will be to focus on developing an understanding at each level of the interplay between the forces that influence the behavior of the cell. An important point to emphasize is that while many of the processes and systems studied in biological sciences are extremely complex, many of the fundamental processes and physical forces at play in these systems can be understood at a more basic or "idealized" level and then extrapolated to an "nonideal" or indeed more "real" level. The intellectual process that has been initiated with the discussion of thermodynamics then can be repeated in kind at various levels, until a description of a complex system is accomplished.

A useful approach to the description and understanding of complex systems is to employ model building exercises. This approach is especially useful in complex systems that have many apparent control mechanisms and therefore multiple forces acting to bring an initial state to a final state. As an example, consider the behavior at the interface between a cell and its external environment. The general nature of the cell's interaction with its environment can be described as one in which the cell senses and responds to its environment. This is a two-way interaction, in which the environment modifies the cell and then in many cases the cell modifies its environment. The physical boundary where much of the cellular - environmental interaction takes place is at the interface between the cell and the environment. Therefore, to understand the biology of the cell interacting with its environment requires an understanding of the interface between the two.

What are the conditions at this interface? Playing a mind game (i.e., building a model) will allow a brief consideration of such a meeting between a cell and its surroundings. Imagine first being able to travel among the molecules and spaces in and around the cell so as to cause no perturbation, in other words, to be a completely casual and objective observer at the molecular level. An overview of the trip shows that surrounding the cell (at a sufficient distance) will be a homogeneous aqueous solution with multiple components that include small macromolecules (this trip will be with the bulk solution at equilibrium conditions). Approaching the cell, the interface is entered and then the cell membrane is encountered. The cell membrane is a nonaqueous phase constructed of lipid molecules containing both polar and nonpolar regions and arranged in a bilayer with the hydrophobic portions internalized. This lipid phase has associated with it numerous components, mainly lipids, but in addition many components are proteins and carbohydrates. Finally, as the membrane is penetrated, the intracellular space is encountered. This is a sea of membranes and membrane-bound vesicles in an otherwise aqueous environment. The stage having been set, the mind trip can begin.

Starting far from the cell, in the homogeneous aqueous solution, several observations can be made immediately. Most striking is the observation that regardless of direction looked toward, the structure of the solution is uniform. This statement does not imply that there is no variation in the structure of the water since it is clear that there will be interactions and disturbances in the "bulk" water structure when other components are dissolved in it. Just what the alterations in solvent structure are will be the subject of extensive consideration in the next several chapters, but for now it is sufficient to note that the overall collection of molecular associations are uniformly distributed. The second observation that can be made is that there is no electric field operant. This is in keeping with the notion that nature abhors an uncompensated charge separation, and therefore all the potential charges in the solution are carefully accounted for and balanced. Both of these observations make perfect sense if viewed in the terms of the simplified partial differential equation, since at equilibrium ΔG will be zero and differences in chemical potential and charge will also be zero, leading to the conditions noted.

Because of the completely objective nature of the special vessel making these observations, one other condition can be appreciated. It will be noted that all the molecules making up this solution are in constant motion. This motion is the random activity of Brownian motion. However, in spite of all the motion, when a single particle or group of particles is observed for a reasonable length of time, it will be found that there is no net motion. In other words, if taken on an appropriately long time scale (thermodynamic time), no observable flux or net motion results. This condition would not be the case if a new component were added or an electric field were to be imposed across the system under observation. In these cases, diffusion or conduction would take place. Such a system must be considered in nonequilibrium terms. These topics will be considered in the upcoming chapters.

In summary, the conditions associated with this external aqueous solution are (1) there is a uniform, though not necesarily continuous, structure to the solution, (2) there is electrical neutrality and no electrical forces due to charge separation, and (3) there is random Brownian motion but no net movement of molecules from one point to another.

As the trip now moves toward the cell, it would be worthwhile to consider that the structure of bulk water or of solutions, to be discussed in the following chapters, is a reasonable description as long as the point of observation is completely surrounded by water. What will be the result when the solvent phase ends? In other words, at the boundary of the phase, are the structure and behavior of the molecules comprising that phase the same as they are in the bulk solution? Just a moment's reflection on this should make it clear that since the structure of the solution phase is carefully described as a three-dimensional structure, at any boundary, including those associated with a surface structure such as a membrane, the solution phase structure would have to be altered in at least one dimension. The same considerations of course hold for the structure of the second phase, whose three-dimensional structure ends at the boundary that ends the solution phase structure. At this point (the interface), the structure, and hence the behavior, of each of the phases will be different from that of the bulk phases. Consequently, at the interface between the aqueous phase and the cell membrane, the character of the interactions will be dominated by boundary properties and not bulk properties.

At this interface between two phases, named the interphase, the different chemical components will have the capacity and tendency to interact with each other, driven by both forces from the altered boundary conditions as well as forces deriving from the disequilibrium caused by association with a new phase. Interactions between the components of the phases lead to alterations in surface energy and the surface tension of the phases. Furthermore, at the interface, adsorption of various molecules preferentially to one phase or another will lead to changes in the driving forces across the interface. The cell membrane is especially interesting because it has qualities of both a hydrophobic gel phase and a solid surface. It is a natural process that there will be a preferential adsorption of certain molecules from one phase to the other at the interface. As this occurs, imbalances are created that lead to the generation of driving forces back toward equilibrium. It is important to recognize that the creation of the interface then has generated a nonequilibrium environment.

Another force at this particular interface between cell and external aqueous environment must be considered. As the observation vessel enters the interphase, it clearly registers the presence of an electric field. Where does the electrical potential field have its origin? In fact, there are several sources of the necessary charge separation. Many of the molecules of biological significance are charged, and when a specific adsorption of these molecules occurs, a charge is accumulated in one place at the expense of another and a charge separation occurs. As will be seen, this universal effect at an interface leads to an electrified region extending into both phases. Therefore, the interphase region is electrified. The membrane itself contributes to the electric field because it is comprised of functional groups that are charged at physiologic pH. These charges are physically associated with the membrane and are not free to move, and hence the membrane acts like a metallic electrode with a charge imposed on it by an electrical source. A further source of charge separation across the membrane is the unequal partition of charge-carrying molecules on opposite sides of the membrane. Charge separation occurs because large polyelectrolytes (nucleic acids, proteins, and glycoproteins) are confined to one side of a membrane and because the cell actively partitions ions across the membrane. Any molecule with charge or capable of being polarized (i.e., having a charge separation induced in it) will therefore behave differently in the electrified environment of the interphase than it would in the bulk solution. Since it was a starting premise that understanding the biology of interactions of the cell with its environment depends on sensing and manipulating events near the interface, the radically different environment of the interphase compared to the bulk solution will be important to comprehend.

Knowledge of the structure and behavior of bulk water and the basics of equilibrium thermodynamics will provide a solid grounding on which the exploration of more complex phenomena can be built. The principles discussed in this section will be constantly used as an understanding is gained of the behavior of concentrated and nonideal solutions, multiphasic systems, nonequilibrium processes, and finally the molecular forces that are operant at the cellular level.

Part II
The Nature of Aqueous Solutions

CHAPTER 9

Water: A Unique Structure, A Unique Solvent

9.1. Introduction

Water is a unique compound whose physical properties set it apart from every other liquid that has been described to date. On what evidence is this statement based?

1) Water has one of the highest dielectric constants known (Table 9.1-1). Water is therefore one of the most polar of solvents, with the consequence that electrically charged molecules or particles are easily separated in its presence. Ionizations take place readily because the coulombic forces between opposite charges are weakened and more easily broken. These forces are proportional to $q^+q^-/\epsilon r^2$, where ϵ is the dielectric constant, and q^+ and q^- are the respective cationic and anionic charges being separated.
2) Water has a very high heat capacity; that is, a large amount of heat is needed to raise the temperature by one degree Kelvin. In biological systems, this is an advantage, since the temperature of a cell undergoing moderate metabolic activity is unlikely to rise significantly. It has

Table 9.1-1. Dielectric constants of selected compounds.

Substance	Dielectric constant (at 298 K)
Water	78.5
CH_3OH	32.6
CH_3CH_2OH	24.0
H_2S	9.3
C_6H_6	2.2
CCl_4	2.2
CH_4	1.7
Air	1.000059
Mica	5.4
Polystyrene	2.5

Table 9.1-2. Heats of vaporization for selected substances.

Substance	Heat of vaporization ($kJ\ mol^{-1}$)
Water	40.7 @ 373 K
Acetic acid	41.7 @ 391 K
Ethanol	40.5 @ 351 K
Hexane	31.9 @ 341 K

been speculated that if the heat capacity of water were not so high and warm ocean currents could not carry heat so efficiently, aquatic life would be profoundly affected and the oceans would be devoid of most fish. For example, the Gulf Stream is a stream of warm water ½ km deep and 200 km wide that carries an energy equivalent to 160 billion kilograms of coal burned per hour.

3) Water has a high heat of vaporization (Table 9.1.2) as do other liquids capable of hydrogen bonding, such as ethanol and acetic acid. In contrast, nonhydrogen-bonded liquids such as hexane have lower heats of vaporization. This is the reason perspiration is such an effective method of cooling the body.
4) Liquid water has a higher density than ice at standard temperature and pressure. There is a positive volume change upon freezing, which means that ice floats. If ice did not float, lake fish could not survive in lakes because the water would freeze from the bottom up. Furthermore, there is an insulating effect of the ice layer without which lakes would be too cold (or completely frozen) to support life.
5) Water has a high surface tension. This requires biological organisms to use detergent-like compounds to modifiy the surface tension. Lung surfactants are an example of these detergents. Lung surfactants significantly decrease the work needed to open the alveolar spaces and allow efficient respiration to take place. The absence of these surfactants leads to severe respiratory disease and often death.
6) Water has a higher conductivity than might be expected on the basis of its ionization constant alone. The conductivity of ice is also high even though its ionization constant is 1000-fold lower. This has been attributed to the presence of an ordered structure which allows the electric charge to be carried by a modified "brigade" mechanism.

Why does water behave so uniquely in comparison to other solvents such as those listed in Table 9.1-2? The main explanation of these singular properties is the presence of extensive hydrogen bonding between water molecules. The resulting structure is the subject of the remainder of this chapter.

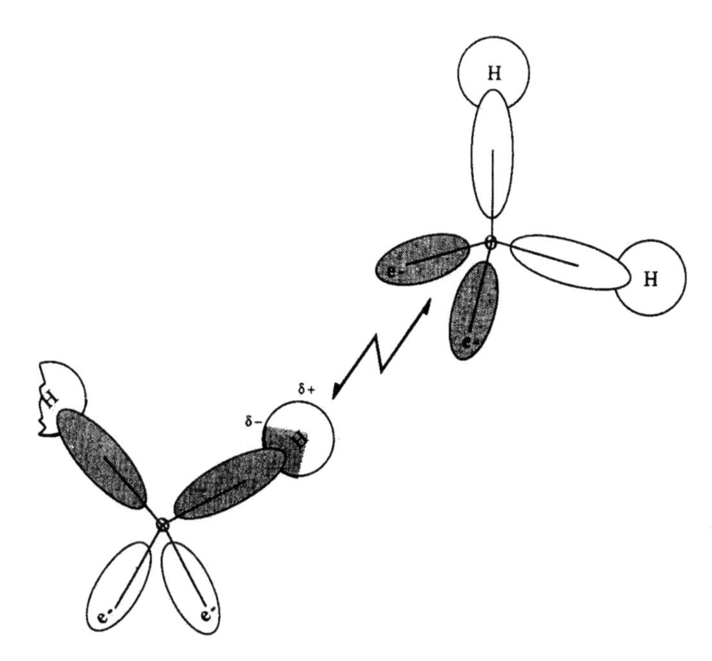

Figure 9.2-1 Molecular orbital arrangement that leads to formation of a hydrogen bond, showing the partial positive charge on the hydrogen atom due to the strong electronegativity of the covalently bonded atom. This partial positive charge interacts with the electron density of the unbonded pair of electrons on the second electronegative species.

9.2. Hydrogen Bonds in Water

Hydrogen bonds are formed by hydrogen atoms located between two electronegative atoms. When a hydrogen atom is covalently attached to a highly electronegative atom, such as an oxygen or a nitrogen, it takes on a partial positive charge due to the strong electronegativity of the oxygen or nitrogen atoms. This electron-deficient hydrogen will then be attracted to the free electron pairs that reside on other oxygens or nitrogens, and the hydrogen bond is thus formed (Figure 9.2-1).

The hydrogen bond is a relatively weak bond of between -20 and -30 kJ mol^{-1}. The strength of the bond increases with increasing electronegativity and decreasing size of the participating atoms. Thus, hydrogen bonds can exist in a number of compounds, not just in H_2O. The study of hydrides, amides, hydrogen halides, hydrogen cyanide, and ammonia has helped to clarify the nature of the hydrogen bonds which link compounds of these types into linear polymers. This text will however be concerned only with hydrogen bonds in water, in aqueous solutions, and within the solutes in such solutions. Hydrogen bonds arise because the

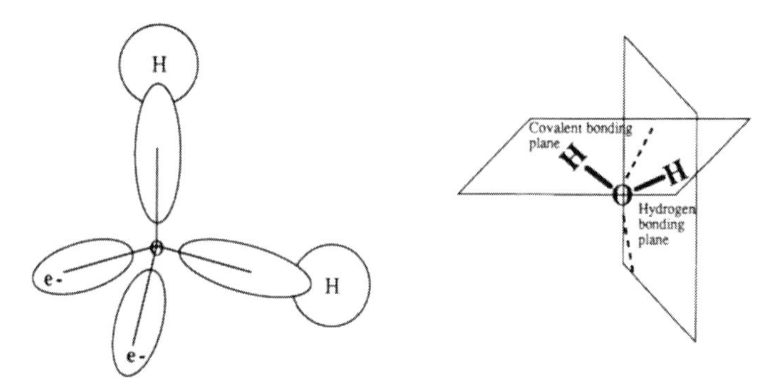

Figure 9.2-2 Schematic of the molecular orbitals of water; the axis is perpendicular to the page.

structure of the water molecule favors such interactions, as its molecular orbital representation indicates (Figure 9.2-2).

In the case of water, hydrogen bonding is strongly favored because each proton that is covalently bonded to the strongly electronegative oxygen atom finds an unbonded electron with which to interact in a one-to-one relationship. Because of this one-to-one proton-to-electron relationship, the orbital structure dictates that each oxygen atom on the average is involved in four bonds to hydrogen atoms. Two bonds are covalent and approximately 0.10 nm in length, and two are non-covalent hydrogen bonds and about 0.20 nm in length (Figure 9.2-3). These four bonds are arranged in a tetrahedral array around the oxygen atom at angles of approximately 104°, as first proposed by Bjerrum (Figure 9.2-4).

X-ray diffraction of liquid water shows a major peak at 0.28 nm which corresponds to the distance between oxygen atoms in the tetrahedral hydrogen-bonded water structures, confirming the existence of the predicted hydrogen-bonded structure. The hybrid orbitals that will allow

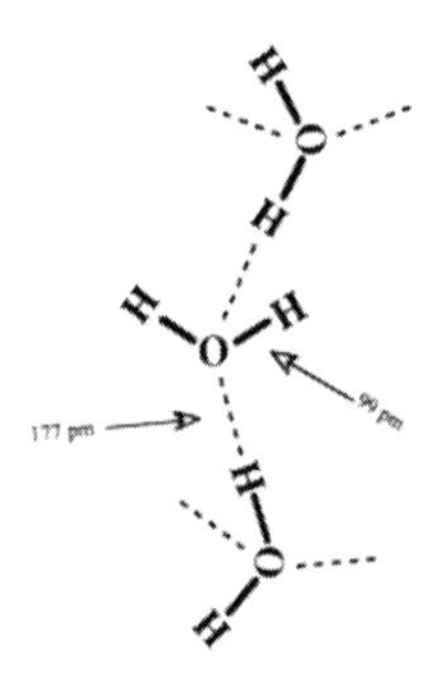

Figure 9.2-3 Water molecules are arranged so that each oxygen has a pair of covalent bonds and a pair of hydrogen bonds.

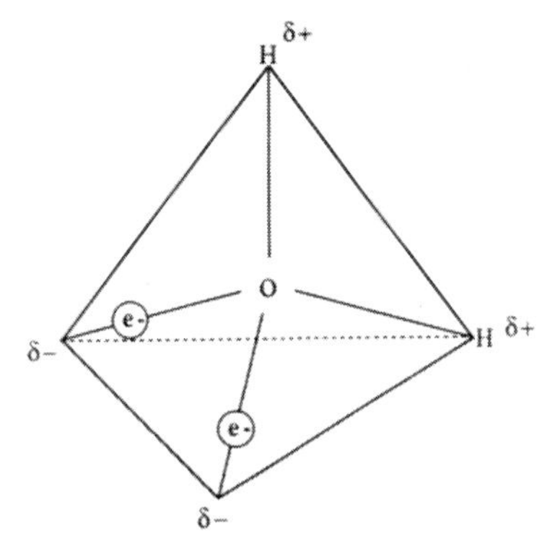

Figure 9.2-4 Tetrahedral bonding arrangement proposed by Bjerrum, where $\delta+$ is a partial positive charge and $\delta-$ is a partial negative charge.

formation of hydrogen bonds between three atoms found on two separate molecules are expected to be colinear. While Pauling's original description of the hydrogen bond stipulated such colinearity, it is now known that a bend of up to 10° can be tolerated before the energy of interaction becomes so weak that the hydrogen bond breaks.

Although the unusual properties of H_2O were well known, experimental evidence of the extensive existence of hydrogen bonds came when spectral techniques, which permitted comparison of normal—OH bonds (e.g., in alcohols) with—OH bonds in water, were developed. These studies were largely based on the effect of hydrogen bonds on vibrational transitions, that is, on movements of atoms with respect to others to which they are covalently linked. The energies associated with such transitions fall in the infrared spectral region. If molecules which do not interact with each other, such as molecules of water vapor at very low pressure, undergo the normal vibrational modes shown in Figure 9.2-5A–C, then the corresponding infrared spectra of liquid water would show each of these vibrational modes, especially deformation, to be perturbed if the water molecules participate in extensive mutual hydrogen bonding (Figure 9.2-5D–F). These types of studies originally were made difficult by the very broad infrared bands associated with the H_2O molecule but became possible with the availability of D_2O, which permitted determination of the shifts in absorbance peaks caused by hydrogen bonding. Such studies led Pauling to postulate that the hydrogen bond was a major interaction which played a critical role not only in water structure, but also in the structure and function of biological macromolecules.

The importance of hydrogen bonding in water can be emphasized by comparing the properties of H_2O and H_2S. Hydrogen sulfide is a compound which might, by virtue of the location of S below O in the periodic table and its almost identical structure to H_2O, be expected to have similar

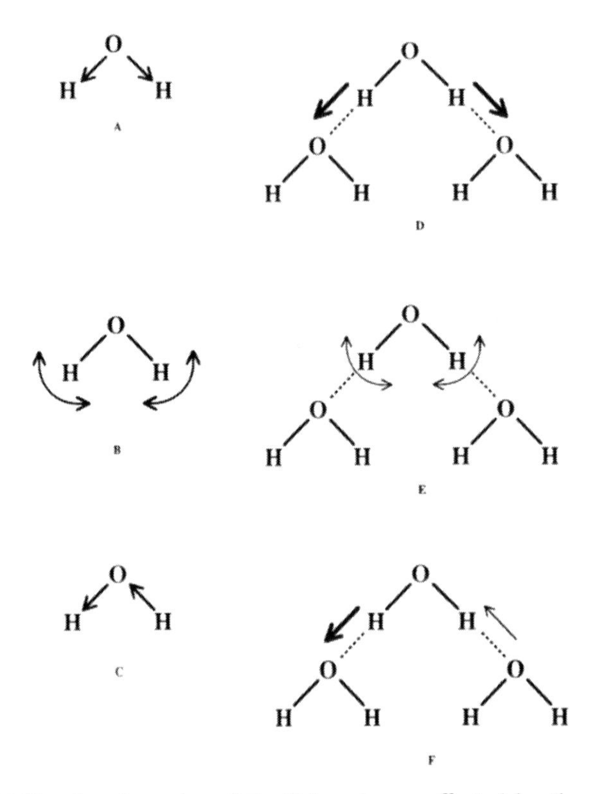

Figure 9.2-5 Vibrational modes of O—H bonds are affected by the presence of hydrogen bonds. A–C are the vibrational motions without hydrogen bonding: (A) the symmetric stretching; (B) the deformation of the bonds; (C) the asymmetric stretching. In D-F hydrogen bonds affect each of these vibrational modes: (D) the symmetric stretching is enhanced and needs less energy; (E) the deformation of the bonds is inhibited and needs more energy to occur; (F) the asymmetric stretching is partly enhanced and partly inhibited.

properties to water (Figure 9.2-6). The O—H bond in H_2O is stronger than the S—H bond in H_2S, since the respective lengths of these bonds are 0.099 nm and 0.134 nm (Figure 9.2-6). There is also no hybridization of atomic orbitals in H_2S. The tetrahedral angle between bonds in H_2O

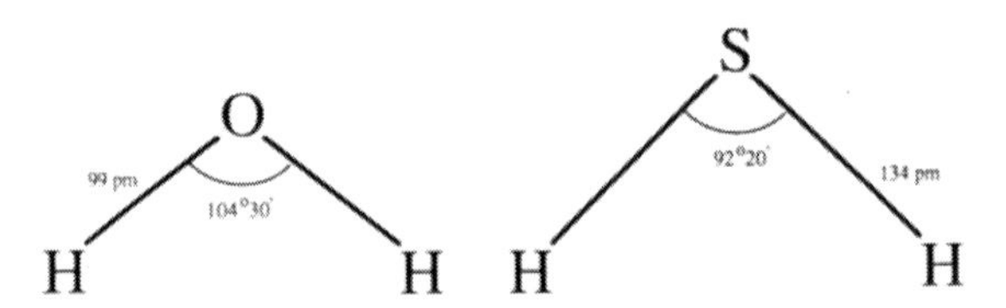

Figure 9.2-6 Structures, bond lengths, and angles for H_2S and H_2O.

Table 9.2-1. Comparison of thermodynamic parameters of water and hydrogen sulfide.

Property	H_2O	H_2S
$T_{solid-liquid,\ 1\ atm}$	273 K	190.1 K
$T_{liquid-gas,\ 1\ atm}$	100 K	−59.6 K
$Density_{solid}$ @ T_m	0.9998 kg dm^{-3}	1.80 kg dm^{-3}
$C_{p,\ liquid}$	76.02 J mol^{-1} K^{-1}	36.1 J mol^{-1} K^{-1}
$C_{p,\ gas}$	36.44 J mol^{-1} K^{-1}	41.56 J mol^{-1} K^{-1}
$\Delta H_{solid-liquid,\ 1\ atm}$	6.003 kJ mol^{-1}	2.386 kJ mol^{-1}
$\Delta H_{liquid-gas,\ 1\ atm.}$	40.656 kJ mol^{-1}	18.747 kJ mol^{-1}
Conductivity at $T_{liquid-gas,\ 1\ atm}$	4×10^{-10} Ω^{-1} m^{-1}	1×10^{-13} Ω^{-1} m^{-1}

is 104°30′, whereas the angle in H_2S is only 92°20′, an angle which cannot fit into a tetrahedral array and which prevents the formation of any further bonds to the S atom. Both molecules are polar, but the dipole moment for H_2O is considerably larger (1.8 debye for H_2O, 1.1 debye for H_2S). As Table 9.2-1 indicates, there are significant differences between the various thermodynamic parameters of these two compounds, which in the absence of hydrogen bonds, might have been expected to be rather similar. The main general observation that can be made about the properties and the thermodynamic variables given in Table 9.2-1 is that more energy is required to heat, melt, or vaporize one mole of H_2O than one mole of H_2S. This is attributed to the strong tendency of H_2O to form hydrogen bonds with other water molecules, while H_2S does not form hydrogen bonds.

9.3. The Structure of Crystalline Water

While the proposed structures for liquid water are still the subject of controversy among physical chemists, there is agreement upon the most probable structure for **crystalline water** (ice) at 1 atm pressure and 273 K, shown in Figure 9.3-1. The four bonds point at the corners of a tetrahedron as Bjerrum had predicted, resulting in tetrahedral arrays of hydrogen-bonded water. The cross section of these bonded arrays creates a hexagon as seen in Figure 9.3-2. The structures are reminiscent of the chair configuration found in cyclic six-carbon hydrocarbon rings.

The hexagonal arrays of H_2O are quite large with large amounts of empty space; the hexagons in ice as depicted in Figure 9.3-2 are empty. These spaces within the hexagons could be filled by small molecules such as monomolecular water (Figure 9.3-3) or other noncharged small molecules, even if they are hydrophobic. Under high pressure, at low temperatures, an entire second hexagonal array within the spaces of the first can be created. Any array having molecules interspaced within the or-

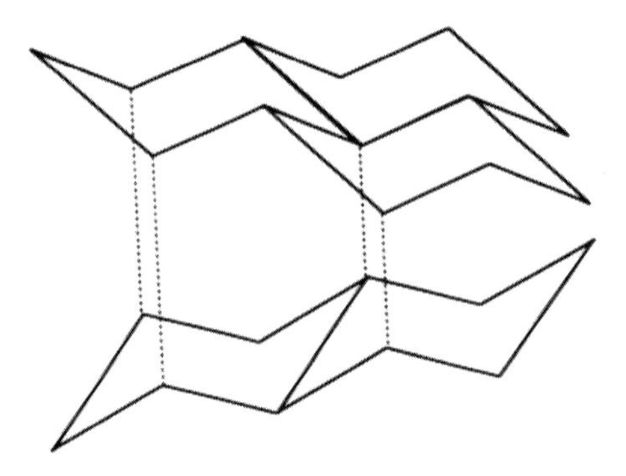

Figure 9.3-1 Structure of ice at 1 atm and 273 K.

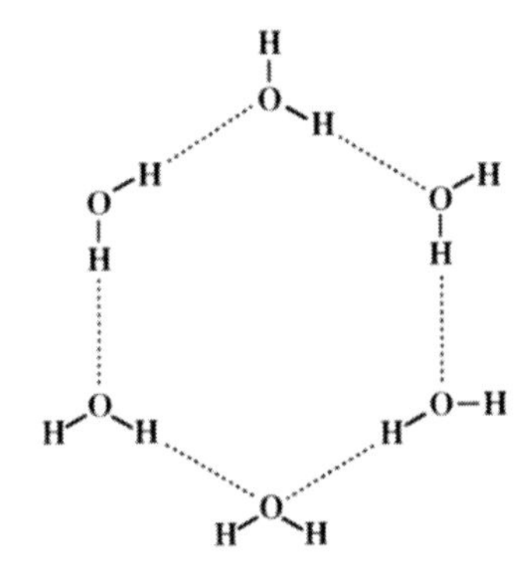

Figure 9.3-2 Cross section of the tetrahedral array shows the hexagonal structure of ice.

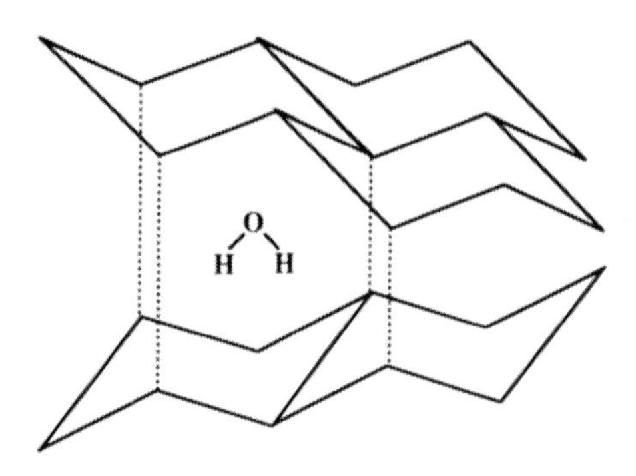

Figure 9.3-3 A small molecule can fit inside the spaces of the hexagonal arrays of water.

dered structure will have a higher density than the comparable structure with empty spaces. In other words, H_2O molecules can penetrate into the hexagonal structure of ice under conditions of greater pressure, or into the less regular, but still hexagonal, structures of liquid H_2O. Figure 9.3-3 represents one of a large number of possible interpenetrating H_2O structures.

9.4. Theories of the Structure of Liquid Water

A rigorous discussion of the various theories of liquid water structure is beyond the intended scope of this text but is well covered elsewhere, notably in *Water, A Comprehensive Treatise*, edited by Franks (see Appendix I). There is still controversy as to the actual structure of liquid water. A number of models of the structure of liquid water have been proposed, differing largely in the mathematical treatments. Most of them treat liquid water as a mixture of distinct states of organization or as a continuum of structures. All models share the feature of being based upon a high degree of hydrogen bonding between water molecules and take into account the strong dependence of the hydrogen-bond energy on the configuration of the bond. If there is bending of the hydrogen bond so that the atoms forming the bond are not colinear, then there is a loss of bond energy. Consideration of the effects of the interaction of the water dipoles with one another is also a general feature of the theories.

This hydrogen-bonded structure leads to a bulk water state with a very high entropy and to a much lower entropic (and therefore less favorable) state when single molecules are isolated from the bulk. The high entropy of water is a result of the equivalence of the hydrogen bonds throughout the bulk phase. Because of this equivalency, there is an enormous degeneracy in the possible energy states that a particular water molecule may occupy throughout the bulk phase. As a result of this degeneracy, a large number of microstates, and hence a large entropy, can be postulated. In entropic terms then, the hydrogen bonding in water does not restrict choice, but rather maximizes choice for the water molecules. In other words, a particular water molecule in the bulk structure cannot be identified with certainty. It is important to recognize the unique aspect of this entropic effect, because, when water molecules are removed from the bulk, or restricted in their ability to associate with the bulk, the possible number of microstates is sharply decreased, and the entropy of the system falls. This effect will be important in biological solvent–solute interactions and will be discussed further in coming chapters.

The salient features of these models can be explained by considering the behavior of proton conduction through a brigade-type mechanism as described by Klotz (cf. Section 18.6). Since all water molecules are identical, the hydrogen bonds in an array can be exchanged to include a new water molecule with virtually no loss in energy, as long as the net number of bonds and their environment remain constant. In the Klotz model, a specific proton does not need to move physically in order to propagate a current. Instead, hydrogen bonds break and then reform successively like a group of dominoes falling, with the net result being the apparent net movement of a proton. This unique capability accounts for the high electrical conductivity of ice even though the degree of ionization is 1000-fold lower than that of liquid water ($pK_{ice} = 10$, $pK_{liquid\ water} = 7$). The structure of ice and water facilitates the charge transfer, and the conductivities of ice and of liquid water are hence almost equal. The case of proton conduction will be considered in more detail in Section 18.6.

The Klotz model proposes a single class of H_2O molecules. This means that all the molecules are considered to behave identically and to have identical thermodynamic properties. A continuum of a hydrogen-bonded species exists in which bonds are constantly being made and broken between indistinguishable molecules. This type of structural model satisfactorily accounts for the observation by nuclear magnetic resonance spectroscopy that there are resonance peaks only for a single type of hydrogen-bonded species. Mixtures of defined clusters or aggregates with nonhydrogen-bonded species (i.e., two different types of H_2O, bonded and nonbonded) are not compatible with the data. The flickering cluster of Klotz's model is characterized by the constantly changing association and reassociation of groups of 50 to 70 water molecules. The time of association is very short in the liquid state, approaching 10^{-12} s, and shortens as the temperature increases.

The debate over the various models of liquid water structure is ongoing. For the purposes of this text, it is sufficient to regard all models as essentially "mixture" models, in which the structure of water is treated as a dynamic, fluctuating, assembly of water molecules at different states of hydrogen bonding. The differences between the models lie in the size, extent, and degree of deformation of the hydrogen-bonded component(s) of the mixture, since, as Franks has shown, there is cooperativity between hydrogen bonds.

There is greater Brownian motion in liquid water, and consequently, there is less organization and greater translational rearrangement than in ice. Hydrogen bonds in liquid water are weaker than those in ice because the interaction between adjacent hydrogen bonds is statistically more likely to be absent, rendering each individual hydrogen bond weaker. The nature of pure water and its properties have attribution to the formation of hydrogen bonds. A single kind of hydrogen bond between identical H_2O molecules is implicated in the properties of crystalline and liquid water. As will be seen in the following chapters, the role that this structured water plays in solvent–solute and solute–solute interactions is significant and important to understand.

CHAPTER 10

Introduction to Electrolytic Solutions

10.1. Introduction to Ions and Solutions

Ionized chemical species are common in all biological systems, and therefore it is crucial to attain a solid understanding of how ions interact with one another and with their solvent. No natural, and only occasional pharmacologically relevant solutions, are simple aqueous solutions of ions. Yet knowledge of the solute and solvent interactions in homogeneous electrolyte solutions is an important first step toward ultimately understanding the role of ions in biological systems. It is recalled from Section 7.2 that ideal solutions fulfill the requirements of Raoult's law; that is, the vapor pressure of each component is proportional to the mole fraction over the entire range of concentrations. This has the physical meaning that the solution's components have neither solute–solute nor solvent–solute interactions. A moment's reflection will suggest that the presence of electrically charged species in a polar-structured solvent such as water will not satisfy these "isolationist" requirements. The nature of these interactions will be the first focus of this chapter.

Ions are, by their nature, charged chemical species because of the imbalance between the positive charge derived from protons in the nucleus and the negative charge residing in the electron clouds of the atom or molecule. Because they are charged, ions are capable of carrying an electric current, which, as already defined, is simply the transfer of charge from one point of potential to another. Ions may do work by moving down an electrical potential gradient and may also generate potential gradients through the separation of charges:

$$\text{Work} = -EQ \tag{10.1-1}$$

Work is given in joules when the charge, Q, is in coulombs and the potential gradient, E, is in volts. Ions, because their nature is intrinsically electrical, will obey the same rules that govern all other electrical processes. While it is not the purpose of this chapter or book to fully discuss the physics of electricity, a brief review will be of value.

10.1.1. The Nature of Electricity

Electricity is concerned with the behavior of charges either at rest or in dynamic motion. The study of the behavior of resting charges is called **electrostatics.** When charges are in motion, electrodynamic phenomena are described by the laws of **electromagnetics.** Transport and behavior of charge in chemical systems is the basis of **electrochemical** studies. The identity of the charge carrier depends on the charge-carrying environment; for example, in metal conductors charge is carried predominantly by electrons whereas ions carry the majority of the charge in aqueous solutions. The study of interactions of ions with their environment is called **ionics**, the study of electron behavior is **electronics,** and the study of charge transfer across an interface (an electrode) is the study of **electrodics.** At this point, the focus will be on electrostatic behavior.

If charges are stationary and are separated in space, then the system is an **electrostatic system** and can be treated in the following fashion. Two charges, q_1 and q_2, are separated a distance r, with an electrical force resulting. According to **Coulomb's law**:

$$F = k\frac{q_1 q_2}{r^2} \tag{10.1-2}$$

The force produced is described with the point of reference being q_1 and extends between q_1 and q_2. If the charge on both particles is the same, whether positive or negative, the force between the two objects will be repulsive. An attractive force between the particles will result if the two objects possess charges of opposite sign. By convention, the force is said to be directed from the positive toward the negative body. The value of k depends on the units used. Coulomb's law as written in Equation (10.1-2) has $k = 1$, q is given in esu (electrostatic units) and r in centimeters; the force described is in dynes. In the SI system, k is equal to $1/4\pi\epsilon_o$, changing Equation (10.1-2) into the following:

$$F = \frac{q_1 q_2}{4\pi\epsilon_o r^2} \tag{10.1-3}$$

where ϵ_o is the permittivity of free space and has a value of 8.854×10^{-12} $C^2\,N^{-1}\,m^{-2}$, q is in coulombs, r in meters, and force is in newtons.

An electrical potential field, E, is generated by the coulombic force described above acting on a test object or charge, q_o, which is introduced at some distance, X, from the set of point charges:

$$E = \frac{F}{q_o} \tag{10.1-4}$$

The electric field is a vector quantity of force per unit charge. In SI units, this field is given as newtons/coulomb. Convention dictates that the vector of this field points in the direction toward which a positive test charge will be pushed. In a vacuum, the force that exists between two charged objects depends only on the distance separating the objects and the charge on the objects that are separated. It can be shown however that if various nonconducting materials are inserted in the space between the charges, the force per unit charge will be decreased. These materials are called **dielectrics**, and the relative measure of their electrical force attenuating properties is called the **dielectric constant** and is given the symbol ϵ. Since a vacuum has a dielectric constant equal to unity, the constants of all dielectric materials are greater than unity. Equation (10.1-3) is rewritten

$$F = \frac{q_1 q_2}{4\pi\epsilon_0 \epsilon r^2} \tag{10.1-5}$$

where ϵ is the dielectric constant.

10.2. Intermolecular Forces and the Energies of Interaction

Except in the case of nuclear chemistry, all interactions between molecules are electronic in nature. For the student of chemistry, the most commonly considered reactions between molecules are those where electrons are actually shared, leading to covalent bonding interactions. However, many of the forces acting between molecules of concern in this text do not involve the breaking and reformation of covalent linkages between molecules. It is worth reviewing briefly the nature and energy of the noncovalent forces acting between molecules. Many of the intermolecular forces are derived from electrostatic interactions similar to those just discussed.

Ion–ion interactions like those found in a pure electrolyte such as KCl are similar to the interactions discussed above. Two charged particles exert an interactive force on one another. The potential energy of interaction, U, is derived in Appendix 10.1 and is given here as

$$U_{\text{i-i}} = \frac{q_1 q_2}{4\pi\epsilon_0 \epsilon r} \tag{10.2-1}$$

A graph of the variation in $U_{\text{i-i}}$ with respect to both distance and dielectric constant is given in Figure 10.2-1.

Ion–dipole interactions occur when one of the molecules is an ion and the other a **dipole**. Dipoles are molecules that carry no net charge yet have a permanent charge separation due to the nature of the electronic distribution within the molecule itself. To illustrate the concept of the dipole, consider the structure of the two molecules water and carbon dioxide. Figure 10.2-2 illustrates this discussion. Oxygen is more strongly electronegative than either hydrogen or carbon, and the shared electrons in these molecules will spend more time with the oxygen atoms. The

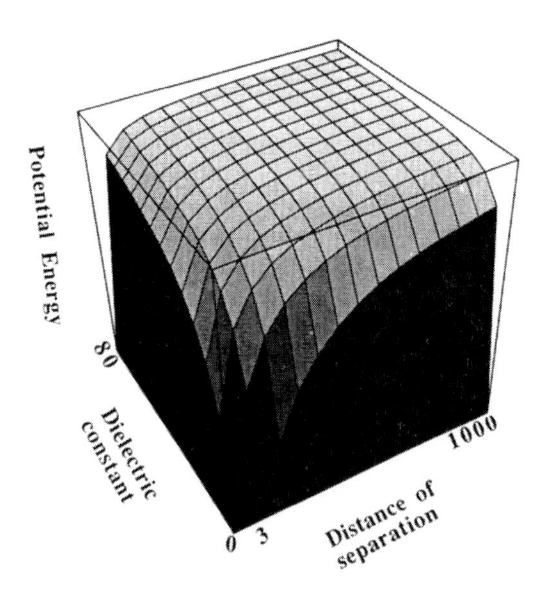

Figure 10.2-1 The potential energy of interaction between two ions based on electrostatic considerations only. The graph shows the relationship between distance of charge separation and dielectric constant and the interactional energy.

result of this preference for the oxygen atom by the negatively charged electrons will result in a partial negative charge on the oxygen atoms, while the companion carbon or hydrogen atoms will carry a partial positive charge. This charge asymmetry in the molecule is given a magnitude

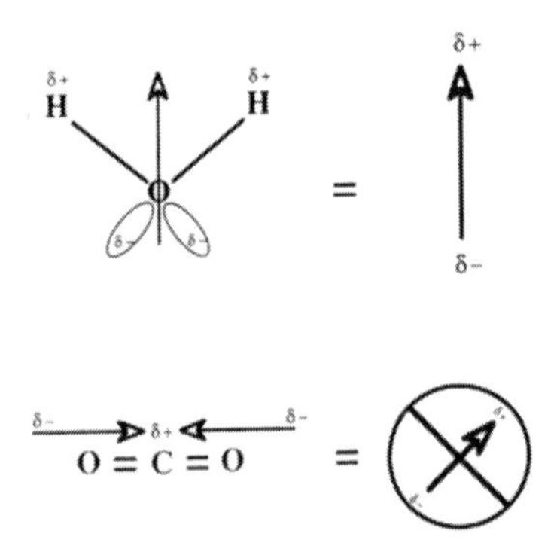

Figure 10.2-2 H_2O has a dipole moment because the vector addition of the electrical forces derived from its partial charge separation leads to the creation of an electric dipole. On the other hand, carbon dioxide has no electric dipole behavior (and hence no dipole moment), because the symmetry of the molecule leads to a cancellation of the electrical vectors generated from its partial charge separations.

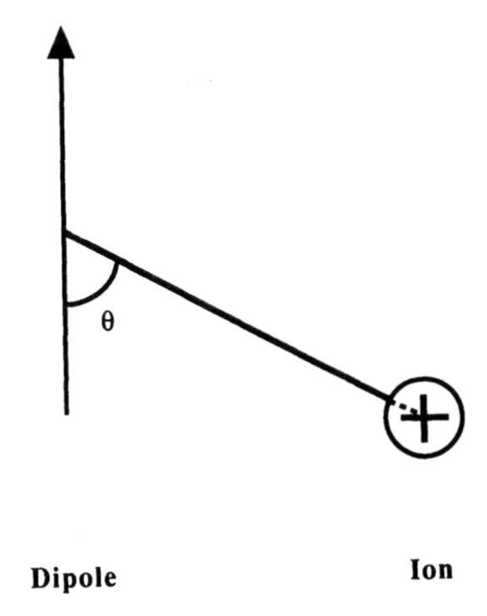

Figure 10.2-3 Schematic showing cos θ for the calculation of ion–dipole interactions.

and direction by calculating the **dipole moment,** μ. The dipole moment is given in **debyes** and is found by multiplying the separated charge times the distance of separation. The existence of a dipole moment requires an asymmetric molecule. As Figure 10.2-2 shows, H_2O has a dipole moment of 1.85 debye and CO_2 has no dipole moment, because H_2O is asymmetric and CO_2 is symmetric. In CO_2, the electric vectors derived from the partial charges cancel one another, and no dipole moment is measured. On the other hand, the H_2O molecule is bent, and the electric vectors are added, resulting in the formation of an **electric dipole**. The electric dipole is treated like any other charge, although the orientation of the dipoles (cos Θ) must be taken into account in calculating the ion–dipole force (Figure 10.2-3). Ion–dipole interactions will be discussed at length later in this chapter. Graphs of the variation in $U_{\text{i-d}}$ with respect to both distance and dielectric constant and distance and cos θ are given in Figure 10.2-4. $U_{\text{i-d}}$ is calculated as

$$U_{\text{i-d}} = \frac{q_1\mu_2 \cos\theta}{4\pi\epsilon_o\epsilon r^2} \qquad (10.2\text{-}2)$$

Even molecules that have no net dipole moment, such as carbon dioxide or carbon tetrachloride, can have a transient dipole induced in them when they are brought into an electric field. Electronic interactions of this type are called **ion–induced dipole** interactions, and the potential of the interaction will depend on the ability of the neutral molecule to be induced into a dipole, that is, its **polarizability**, α. Polarizability will be a topic of discussion later in this chapter.

$$U_{\text{i-id}} = \frac{q^2\alpha}{8\pi\epsilon_o\epsilon^2 r^4} \qquad (10.2\text{-}3)$$

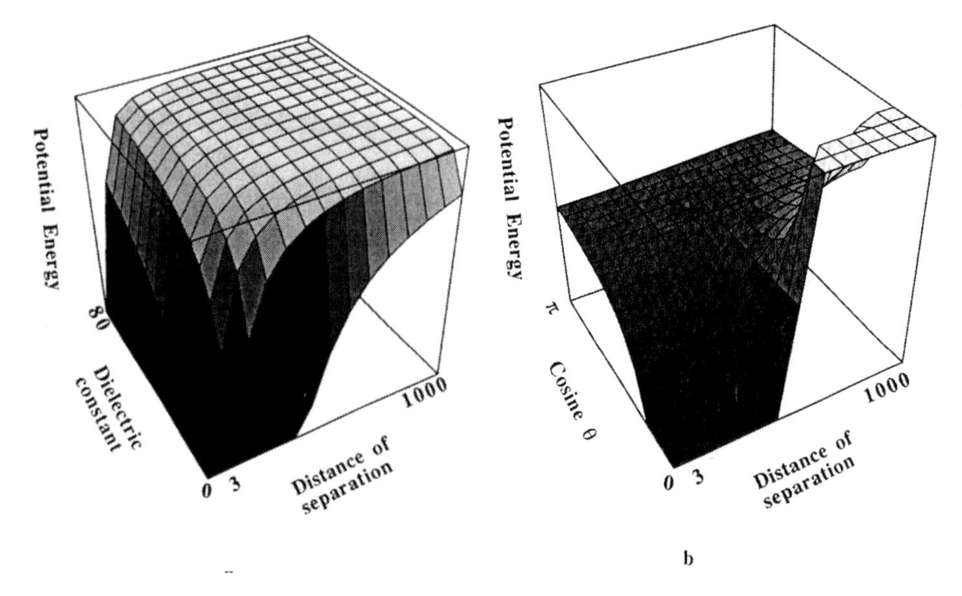

Figure 10.2-4 Energy of interaction for ion–dipole interactions. In the graph on the left, the potential energy is plotted against the distance of separaation and dielectric constant; in the graph on the right, the potential energy is plotted against the distance of separation and cos θ.

Graphs of the variation in $U_{\text{i-id}}$ with respect to both distance and dielectric constant, and distance and α are given in Figure 10.2-5.

When molecules are in very close proximity, attractive forces called **van der Waals** interactions occur. Van der Waals forces are all cohesive forces that vary with respect to distance as $1/r^6$. It is generally useful to subdivide the van der Waals interactions into three types, all derivable from electrostatic considerations. These include permanent dipole–dipole and permanent dipole–induced dipole forces and induced dipole–induced dipole interactions. Van der Waals forces are considered to be the molecular explanation for the cohesive energies of liquids and are similar in magnitude to the enthalpies of vaporization of most liquids, approximately -41.84 kJ mol^{-1}. They can become dominant in reactions where close fitting and proximity are important.

Dipole–dipole interactions occur when molecules with permanent dipoles interact. In dipole–dipole interactions, the orienting forces acting to align the dipoles will be countered by randomizing forces of thermal origin. These terms are reflected in Equation (10.2-4). This equation indicates that dipole–dipole interactions are sensitive to both temperature and distance:

$$U_{\text{d-d}} = -\frac{2}{3}\left(\frac{\mu_1^2\mu_2^2}{kT(4\pi\epsilon_o)^2 r^6}\right) \tag{10.2-4}$$

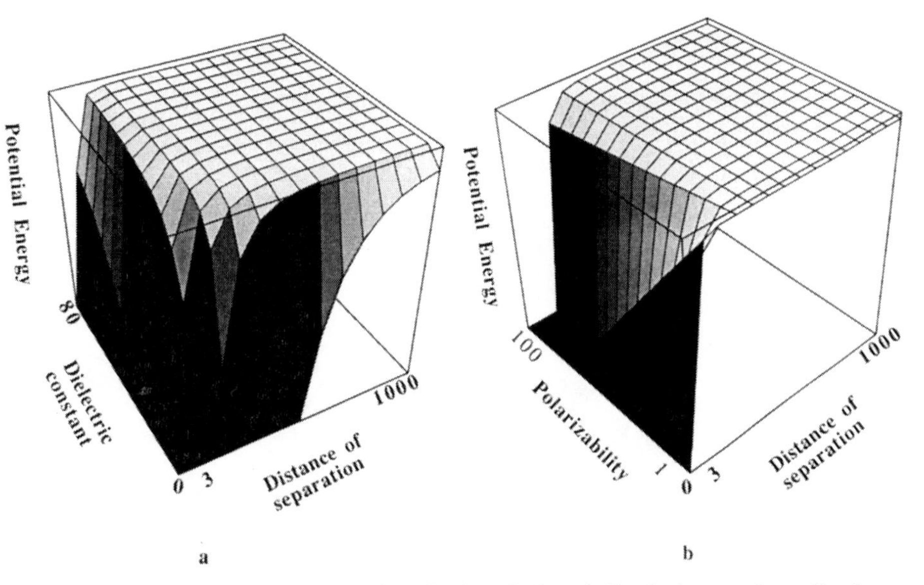

Figure 10.2-5 Energy of interaction for ion–induced dipole interactions. In the graph on the left, the potential energy is plotted against the distance of separation and dielectric constant; in the graph on the right, the potential energy is plotted against the distance of separation and the polarizability, α.

A graph of the variation in $U_{\text{d-d}}$ with respect to both distance and temperature is given in Figure 10.2-6.

Permanent dipoles can induce a dipole moment in a neutral molecule in a fashion similar to that discussed above for ion–induced dipole interaction. This **dipole–induced dipole** interaction depends on the polarizability of the neutral molecule but is not sensitive to the thermal randomizing forces. This is reflected in Equation (10.2-5):

$$U_{\text{d-id}} = -\frac{2\mu_1^2\alpha_2}{16\pi^2\epsilon_o r^6} \tag{10.2-5}$$

A graph of the variation in $U_{\text{d-id}}$ with respect to both distance and temperature is given in Figure 10.2-7.

The dipole moment is a time-average measurement. If a snapshot were taken of any neutral molecule at an instant of time, there would be a variation in the distribution of the electrons in the molecule. At this instant, the "neutral" molecule would have a dipole moment. This instantaneous dipole is capable of inducing an instantaneous dipole in another neutral molecule; thus are born **induced dipole–induced dipole** forces. These forces are also called **London** or **dispersion** forces. Dispersion forces fall off very rapidly with distance but can be quite significant for molecules in close proximity. The deformability of the electron clouds,

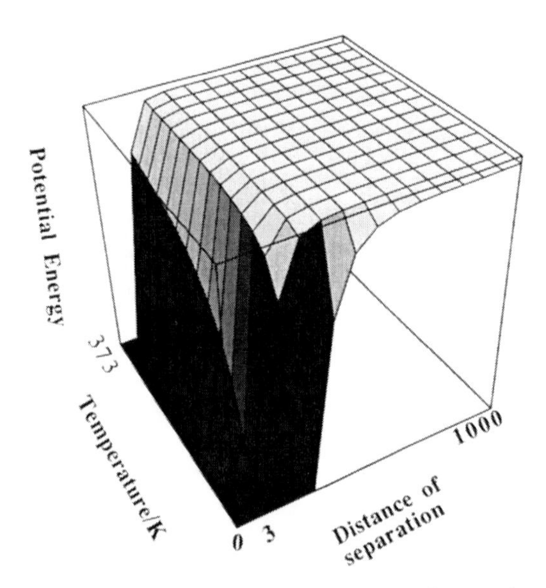

Figure 10.2-6 Energy of interaction for dipole–dipole interactions. The potential energy is plotted against the distance of separation and temperature.

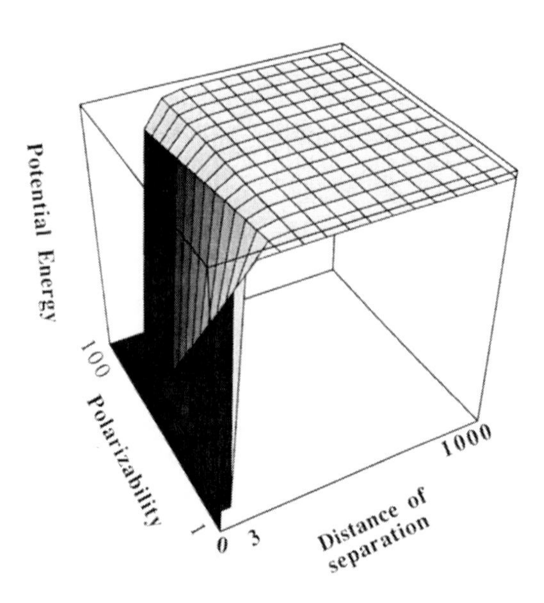

Figure 10.2-7 Energy of interaction for dipole–induced dipole interactions. The potential energy is plotted against the distance of separation and the polarizability of the neutral molecule.

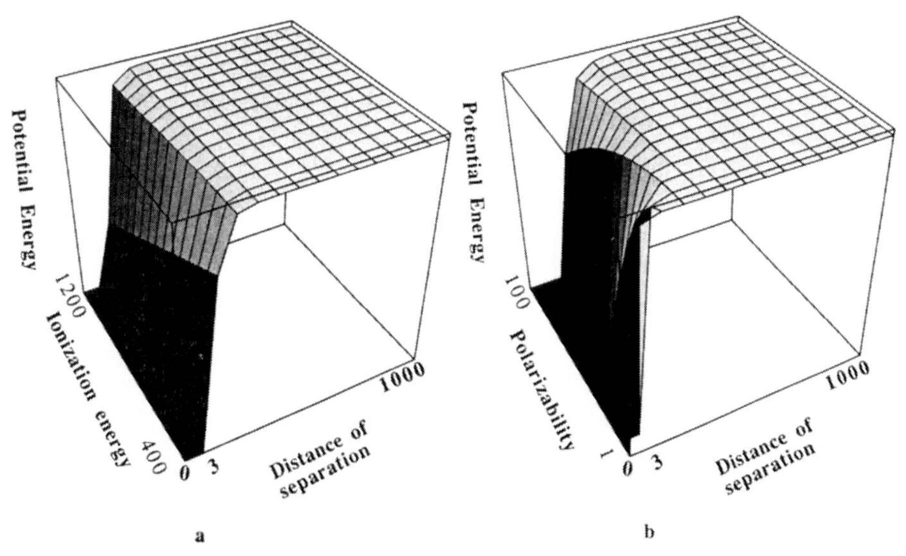

Figure 10.2-8 Energy of interaction for induced dipole–induced dipole interactions. In the graph on the left, the potential energy is plotted against the distance of separation and $f(I)$; in the graph on the right, the potential is plotted against the distance of separation and the polarizability, α.

as reflected by α, is obviously important in these interactions. The rigorous calculation for dispersion forces is quite involved, but an adequate approximation can be written:

$$U_{\text{id-id}} = -\frac{f(I)\alpha^2}{16\pi^2\epsilon_o r^6} \tag{10.2-6}$$

The term $f(I)$ is a function of the ionization energies of the two molecules and is equal to

$$f(I) = \frac{3I_1I_2}{2(I_1 + I_2)}) \tag{10.2-7}$$

Graphs of the variation in $U_{\text{id-id}}$ with respect to both distance and $f(I)$ and distance and polarizability are given in Figure 10.2-8.

The overall interactional energy for van der Waals forces can be written in terms of these three electrostatically derived interactions:

$$U_{\text{vdW}} = -\frac{2}{3}\left(\frac{\mu_1^2\mu_2^2}{kT(4\pi\epsilon_o)^2 r^6}\right) - \frac{2\mu_1^2\alpha_2}{16\pi^2\epsilon_o r^6} - \frac{\mathrm{f}(f(I)\alpha^2}{16\pi^2\epsilon_o r^6} \tag{10.2-8}$$

The **attractive interactional energy** represented by the van der Waals forces is often written in the form:

$$U_{\text{interaction}} = -\frac{A}{r^6} \tag{10.2-9}$$

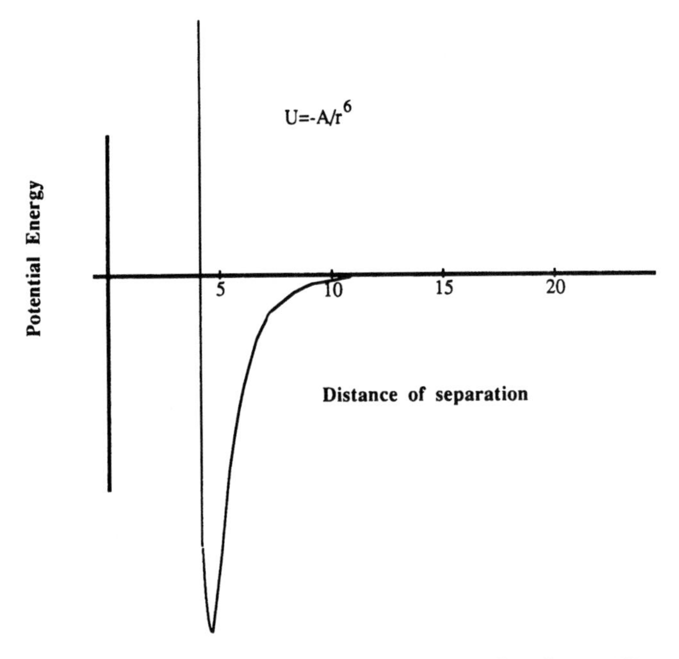

Figure 10.2-9 Energy diagram of van der Waals attraction forces, $U = -A/r^6$.

where A is a different constant for each different molecule. A graph of the van der Waals interactional energy would show that the attractive force increases rapidly as molecules get closer until the molecules actually contact one another, at which point the interactional energy goes instantly to infinity (Figure 10.2-9). In fact, before the molecules try to occupy the same space, another event occurs that is a repulsive force. This is a repulsion that occurs between the electron clouds of the molecules as they approach one another. This electron repulsion must be added into the formula for interactive energy, Equation (10.2-9), and the interactive energy is then usually written in the form of the **Lennard–Jones potential**:

$$U_{\text{interaction}} = -\frac{A}{r^6} + \frac{B}{r^{12}} \qquad (10.2\text{-}10)$$

A graph of the Lennard–Jones potential is shown in Figure 10.2-10.

The **hydrogen bond** is of paramount importance in aqueous systems and has been described in some detail already (cf. Section 9.2). The hydrogen bond is a low-energy bond, the bond energy being on the order of -20 to -30 kJ mol^{-1}.

10.3. The Nature of Ionic Species

If ions can behave as carriers of electric charge, then they must be free to move in the presence of an electric field. How can a system that contains mobile ions be produced? There are several possible methods

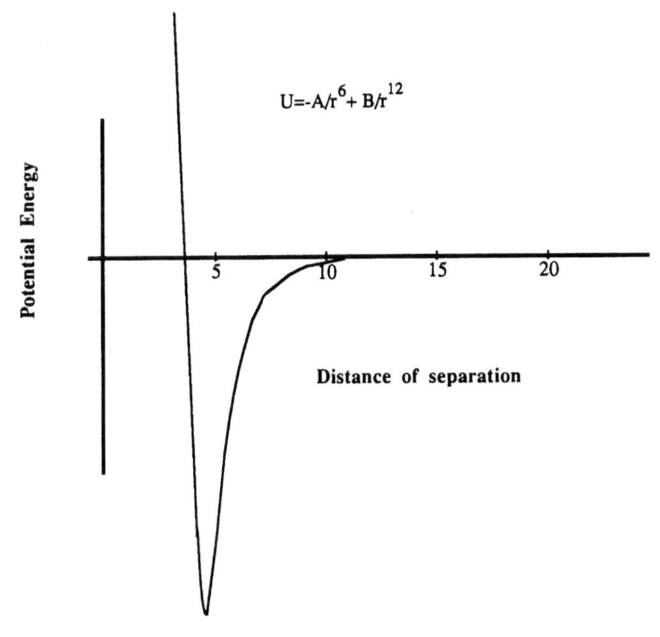

Figure 10.2-10 Energy diagram of a Lennard–Jones potential, $U = (-A/r^6) + (B/r^{12})$.

for producing these mobile ions. Each method reflects the chemical nature of the parent electrolyte. It is worth noting that the term **electrolyte**, as used in biomedicine and electrochemistry, refers to both the ionically conducting medium and to the substances that produce it when they are dissolved or liquefied. A specialized form of electrolyte is the **colloidal electrolyte**, a compound that gives rise to at least one ionic macromolecular species which becomes the ionic conductor.

Materials, such as salts (e.g., sodium chloride), will conduct a current in the solid state whether or not they are in solution. It would be a reasonable inference that such materials must be comprised of ions in their pure state. Such electrolytes are named **ionophores**, or **true electrolytes**. A common nomenclature also calls such electrolytes **strong electrolytes** because when dissolved in water they conduct strongly. This is in contrast to substances called **weak electrolytes**, which do not give rise to high conductances when dissolved in water (see below). This historical nomenclature can lead to some problems in nonaqueous systems. In this book, the term true electrolyte will be used. True electrolytes in the solid form are comprised of a lattice of positive and negative ions arranged in highly ordered and repeating structures. If a crystalline true electrolyte is heated, the translational motion of the ions will increase. A point is reached where the lattice structure is disrupted and the crystal melts. This liquid salt is a **liquid ionic conductor**, because the ions carrying the charge are now free to move in a liquid phase. A true electrolyte therefore is not required to dissolve in a solvent to become a liquid ionic conductor. However, the temperatures necessary to generate a liquid ionic conductor from a pure salt are significant (500 to 1300 K) and certainly not compatible with life on this planet. When a true electrolyte is dissolved in an aqueous solvent, the ions are displaced from their lattice and become associated with the solvent molecules. A liquid conducting solution is generated that does not require the addition of heat to dissociate the ions from their lattice. This event leads to questions about the interactions between the ions and the water molecules that allow the ions to leave the ion lattice structure and carry charge in the aqueous solvent. Indeed, this is the intuitive process that should previously have suggested that there will be solute–solvent interactions in ionic solutions, at least those composed of true electrolytes, such as salts.

When a similar analysis is applied to certain other materials, for example, an organic acid, such as butyric acid, it will be found that in the pure state there is little electrical conduction. This lack of conduction occurs because the carboxylic acids are not ionized in their pure forms. Without ions, no charge transfer can occur. There is a significant difference between the liquid form of a true electrolyte and the liquid form of these organic acids. When an organic acid such as butyric acid is dissolved in water however, an interesting change in behavior takes place; the solution of the acid in water becomes a liquid ionic conductor. By definition, the butyric acid is an electrolyte since it gives rise to a liquid ionic conductor, but, clearly, a different set of events has taken place in this scenario compared to the case of the true electrolyte. The difference is that butyric acid and other carboxylic acids in solution with water will undergo dissociation into an acid–base pair, both of which are ionized and therefore will carry an electric charge. Electrolytes such as these, whose ionic character depends on a chemical reaction to generate their ionic nature, are called **potential electrolytes**. It is evident that a potential electrolyte must interact with its solvent, since it depends on this solvent–solute interaction to become a conducting solution. Historically, electrolytes that gave rise to limited conductance when dissolved in water were called weak electrolytes. This is a solvent-specific classification (a "weak" electrolyte in water may give rise to "strong" electrolyte conducting properties in another solvent), and the preferred term of potential electrolyte will be used in this book. What are the nature of the solute and solvent interactions for both true and potential ions in aqueous solutions? Understanding these interactions will be the task for the remainder of Part II.

Appendix 10.1. Derivation of the Energy of Interaction Between Two Ions

The interactional energy between two ions can be found simply. The starting point is that there is an electrostatic interaction between the two ions given by Coulomb's law:

$$F = \frac{q_1 q_2}{4\pi\epsilon_o \epsilon r^2} \tag{A10.1-1}$$

The energy of the interaction is found by calculating the electrical work necessary to bring a total charge q_1 in infinitesimal increments dr, until it rests a distance r from charge q_2. This work is calculated from the equation that defines work:

$$w = U_{\text{i-i}} = \text{Force} \times \text{Distance} \tag{A10.1-2}$$

$$\int dw = -\int_{\infty}^{r} \frac{q_1 q_2}{4\pi\epsilon_o \epsilon r^2}\, dr \tag{A10.1-3}$$

The constants can be removed from the integral:

$$\int dw = -\frac{1}{4\pi\epsilon_o \epsilon}\int_{\infty}^{r} \frac{q_1 q_2}{r^2}\, dr \tag{A10.1-4}$$

which gives the result

$$U_{\text{i-i}} = \frac{q_1 q_2}{4\pi\epsilon_o \epsilon r} \tag{A10.1-5}$$

which is the result of interest. When two oppositely charged particles are brought together, the sign of the interactional energy will be negative, indicating that there is an attractive force between them.

CHAPTER 11
Ion–Solvent Interactions

11.1. Understanding the Nature of Ion–Solvent Interactions Through Modeling

11.1.1. Overview

Complex systems are often understood by making a number of simplifying assumptions, building a model based on these assumptions, and then testing the model against the real world in the laboratory. Subsequent modifications can be made in the model depending on the empirical results. In building such model systems, the thermodynamic formulations already developed can be invaluable. The model building exercise that will follow is a reasonably detailed examination of the forces that exist between ions and a solvent and will lead to an understanding, in molecular terms, of the deviations from the ideal, characterized empirically by the activity coefficient. Initially, a very simple system that neglects the reader's knowledge of the molecular structure of water will be examined to point the way toward a more detailed examination of ion–solvent interactions. It will be found that, to a crude approximation, the behavior of ion–solvent interactions can be described in terms of simple electrostatics. As further detail about the solvent and its interaction with ions is sought however, it will become apparent that the structure of water has important effects in modifying the electrostatic environment. In considering the dielectric constant and the Kirkwood equations, the effects of water structure on the electrostatic forces are described in some detail, an exercise that is important when considering the more complex relationships in cellular systems.

In studying this section, the reader should endeavor both to gain a reasonable knowledge about the relationship of an aqueous solvent to ions and the charged portions of molecules and also to begin to master the techniques of theoretical model building as a method of scientific research.

11.1.2. The Born Model

For a first approximation of ion–solvent interactions, a set of assumptions will be set forth that may seem too simplistic to the reader, but there is value in understanding the model that can be built with these assumptions. Simple models are usually most appreciated when the formulation becomes more accurate and also more mathematically complex. The system for first consideration will be one proposed in the early part of this century by Born and is called the **Born model**. The model is based on two assumptions (Figure 11.1-1). First, it is assumed that the ion may be represented as a rigid sphere of radius r_i and charge z_ie_o, where z_i is the charge number and e_o is the charge on an electron. The second assumption is that the solvent into which the ion is to be dissolved is a structureless continuum. It will be these assumptions that will be tested when the model's predictive power is assessed in an empirical experiment. A priori then, this model should permit an understanding of the relationship between an ion and its solvent, in this case water.

The ultimate picture of this model is the formation of a solution of ions, represented as rigid spheres of a particular charge, by floating them in a continuum. Where do these charged spheres come from? The source of the ions is really a matter of convenience, and it is most convenient to consider the problem as the movement of a charged sphere (the ion) from a vacuum into the continuum. Now the attention paid earlier to

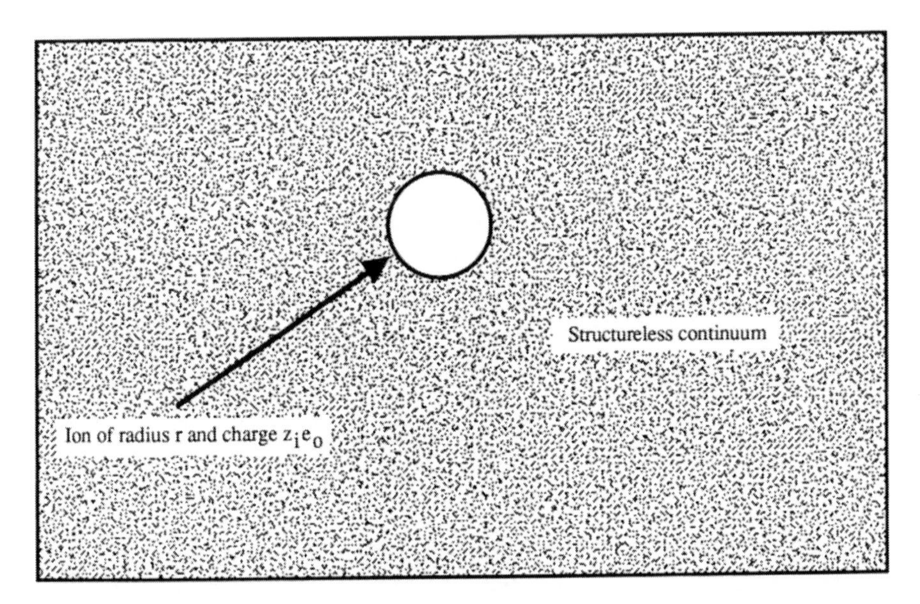

Figure 11.1-1 The basic assumptions on which the Born model is based: (1) that the ion under consideration can be represented by a sphere of radius r_i and charge z_ie_o, and (2) that the solvent into which it is dissolved is a structureless continuum.

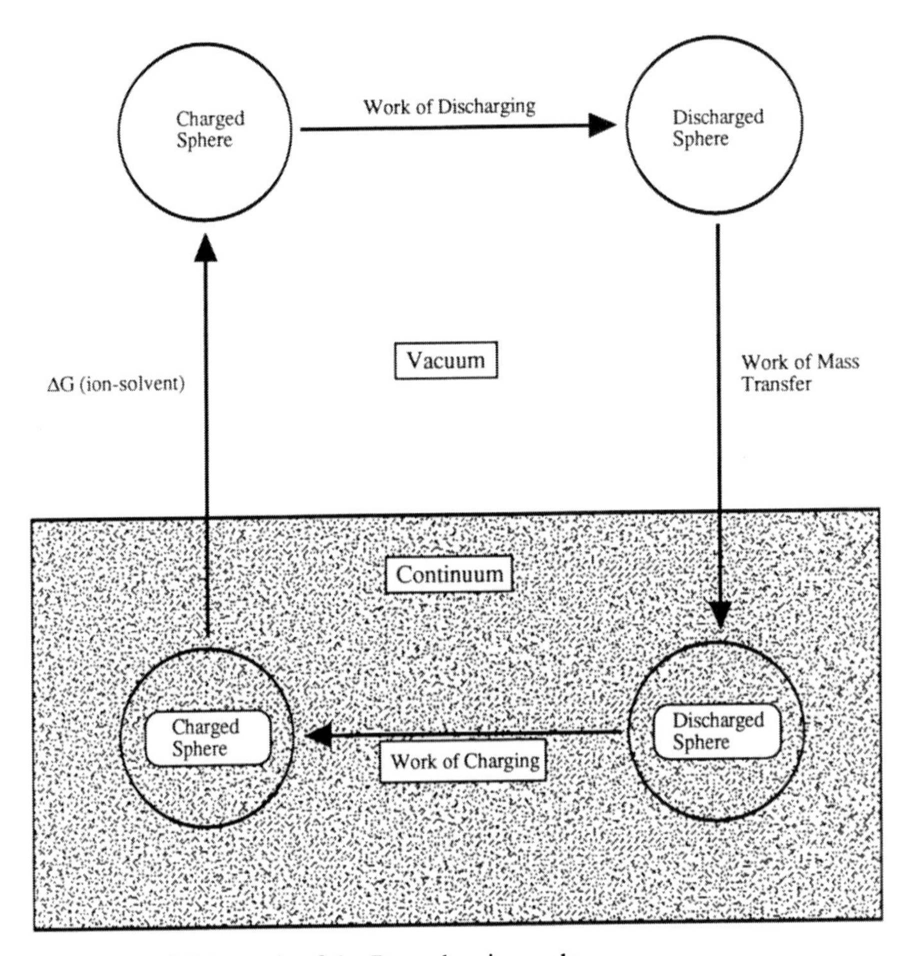

Figure 11.1-2 Schematic of the Born charging cycle.

the concept of a thermodynamic cycle will be valuable. If the work done each step of the way in the movement of the charged sphere from vacuum to continuum and back to vacuum is considered, a cyclic integral will result. Since the entire nature of the ionic character is to be given in this model by the fixed charge on a rigid sphere, the work functions and ion–solvent interactions will be defined strictly by electrostatic forces. Since the goal is to write a thermodynamic cycle that is convenient to solve, the following cycle is proposed (Figure 11.1-2):

1) Start with a charged sphere in a vacuum.
2) Discharge the sphere so that it is electrically neutral; this will be a work function of discharging.
3) Once the sphere is discharged it can now be freely slipped into the continuum, since in this model the only interactions are those that

are due to electrostatic forces; without charge on the sphere, no interactional force will exist. Although equal to zero, this is the work of mass transfer.
4) Once inside the continuum, the sphere is charged back up to its original charge, $z_i e_o$. This will be the work of charging.
5) Since the goal is to design a thermodynamic cycle, the charged sphere is now removed from the continuum and brought back to its starting place in the vacuum. The work necessary to do this is the free energy of the ion–solvent interaction, $\Delta G_{\text{ion-solvent}}$. Since the work to remove the sphere from the continuum into a vacuum is equal to, but opposite in sign to, the work necessary to move the sphere into the continuum, the free energy is $-\Delta G_{\text{ion-solvent}}$ $(-\Delta G_{\text{i-s}})$.

This cycle can be written out as follows:

$$w_{\text{discharge}} + w_{\text{transfer}} + w_{\text{charging}} + (-\Delta G_{\text{i-s}}) = 0 \qquad (11.1\text{-}1)$$

Since $w_{\text{transfer}} = 0$ (cf. step 3 above), this can be algebraically manipulated:

$$w_{\text{discharge}} + w_{\text{charging}} = \Delta G_{\text{i-s}} \qquad (11.1\text{-}2)$$

If the electrostatic work problems can be solved, this model will provide results in terms of ΔG, which is advantageous, since it is already clear that by measuring enthalpies and calculating entropies, an empirical check of the model and its assumptions can be made. Thus, the stage is set for the solution to this problem.

The work represented by the first term, $w_{\text{discharge}}$, occurs in a vacuum and can be determined by solving the problem of charging a sphere in vacuum. This answer is found by bringing infinitesimal bits of charge from infinity to the sphere until the sphere has the correct charge. The solution to this problem is

$$\int dw = \int_0^{z_i e_o} \psi \, dq = w_{\text{charging}} = \frac{(z_i e_o)^2}{8\pi\epsilon_o r_i} \qquad (11.1\text{-}3)$$

(See Appendix 11.1 for the derivation.) Since the work of discharging is just the opposite of the work of charging:

$$w_{\text{discharge}} = -\frac{(z_i e_o)^2}{8\pi\epsilon_o r_i} \qquad (11.1\text{-}4)$$

The work of charging the sphere in solvent is similar, except that the presence of the solvent, no matter what its structure, requires consideration of its dielectric constant, ϵ. The physical treatment of the alteration in the electric field was discussed in Section 10.1 and, in this problem of structureless solvent, is adequate for understanding what is to follow. A molecular-structural treatment of the dielectric constant will be forthcoming.

Earlier, it was shown that the electrostatic force can generally be written

$$F = \frac{q_1 q_2}{4\pi\epsilon_o \epsilon r^2} \tag{11.1-5}$$

where in a vacuum ϵ is equal to 1. When a medium other than a vacuum exists between the two point charges, the force will be less depending on the medium's ability to modify the electrostatic force. In this case, the dielectric constant of the continuum (solvent) is chosen. The following equation for the work of recharging the ion in the solvent is obtained:

$$w_{\text{charging}} = \frac{(z_i e_o)^2}{8\pi\epsilon_o \epsilon r_i} \tag{11.1-6}$$

It is now possible to write the equation describing the free energy of the ion–solvent interaction:

$$\Delta G_{\text{i-s}} = w_{\text{discharge}} + w_{\text{charging}} \tag{11.1-7}$$

Substituting the terms from Equations (11.1-4) and (11.1-6) gives

$$\Delta G_{\text{i-s}} = -\frac{(z_i e_o)^2}{8\pi\epsilon_o r_i} + \frac{(z_i e_o)^2}{8\pi\epsilon_o \epsilon r_i} \tag{11.1-8}$$

Rearranging algebraically gives

$$\Delta G_{\text{i-s}} = -\frac{(z_i e_o)^2}{8\pi\epsilon_o r_i}\left(1 - \frac{1}{\epsilon}\right) \tag{11.1-9}$$

This result is for a single ion. To make experimental analysis more practical, it is useful to describe ΔG on a per mole basis:

$$\Delta G_{\text{i-s}} = -N_A \frac{(z_i e_o)^2}{8\pi\epsilon_o r_i}\left(1 - \frac{1}{\epsilon}\right) \tag{11.1-10}$$

What has now been derived is an expression that defines the energy of interaction between a solvent and an ion in the system. The equation shows that the energy of interaction will depend on the charge on the ion and its radius alone, assuming that the dielectric constant does not change under the conditions of the experiment. Since the concern is to study a model of aqueous electrolyte solutions, the experimentally known dielectric constant for bulk water, 80, will be used for the calculation of the ion–solvent interactional energy in the Born model.

The point of this exercise is to determine if a model of the interaction of rigid charged spheres and a structureless continuum has any validity as a description for aqueous electrolyte behavior. The next step is to define a protocol that can be used to compare the theoretical behavior of the ions to empirical results derived from experiment. The most practical experimental approach is to measure the enthalpy of the ion–solvent interaction. ΔG is related to enthalpy through the following relationship:

$$\Delta G = \Delta H - T\Delta S \tag{11.1-11}$$

The enthalpy is

$$\Delta H_{\text{i-s}} = \Delta G_{\text{i-s}} + T\Delta S_{\text{i-s}} \tag{11.1-12}$$

Before the model can be compared to experiment, some expression for $\Delta S_{\text{i-s}}$ must be found. Earlier (see Equation 5.3-10), it was shown that

$$-\Delta S = \frac{\partial \Delta G}{\partial T} \tag{11.1-13}$$

Therefore, $\Delta S_{\text{i-s}}$ can be found by differentiating $\Delta G_{\text{i-s}}$ with respect to temperature (T). It can be shown experimentally that the dielectric constant varies with temperature, and therefore it must be treated as a variable in the differentiation:

$$\Delta S_{\text{i-s}} = -\frac{\partial \Delta G_{\text{i-s}}}{\partial T} = N_A \frac{(z_i e_o)^2}{8\pi\epsilon_o r_i} \frac{1}{\epsilon^2} \frac{\partial \epsilon}{\partial T} \tag{11.1-14}$$

Finally, solving for $\Delta H_{\text{i-s}}$:

$$\Delta H_{\text{i-s}} = -N_A \frac{(z_i e_o)^2}{8\pi\epsilon_o r_i} \left(1 - \frac{1}{\epsilon} - \frac{T}{\epsilon^2} \frac{\partial \epsilon}{\partial T}\right) \tag{11.1-15}$$

The solutions of this equation are a series of straight lines that depend on the magnitude of charge, z_i, of the ion and the ionic radius. An example of the solution is shown in Figure 11.1-3. What is necessary now is to compare the theoretical values to enthalpies of solvation. It is worth considering the axes of this graph. What in the empirical world corresponds to the radius of the rigid Born sphere? How can the enthalpy of solvation for a single ion be measured? The first question is fairly easily answered. Since this approach supposes that the ion can be represented as a rigid sphere, crystallographic radii of ions obtained from X-ray crystallography will be used. The second question requires a somewhat more complicated answer.

How could a solution of a single ionic species be made in order that the heat of solvation could be measured? Figure 11.1-4 illustrates this problem. Consider the events that would occur as this solution of single ions is made. Initially, the solvent has a neutral electrical charge. Now ions of a single type are added to the solution. As the ions are added, an unbalanced charge is added to the solution and an ever-growing coulombic force is generated. As each additional ion is brought to the solution, an increasing amount of energy is required to provide the work necessary to move the ions into solution. The difficulty with this scenario is that the energy measured in the production of this solution represents both the coulombic work of getting the ions to the solution as well as the energy of solvation. Such an approach will not generate the enthalpy number that is being sought.

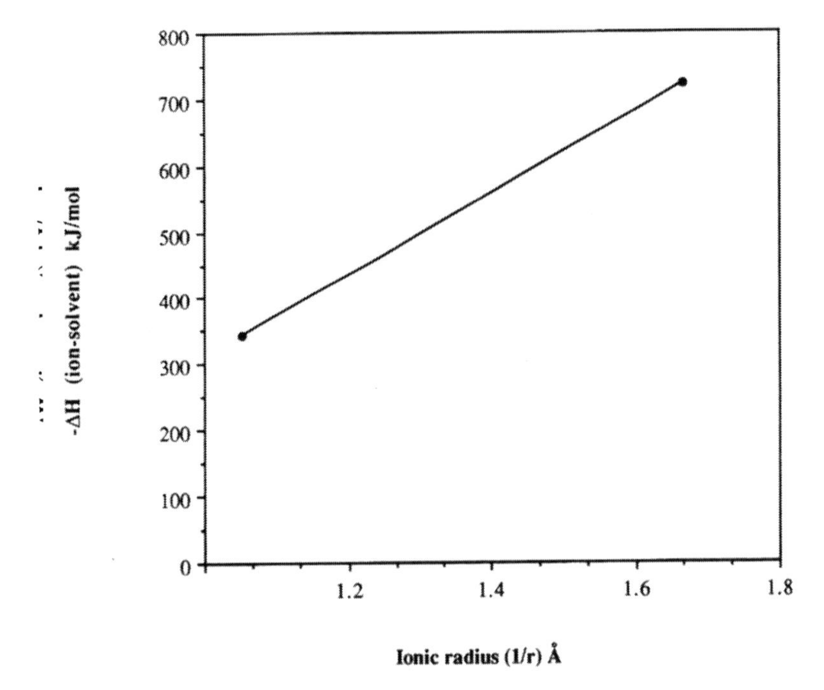

Figure 11.1-3 ΔH of interaction for ions of charge z_i as calculated from the Born model.

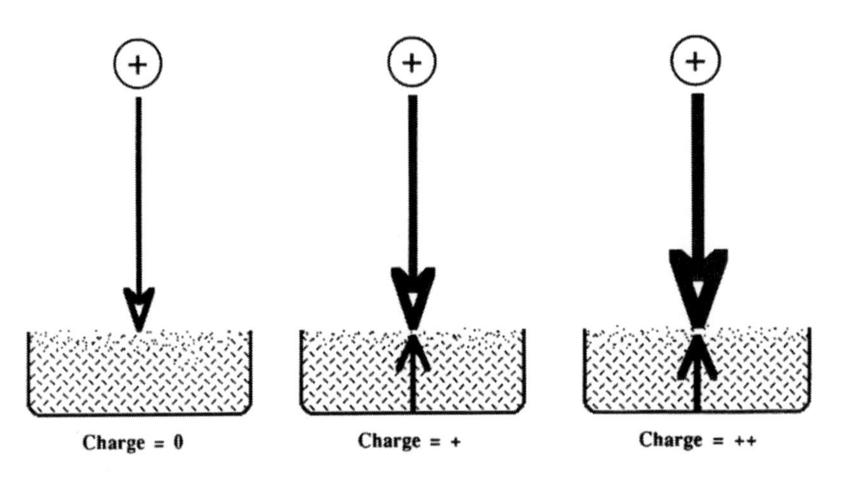

Figure 11.1-4 Illustration of significant difficulties encountered while making a solution of one ionic species.

Beyond the theoretical problems with this approach of adding a single unbalanced charged species to a solution, the usual practical approach to making an electrolytic solution is to dissolve a salt in the appropriate solvent. For example, if a NaCl crystal is being used, the positively charged sodium cations are added with an equal number of negatively charged chloride anions. This solution maintains electrical neutrality, so the enthalpy measured will represent the solvent–solute interaction and will contain no added term that represents work done in overpowering an electric force field. But while the generation of an electrolyte solution by the addition of a salt has eliminated the electroneutrality problem, the solution created now has two species of ions instead of the single species that has been modeled. This problem in obtaining information pertaining to the thermodynamics of a single ion in solution is an important one to solve because many questions in biological systems are concerned with the effects of single ions on a cellular system. Through the use of a relative scale of enthalpies, thermodynamic information about single ion enthalpies can be indirectly derived with reasonable confidence. This indirect method involves using the enthalpies of solvation from a series of salts, each of which shares a common ion, and attributing the differences in enthalpy to the single ion that is not commonly held. For example, the enthalpy of solvation for the single ion sodium could be determined if the enthalpies of solvation for NaCl and KCl were used. Methodology such as this therefore allows the enthalpies of solvation for a single ionic species to be considered. The relative differences found are generally meaningful only if they are linked to some relative point on a scale of other ions. This requires the assumption that the enthalpy of solvation of a salt can be found in which exactly one half of the enthalpy change can be assigned to the solvation of the cation and the other half to the solvation of the anion. Typically, KF is the salt chosen for this assumption because of the virtually identical crystallographic radii of the K^+ and F^- ions. It will appear to the astute reader that there is an element of circularity to this reasoning. That is a correct impression. The problem with obtaining the experimental heats of solvations for single ions is that at some point numerical assignment depends on an assumption based on some theoretical grounds.

Now comparison of experimental data with the theoretical solutions of the Born model can be undertaken. In Figure 11.1-5 the enthalpies of solution for a number of monovalent ions of varying crystallographic radii are compared to the Born solution for the rigid spheres of the same charge and the same radius. While trends between experiment and the model match, there are clearly major discrepancies between the model and the real world. Historically, with empirical corrections to the Born model greater agreement between the experimental and theoretical curves could be achieved. These corrections took two approaches. It can be shown that by simply adding 10 pm to the crystallographic radii of the

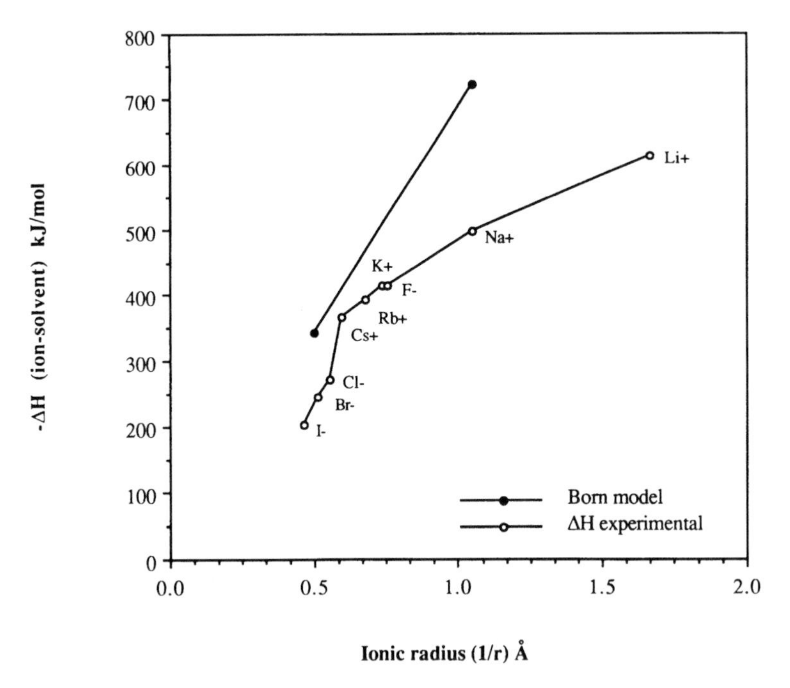

Figure 11.1-5 Comparison between the experimental and theoretical values of the enthalpies of hydration for monovalent ions.

positive ions and 85 pm to the negative ions, the experimental data can become nearly linear, in agreement with the model. It is also possible to consider that the use of the dielectric constant of 80 for the continuum may be incorrect. If, for example, there is an interaction between the charged sphere and the substance of the continuum, could the dielectric constant be changing? These corrections of course strike directly at the heart of the very assumptions on which the hypothesis was originally based. Both suggest that there is something discontinuous about the solvent and something wrong with simply using the crystallographic radii of ions for the size of the rigid sphere. Indeed, both suggest that there are structural considerations that must be included in an examination of the interaction between the ion and its aqueous solvent.

11.2. Adding Water Structure to the Continuum

In the last chapter, the structure of molecular water in its bulk form was discussed. It is important to reemphasize several features. Water is a polar molecule with a dipole moment of 1.85 debye. The negative end of this dipole is oriented toward the oxygen atom and the positive end

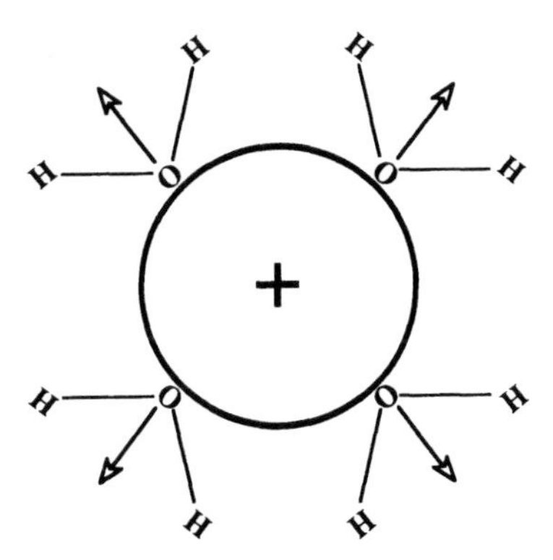

Figure 11.2-1 The interaction of a charged ion with unassociated water molecules. There is an orientation of the water molecules,which behave as dipoles, because of an ion–dipole interaction.

toward the hydrogen atoms. The molecule is ideal for forming hydrogen bonds, and in bulk water these bonds are extensively formed with adjacent water molecules. Because of this hydrogen bonding nature, ice, the crystalline form of water, forms an extensive lattice structure comprised of a puckered hexagonal array of water molecules. While liquid water does not have as highly ordered a lattice structure, significant portions of liquid water maintain the ordered tetrahedral clustering characteristic of ice. At any particular time, water molecules that are not part of the lattice are free to move in and out of the interstitial spaces formed by the structure. The local arrangement of water molecules is in constant flux as some free molecules reassociate with the lattice, and some lattice molecules break off to become free in the interstitial spaces. Many of the thermodynamic qualities of water depend on the hydrogen bonding nature and the structure of the bulk water. What is the effect on the water structure and hence on the properties of the solution when ions (and later macromolecules) are added?

For simplicity's sake, consider what will happen if an ion is added to water molecules that are all free and not in a hexagonal array structure. Figure 11.2-1 illustrates this case. The ion is charged and is the center of an electrical force that emanates symmetrically from the ion. This electrical force will seek to orient any other electrically polar object near it in such fashion that like charges will be pushed away and unlike charges will be attracted. The free water molecules near the ion, being electrically polar, will be subject to this orienting force. If the ion is positively charged, the negative end of the water molecule will be turned toward the ion; if the ion is negative, the water will orient its positive end toward the ion. Consider now the case of structured liquid water. As the ion now exerts its reorienting force on the water molecules, the water molecules are subject to ion–dipole forces that may tear water molecules out of the lattice structure as it orients them toward the ionic point charge. What is different about this model with structured water as the solvent in comparison to the Born model? A principal difference is that the continuum model considered all ion–solvent interactions to be simply electrostatic in nature, while this structured water model considers the interaction to be ion–dipole in nature.

11.3. The Energy of Ion–Dipole Interactions

Understanding ion–dipole interactions requires a sense of the energy of the interaction, $U_{i\text{-}d}$. More detailed analyses are available elsewhere but a qualitative understanding is satisfactory for this discussion. Dipoles were mentioned earlier during consideration of the nonbonding interactions of molecules. It is important to recognize that the separation of charge in a dipole occurs over a finite distance. This characteristic is reflected in the definition of the dipole moment as being the product of the electric charge times the distance separating the two centers. By convention, the direction of this vector is taken as positive when going from the negative to the positive end of the dipole. The energy of interaction between ion and dipole will be the energy between the dipole center and the ion center. The line, r, connecting these two points makes an angle θ with the dipole (see Figure 11.3-1). The energy of interaction is given

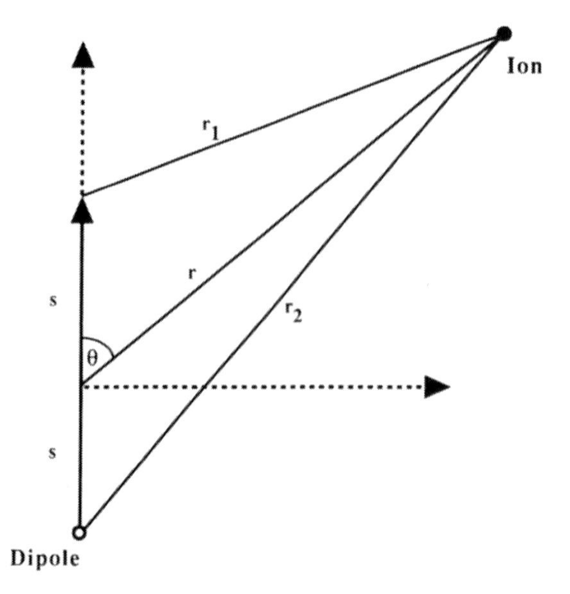

Figure 11.3-1 Geometry for the calculation of $U_{i\text{-}d}$.

by the product of the charge on the ion, $z_i e_o$, and the potential, ψ_r, i.e., $z_i e_o \psi_r$. Finding the potential, ψ_r, due to the dipole can be accomplished if the charges at each end of the dipole are considered separately and then added, according to the law of superposition of potentials. The distance between the charges at each end of the dipole and the ion center however are not equivalent to r but rather to two other radii, r_1 and r_2. Therefore, the potential can be written as

$$\psi_r = \frac{q}{r_1} + \frac{-q}{r_2} \tag{11.3-1}$$

The distances r_1 and r_2 can be found by the Pythagorean theorem. The result has in it a term that includes the distance from each of the ends of the dipole to the midpoint of the dipole. If at this point a very important assumption is made that the distance separating the ends of the dipole from the midpoint is insignificant compared to the distance between the dipole center and the ion center, the following relationship can be ultimately derived:

$$U_{\text{i-d}} = -\frac{z_i e_o \mu \cos\theta}{4\pi\epsilon_o \epsilon r^2} \tag{11.3-2}$$

Clearly, ion–dipole interactions are relatively strong and will depend on the angle of the dipole to the ion, the dipole moment, the charge on the ion, and the distance separating the ion and the dipole. However, as the dipole approaches the ion more closely, the distance r will more closely approximate the ignored distance between the dipole center and the dipole ends. At close proximity, the simplifying assumption of this equation will become unreasonable.

11.4. Dipoles in an Electric Field: A Molecular Picture of Dielectric Constants

Before going back to the question of water structure near an ion, the role of dipoles in an electric field will be considered. An ideal system with which to get an intuitive feel for this subject is a **parallel plate capacitor** (Figure 11.4-1). Using a capacitor to introduce the concepts is useful, because the mathematical descriptions for a parallel plate capacitor are fairly straightforward, and as a thought tool the capacitor lends itself well to the manipulations necessary to understand the topic. Capacitors are so named because of their capacity to store electrical energy as a stored charge. The charge that a given capacitor can store is given by the following expression:

$$Q = C \times V \tag{11.4-1}$$

where Q is the charge in coulombs, C is the capacitance in farads, and

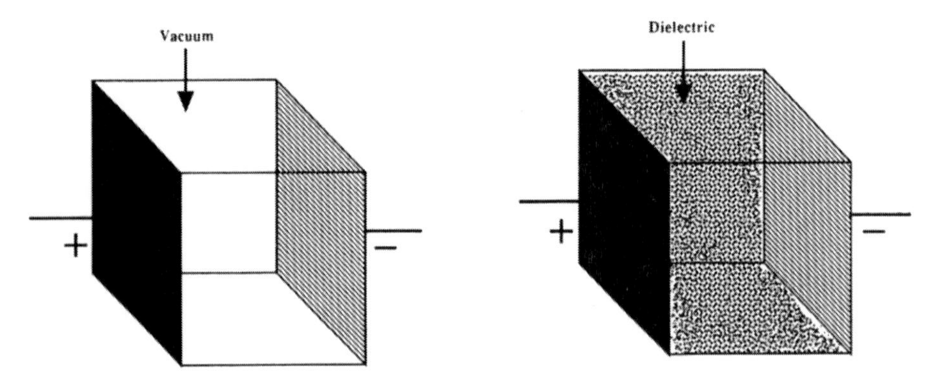

Figure 11.4-1 The parallel plate capacitor can be used as a system to understand dipole behavior in an electric field. On the left, the plates are separated by a vacuum; on the right, the plates are separated by a dielectric material.

V is the potential in volts across the capacitor. In a parallel plate capacitor, the capacitance is given by the equation

$$C = \frac{\epsilon A}{d} \tag{11.4-2}$$

where A is the area of the plates, ϵ is the dielectric constant of the material between the plates, and d is the distance between the plates. Capacitors do not allow current to flow through them because the material separating the plates, called the **dielectric**, is electrically insulating. In a capacitor with an external voltage source applied across it, charge is moved from the battery to the plates. On the plate connected to the negative side of the voltage source, an excess of electrons builds up and this plate becomes progressively more negatively charged. Conversely, the plate connected to the positive side of the battery develops a positive charge as electrons are withdrawn from it. As these charges build up on the plates, an electric force field is generated that ultimately is exactly the same as the voltage of the external source. However, the field direction across the capacitor is opposite to that of the external source. The potential difference falls to zero because of the counterpotential; a point is therefore reached when no more charge can be added and the system is in equilibrium. Figure 11.4-2 illustrates the potential across the capacitor.

The number of fundamental charges necessary to generate the counterpotential is given by Equation (11.4-1). At this point, the voltage or electric field strength across the capacitor is equal to the external force. The capacitor is now disconnected from the circuit. Since the charges on the plates have nowhere to go, the electric field measured across the capacitor will still be the same as the original field strength. If the dielectric is a vacuum, there are no molecules between the plates and the coun-

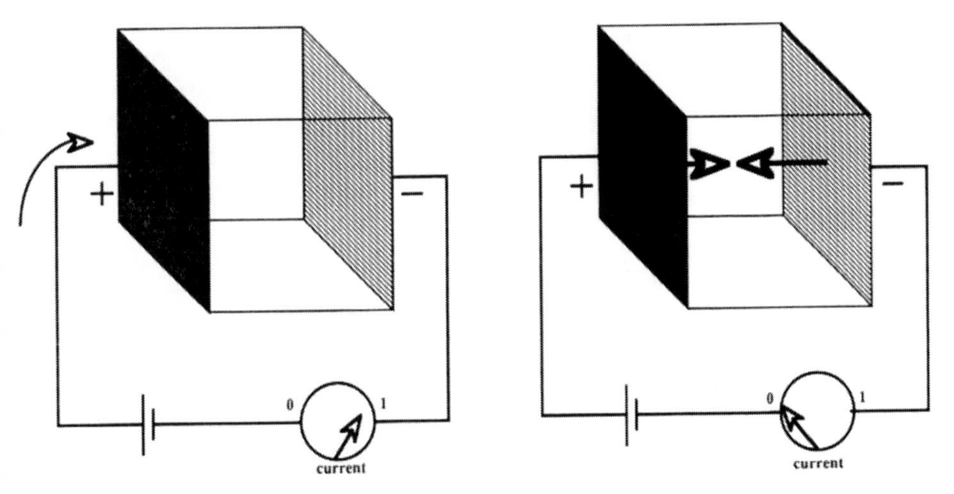

Figure 11.4-2 The buildup of charge on the plates of the capacitor leads to a counter field that exactly opposes the potential of the external source.

terpotential field extends through space as described in the earlier electrostatic problems. What happens however when a dielectric material is now inserted between the charged plates? Inevitably, the measured potential field across the capacitor diminishes, depending on the dielectric material chosen (Figure 11.4-3). This occurs without the loss of charge

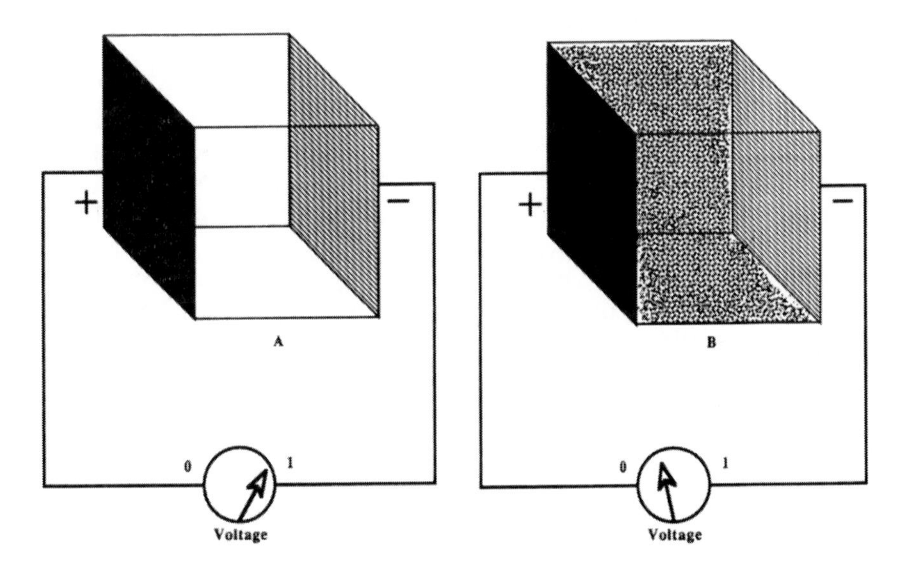

Figure 11.4-3 The measured potential across a charged capacitor is altered by the dielectric constant of the dielectric material separating the two plates.

since the same charge is still on the plates as when the capacitor was initially charged. This means that the capacitance of the parallel plate capacitor has increased solely because the dielectric has changed.

The dielectric constant is derived from a comparison between the capacitance of a parallel plate capacitor using a vacuum as the dielectric material and that of the same capacitor with a different dielectric material replacing the vacuum. The relationship is written

$$\epsilon = \frac{C_{\text{any dielectric}}}{C_{\text{vacuum}}} \tag{11.4-3}$$

What is happening in the dielectric that is changing the measured electrical force across the capacitor? Two cases will be considered: the first will be when the dielectric is comprised of molecules that are permanent dipoles; the second, when the molecules do not have a dipole moment.

In the case where the dielectric is comprised of **permanent dipoles**, how will the dipoles be oriented prior to insertion into the field between the capacitor plates? Measurements on a macroscopic time scale (i.e., microseconds or longer) show that no single overall orientation of the molecules as a group can be ascertained, indicating that the dipoles are oriented randomly. This must be the case since these dipoles comprise a neutral substance that, while made up of polar molecules, is itself apolar. When the randomly arranged dielectric is now placed within the field of the capacitor, the dipoles experience the electrical force and attempt to align themselves with the field. The positive end of the dipole will turn and align itself normal to the negative plate; conversely the negative end of the dipole will align itself normal to the positive plate (Figure 11.4-4).

When a dielectric comprised of molecules without a dipole moment is placed into an electric field, the field will cause the displacement of the electron clouds (negatively charged) away from the nuclei (positively charged), thus inducing a dipole. Because these **induced dipoles** are generated by the field, they are aligned with the field at the instant of generation. They will exist as long as the field remains.

What has been described so far is that an electric field has the ability to align permanent dipoles or induce dipoles along its lines of force. However, these aligned dipole charges are arranged with their charges opposite in orientation to the field. The separated charges of the dipoles themselves will generate an electric field, but one that is opposite to the field of the capacitor. This counterpotential can, by the principle of superposition, be shown to diminish the potential measured across the capacitor. It thus accounts for the increased capacitance of the system.

Are there other forces interacting between the dipoles and the electric field that should be considered? It will be worthwhile investigating this question for it leads back to the original quest for an ion–dipole inter-

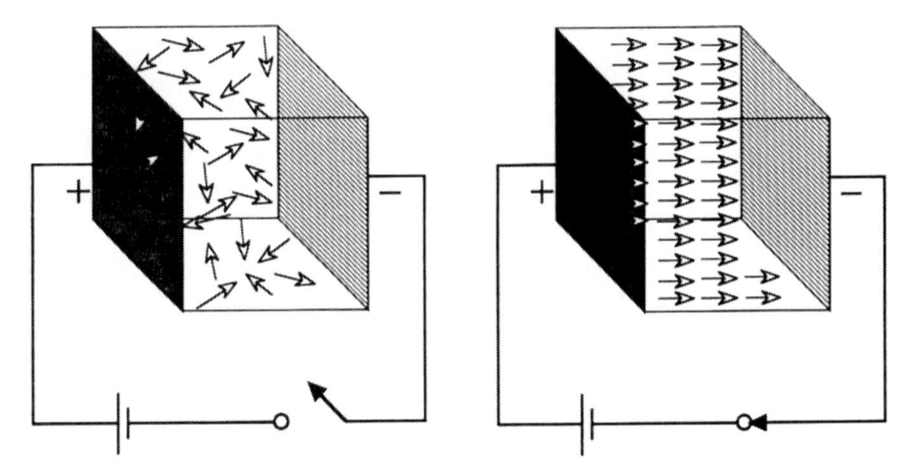

Figure 11.4-4 Alignment of permanent dipoles in the absence and presence of an electric field. In the diagram on the left, there is a random arrangement of dipoles. When the field is switched on, the dipoles attempt to line up in the direction of the lines of electrical force.

action in water solution. There are three further aspects that should be considered:

1) First, what is the ease with which the permanent dipole can be aligned, and how susceptible is the nonpermanent dipole to being induced into an aligned dipole? The measure of these properties is the polarizability of the material.
2) The next consideration is, what are the effects of nonaligning forces on the orientation of the dipoles and on the overall average dipole moment of such dipoles? The primary considerations here are the randomizing thermal effects.
3) Finally, the question of how the structure of the dielectric will respond to the various forces acting on it needs to be examined. Since the primary concern here is with water, the dielectric material of interest will be one containing permanent dipoles. Recognizing that in the liquid state lattice structural considerations will need to be considered, initially it will be simpler to consider each water molecule as free and independent (this corresponds of course to gaseous water, or steam).

The **susceptibility** of a material is the physical property that determines its dielectric constant. When a dielectric is placed in an electric field, there will be two forces acting against each other. The electric field caused by the excess charge, Q, on the capacitor plates either will attempt to align the permanent dipoles or will induce temporary dipoles. Depending on the ease with which these dipoles can be created or aligned, a charge separation, q_{dipole}, will occur. q_{dipole} has a field that is directed counter to the field resulting from the capacitor charge q. An analysis of this problem (Appendix 11.2) gives the result that the electric force, X_{ext}, directed from one plate through the dielectric toward the second plate is

$$X_{ext} = 4\pi (q - q_{dipole}) \tag{11.4-4}$$

q_{dipole} depends on the number of dipoles that can align with the original capacitor-derived field and can be given by an equation that will relate the ease of dipole alignment (α or the **polarizability**), the field that is causing the alignment (X_{ext}), and finally the number of dipoles aligned (n). n is dependent on both α and X_{ext}. The relation is given by

$$q_{dipole} = \alpha X_{ext} n \tag{11.4-5}$$

Combining Equation (11.4-4) with Equation (11.4-5) gives

$$X_{ext} = 4\pi q - 4\pi\alpha X_{ext}\, n \tag{11.4-6}$$

Ultimately, it can be derived that

$$\epsilon - 1 = 4\pi\alpha n \tag{11.4-7}$$

This equation quantifies the idea that the dielectric constant is related to the ability of a material to respond to an electric field by generating a counter force through dipole-based charge separation.

The dielectric constant depends on the number of *aligned* dipoles to generate the counter field. Therefore, the ease of dipole alignment, α, must be balanced against nonaligning or randomizing forces. In the earlier discussion of the Born model, it was noted that the dielectric constant varies with temperature. Thermal effects are a significant randomizing force in dielectrics. Imagine that, as the dipoles swing into alignment with the external field, they are constantly buffeted by thermal collisions that randomly knock them out of alignment. The higher the temperature, the more likely is the chance of a thermal collision. Ultimately, a balance will be struck between the orienting force of the external field and the randomizing force of the thermal collisions. This balance can be written in a form of the **Boltzmann distribution law:**

$$n_o = Re^{-w/kT} \tag{11.4-8}$$

Here, n_o represents the number of dipoles aligned at a particular angle θ to the external field, R is a proportionality constant, and w represents the work done by the molecule in orienting itself with the field:

$$w = -\mu X_{ext} \cos\theta \tag{11.4-9}$$

The average orientation of the dipoles to the field will depend on the thermal energy in the system. As the system becomes hotter and hotter, the dipoles will have a greater tendency to be randomly oriented in spite of the aligning force of the external electric field. By using standard formulas for finding average values, a value for the average dipole moment depending on both field and temperature can be found:

$$<\mu> = \frac{\mu^2 X_{ext}}{3kT} \tag{11.4-10}$$

This formula gives the average dipole moment for a gas dipole. For a number of dipoles, n, the charge can be given by

$$q_p = \frac{n\mu^2 X_{ext}}{3kT} \tag{11.4-11}$$

Thermal effects have a greater disturbing effect on the permanent dipoles than on the induced dipoles. Molecules with induced dipole moments do not move into an orientation but rather are oriented as they are created by the electric field. The charge due to the dipole orientation, q_{dipole}, will be represented by a term containing permanent dipole contributions (q_p) and induced dipole contributions (q_i):

$$q_{dipole} = q_p + q_i \tag{11.4-12}$$

Earlier, it was shown that the polarizability was related to q_{dipole}:

$$\alpha = \frac{q_{dipole}}{nX_{ext}} \tag{11.4-5}$$

Combining Equations (11.4-12) and (11.4-5) gives the following result:

$$\alpha = \frac{q_p}{nX_{ext}} + \frac{q_i}{nX_{ext}} \tag{11.4-13}$$

which is the same as

$$\alpha = \alpha_{orient} + \alpha_{deform} \tag{11.4-14}$$

The first term, α_{orient}, represents the contribution to the total polarizability of the dielectric molecules from the permanent dipoles and the second term, α_{deform}, represents the polarizability due to the ability of the electric field to deform the molecule and induce a dipole. This equation can be rewritten in terms of the dielectric constant:

$$\epsilon - 1 = 4\pi n\alpha_{orient} + 4\pi n\alpha_{deform} \tag{11.4-15}$$

The first term can be rewritten in terms of Equation (11.4-11), giving

$$\epsilon - 1 = \frac{4\pi n\mu^2}{3kT} + 4\pi n\alpha_{deform} \tag{11.4-16}$$

When the temperature rises, the contribution to the dielectric constant from the permanent dipoles grows smaller and eventually becomes negligible, and the entire value of e is accounted for by the induced dipoles in the system.

11.5. What Happens When the Dielectric Is Liquid Water?

Thus far, the considerations related to the interactions of a dipole with an orienting electric field have been confined to dipoles that interact only with the field but not with one another. In other words, there is no structural constraint on the interaction. This scenario only holds for dipoles that exist in dilute gaseous form. How will the structure of liquid water affect the ability of an electric field to orient the dipoles of the water? This question is answered by considering a single water molecule at the center of a tetrahedral array of water. As the electric field attempts to orient this particular dipole, it must orient the entire complex of water molecules. The issue is no longer simply one of the orientation of a permanent dipole countered by the forces of thermal disarray. The concern rather is the problem of aligning the entire array to the electric field. Attention must shift away from the average dipole moment of a single molecule to the average dipole moment of the icelike structure. The average dipole moment of a tetrahedral cluster of water molecules will be the vector sum of the dipoles for each element of the cluster:

$$<\mu_{\text{cluster}}> = \mu + g(\mu \overline{\cos \gamma}) \tag{11.5-1}$$

where g, the number of nearest neighbors, is multiplied by the average of the cosine values between the central molecule's dipole moment (μ) and those of the neighbors. As can be seen, the dipole moment of the cluster will be greater than the electric moment of the single central dipole itself. A relationship between the dipole alignment and the thermal disorientation can also be made, and the following is the result:

$$<\mu_{\text{group}}> = \frac{\mu^2(1 + g\,\overline{\cos \gamma})^2}{3kT} X \tag{11.5-2}$$

where X is the electric field. Proof that this is a general equation that includes the special case of the gaseous dipole derived earlier is left as an exercise for the reader.

Is the electric force that originates at the capacitor plate the only force that is of consequence for the dipole residing inside the cluster of icelike water? The interaction of the free dipoles in a dilute gas is limited because the individual dipoles are too far apart to feel the field emanating from even the next nearest dipole. In the case of a tetrahedral water cluster, the dipoles are atomically close to one another and the electric field experienced by the central dipole will be a sum of the external electric field derived from the capacitor, X_{ext}, and as well a more local field derived from the surrounding molecules, X_{loc}. Therefore, in the case of liquid water, the expression for the orientation polarizability (Equation 11.4-5) does not simply depend on the external field, as was the case for gaseous dipoles. The better expression for the field acting to orient the dipoles will be derived from both X_{ext} and X_{loc}.

It is possible to derive an expression for the electric field operating on a reference dipole or cluster of dipoles through the approach of Onsager. The reference dipole or grouping is considered to be surrounded by other molecules that, to a very reasonable approximation, can be thought of as forming a spherical cavity around the dipole grouping of interest. If the dipole cluster is removed, an empty cavity is the result, and the field inside the cavity (where the dipole would be located) is a resultant of both the external field and the field generated through the partial orientation of the surrounding dipoles. The expression for the local field inside the cavity acting on the dipole can be written as follows:

$$X_{\text{loc}} = \frac{3\epsilon}{2\epsilon + 1} X_{\text{ext}} \qquad (11.5\text{-}3)$$

Since ϵ is always greater than 1, this equation says that the local field, and hence the orienting force on a dipole in a medium such as liquid water, is always greater inside the condensed phase than when the dipole is in a dilute gas phase.

What then is the effect on the dielectric constant of a condensed phase such as water, now taking into account the effects of structure and dipole interactions in this structure? This can be answered by starting with the earlier expression

$$\epsilon - 1 = 4\pi\alpha n \qquad (11.4\text{-}7)$$

This equation was solved by determining a from the expression

$$\alpha = \frac{q_{\text{dipole}}}{nX_{\text{ext}}} \qquad (11.4\text{-}5)$$

This was shown to be equal to

$$\alpha = \frac{q_p}{nX_{\text{ext}}} + \frac{q_i}{nX_{\text{ext}}} \qquad (11.4\text{-}3)$$

Recall that the first term represents the orientation polarizability and the second the deformation polarizability. Earlier results, Equations (11.4-10) and (11.4-11), are combined to give the following expression for q_p:

$$q_p = n < \mu_{\text{group}} > \qquad (11.5\text{-}4)$$

Substituting the result for the average dipole moment of the group (Equation 11.5-2) and the local electric field X_{loc} (Equation 11.5-3) in place of the field X_{ext}, q_p becomes

$$q_p = \frac{n\mu^2(1 + g\,\overline{\cos\gamma})^2}{3kT} \frac{3\epsilon}{2\epsilon + 1} X_{\text{ext}} \qquad (11.5\text{-}5)$$

The q_i term may also be written

$$q_i = n\alpha_{\text{deform}} \frac{3\epsilon}{2\epsilon + 1} X_{\text{ext}} \qquad (11.5\text{-}6)$$

Table 11.5-1. Comparison of the dipole moments and dielectric constants of H_2O and H_2S, and $(CH_3)_2CO$.

Substance	Dipole moment (debye)	Dielectric constant
H_2O	1.85	78.50
$(CH_3)_2CO$	2.90	20.00
H_2S	1.02	9.3

When these substituted equations are used to evaluate Equation (11.4-7), the following is the result:

$$\epsilon - 1 = 4\pi n \frac{3\epsilon}{2\epsilon + 1}\left(\alpha_{\text{deform}} + \frac{\mu^2(1 + g\,\overline{\cos\gamma})^2}{3kT}\right) \tag{11.5-7}$$

This equation is the **Kirkwood equation** for finding the dielectric constant of a condensed medium. This equation takes into account the interactions that arise from the structured nature of the water molecules and also the effect of the localized field rather than the external field in calculating the dielectric constant of a medium such as water. It is clear that the dielectric constant of a medium such as water is profoundly affected by the cluster structure of the associated dipoles. The structural linking of dipoles will increase the number of nearest neighbors and hence increase the dielectric constant.

With the Kirkwood equation in hand, consider again the case of liquid H_2S and H_2O (Table 11.5-1). Recall that H_2S has a dipole moment (1.02 debye) that is approximately one-half that of H_2O and, largely due to a lack of hydrogen bonding, does not have a strongly associated structure like that of water. The dielectric constant of water is approximately 10 times that of H_2S. While not all of the increased dielectric constant is attributable to the association of the dipoles in H_2O, the largest contribution is due to this structure and not to the increased dipole moment. To illustrate this point, consider a liquid that is nonassociating, yet whose molecules have a dipole moment even larger than that of water, $(CH_3)_2CO$. The dipole moment of $(CH_3)_2CO$ is over one and a half times larger than that of H_2O, yet the dielectric constant of $(CH_3)_2CO$ is four times less that of H_2O. Thus, the Kirkwood equation predicts, and experimental data confirm, that it is the strong interactions between molecules in an associated structure that account for the high dielectric constant of water.

Finally, the Kirkwood equation indicates several parameters that are important for the definition of the dielectric constant. The most obvious is g, the number of nearest neighbors to the reference dipole. For liquid water, X-ray studies have indicated that approximately 4.5 water dipoles are coordinated around a reference central water molecule. Additionally, the Kirkwood equation explicitly gives the variation of dielectric constant with temperature. If this equation is used to calculate a dielectric constant for water at various temperatures, the theoretical values are usually within 10% of the experimental measurements.

11.6. Extending the Ion–Solvent Model Beyond Born

By a circuitous route, the argument has now returned to the origin. What is the relationship between an ion and a structured solvent such as water? The ion will again be considered a charged sphere of fixed radius. There are several forces operating in this system, and the relationship of the water molecules to the ions will depend on the balance between these forces. Since the water molecules act as dipoles, there will be electrostatic forces acting between the ions and the dipoles. The strength of these electrostatic forces falls off according to the inverse square law and therefore is formidable near the ion but negligible at some distance from the ion. The hydrogen bonding interactive forces that will hold the water molecules together in icelike clusters are present throughout the bulk of the water. Finally, there are thermal forces acting to disrupt the water structure. The ion–dipole interactions are relatively powerful forces, especially when compared to the hydrogen bonding that maintains the structure of the water. Near the ion, the electrostatic forces could therefore be expected to dominate, and water molecules will be oriented as dipoles to the ion's electrostatic field. These molecules will be torn from their association with other water molecules and will become immobilized or trapped around the ion. So tightly held are these water molecules that along with the ion they become a new kinetic body. The molecules in this solvent sheath are often referred to as **immobilized** or **irrotational** water, because of the complete loss of freedom. At distances far removed from the ion, where the electrically orienting force is insignificant, the predominant forces will be those of the structure-stabilizing hydrogen bonding and the thermal effects that lead to a randomization of the structure. This is a description of the forces acting in bulk water, and indeed there is essentially no effect of the ion's electric field at these distances.

The two extremes are not difficult to comprehend, but what is the structure in the region where there is a balance between the orienting forces of the ion and the structure-forming forces of bulk water? In this region of balanced forces, the water molecules will be sometimes oriented to the ion-derived forces and at other times oriented to the structure of the bulk water. From a structural viewpoint therefore, the time-averaged positions of water molecules in this region will give the appearance of a broken-down or less icelike arrangement than that of the bulk water. The water molecules in this region will be less associated and therefore have different physical properties, such as a decreased dielectric constant, when

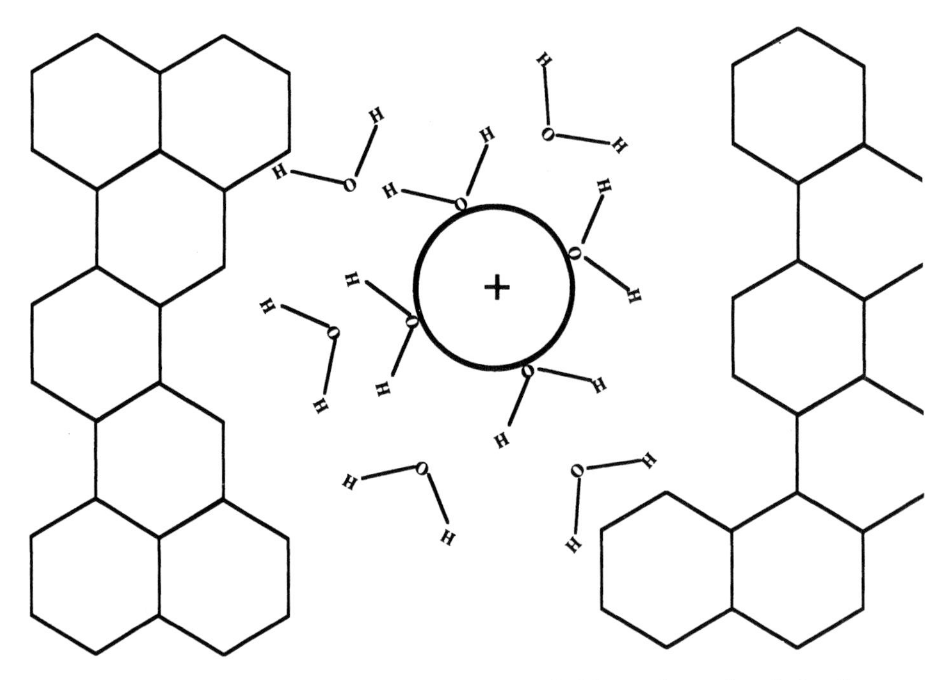

Figure 11.6-1 An ion in water will be surrounded by an irrotational sheath, an intermediate sheath and finally water with bulk properties.

compared to bulk water. On the other hand, the ionic forces acting to orient and fix the water molecules to the ion will not be adequate to make the water in this region a permanent part of the same kinetic entity that the ion and the irrotational water sheath comprise. Consequently, while the irrotational water will move in lockstep with the ion when the ion moves, the water in this secondary sheath will not travel with the ion and its primary hydration sheath.

Solvent interactions with an ion therefore may be considered to occur in three layers (Figure 11.6-1). First, there is a primary hydration sheath comprised of irrotational solvent molecules that move as the ion itself moves. As the distance from the ion increases, a secondary hydration sheath is entered where there is a partial ionic and solvent structural influence. Finally, the third layer is the bulk solvent itself and essentially feels no local force related to the ion.

11.7. Recalculating the Born Model

When the idea of these hydration sheaths is incorporated into an analysis like the Born model, the predicted enthalpies of solution more closely fit the experimental values than those predicted by the Born equation. Such an approach was developed by Bernal and Fowler and by Elay and Evans. In a fashion analogous to the Born model, the heat of solvation, $\Delta G_{i\text{-}s}$, can be calculated by taking a rigid sphere representing an ion in a vacuum and bringing it into the solvent. Since the final structure in the solvent will be an ion surrounded by irrotational water and a secondary hydration sheath, the energy for the events leading to this structure must also be considered. The energy calculations will include enthalpic terms related to both the Born charging of the ion and the energy needed to disrupt the water structure and rearrange the solvent in the solvation sheaths. Furthermore, there will be entropic changes resulting from the alterations in degrees of freedom as the water molecules experience altered structural environments. Because this analysis is only semiquantitative, its focus will remain on the enthalpic changes, and the entropic changes will be ignored. In similar fashion to the treatment of the Born model, a thermodynamic cycle can be constructed to derive an expression for $\Delta G_{\text{ion-solvent}}$.

How does consideration of water structure modify the Born model? A number of changes derive from explicit consideration of the structure of the solvent. Explicit consideration of the structure of the solvent will dictate treatment of the ion as a unit comprised of the ion and its irrotational sheath of hydration water rather than as a simple rigid sphere. Several new terms must be considered as a result of these changes. What are the changes in the new model?

1) Rather than the crystallographic radius, the radius of the ion plus its hydration sheath will be used.
2) The formation of the hydration sheath around the ion will require a specific number of molecules of water, n.
3) The solvent has structure, and therefore the space occupied by the hydrated ion added into the solvent must be explicitly considered. Addition of the hydrated ion will require the removal of enough water molecules to leave space for the hydrated ion. The size of an ion is about the same as that of a water molecule so the volume required for the hydrated ion will be related to the expression $V(n+1)$. In this expression, V is a function of the number of water molecules in the hydration sheath plus one volume (equivalent to a water molecule) for the space occupied by the ion.
4) $n + 1$ water molecules will be removed from the solvent which will both make room for the hydrated ion and provide the molecules necessary to hydrate the ion (which starts in a vacuum). However, for every hydrated ion, one extra water molecule is taken out of the solvent. At some point in the cycle, these molecules must be returned to the solvent.
5) Additional terms must also be added to account for the formation of the hydration sheath around the ion. These terms must include an

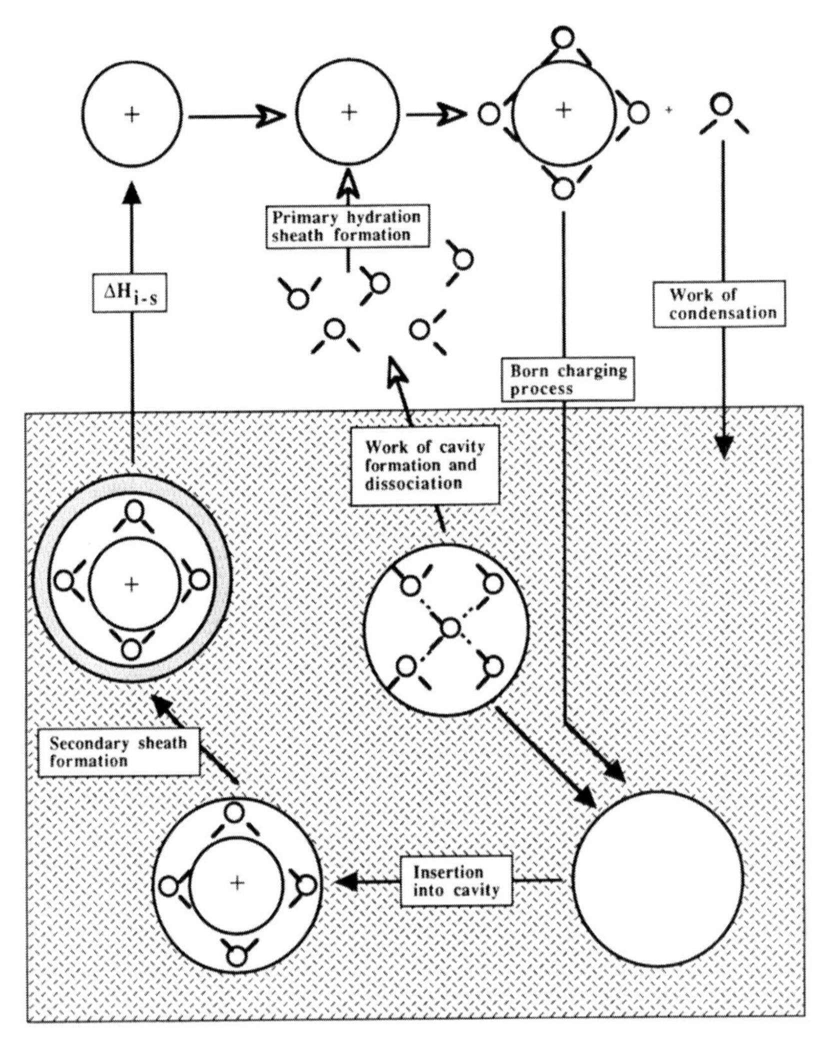

Figure 11.7-1 Schematic of thermodynamic cycle for the refinement of the Born model.

amount of work required to dissociate the $n+1$ water molecules and the work of the ion–dipole interactions necessary to form n molecules into a primary hydration sheath.

6) Finally, the secondary hydration sheath will need to be formed after the ion has been moved into the solvent.

Once all of the terms listed above are considered, then the problem can be treated in a fashion similar to the earlier Born model. Figure 11.7-1 outlines the terms of the cycle described here.

Table 11.7-1. Components for determination of $\Delta H_{(\text{ion-solvent})}$ for the extended Born-type model.

Born charging[a] bond/condensation	Ion–dipole interaction[b]	Hydrogen bonding
$-\dfrac{N_A(z_i e_o)^2}{8\pi\epsilon_o(r_i+2r_s)}\left[1-\dfrac{1}{\epsilon}-\dfrac{T}{\epsilon^2}\dfrac{\partial\epsilon}{\partial T}\right]$	$-\dfrac{N_A n z_i e_o \mu}{4\pi\epsilon_o(r_i+r_s)^2}$	126 kJ mol^{-1}[c]
		84 kJ mol^{-1}[d]

[a]Note that the radius has been changed to reflect the radius of the ion plus its hydration sheath.
[b]Note that the dielectric constant does not appear in this equation because the orientation of the dipoles into the primary sheath is considered to occur in a vacuum.
[c]For negative ions with a hydration sheath of four water molecules.
[d]For positive ions with a hydration sheath of four water molecules.

The enthalpy of ion–solvent interaction for this model can be written as the sum of the various terms just discussed. Table 11.7-1 provides a listing of these terms. As Table 11.7-1 indicates, the heat of interaction will depend on a term reflecting a process similar to the Born charging process derived earlier, and also on a term that describes the ion–dipole interactions resulting from the formation of the primary solvation sheath. The amount of work necessary to remove n water molecules, dissociate them, and condense the extra water molecules as well as the energy of interaction of the primary solvated ion with the secondary solvation sheath must also be known. It turns out that, on a molar basis, this work function is 84 kJ mol^{-1} for positive ions and 126 kJ mol^{-1} for negative ions. These numbers are derived from a consideration of the breaking and reformation of the hydrogen bonds in the water structure as the water molecules are removed, dissociated, replaced, and reoriented in the cycle. If the number of hydrogen bonds broken and then ultimately reformed is added up and multiplied by the energy of hydrogen bond formation (-21 kJ mol^{-1}), these are the values found for a tetrahedral water structure. The difference in energy between the positive and negative ion interaction occurs because of the different orientation of the water molecules in the primary sheath. In the case of the positive ion, the negative pole of the water dipole is closest to the ion, and for the negative ion, the positive pole of the dipole orients to the ion. In the case of the positive ion, there are two fewer hydrogen bonds reformed after the water-removal-ddissociation–reassociation process, while four less hydrogen bonds are reformed in the case of the negative ion. The structural proof of this statement is left as a problem for the reader. The work required for the return of one mole of water for each mole of ion added to the solvent is the **latent heat of condensation**, about 42 kJ mol^{-1}. By adding together the heat of condensation and the energy necessary to break the requisite number of hydrogen bonds in each case, the numbers given in Table 11.7-1 are found.

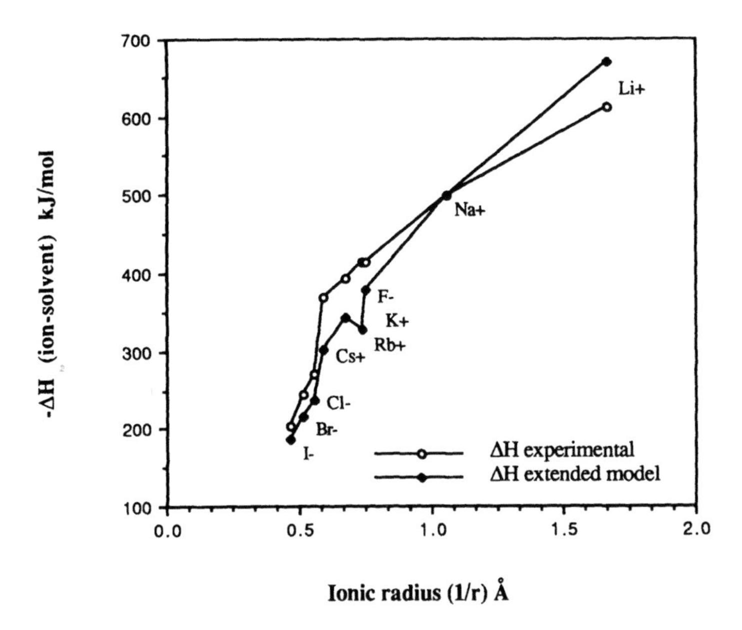

Figure 11.7-2 Comparison of $\Delta H_{(\text{ion-solvent})}$ from experiment and extended Born theory.

These considerations allow the following equations to be described for the ion–solvent interaction in an aqueous solvent, where water is considered a dipole and the coordination number of the water in the primary sheath is four:

For negative ions:

$$\Delta H_{\text{i-s}} = 126 - \frac{4N_A z_i e_o \mu_W}{4\pi\epsilon_o(r_i + r_W)^2} - N_A \frac{(z_i e_o)^2}{8\pi\epsilon_o(r_i + 2r_W)}\left(1 - \frac{1}{\epsilon} - \frac{T}{\epsilon^2}\frac{\partial\epsilon}{\partial T}\right) \quad (11.7\text{-}1)$$

For positive ions:

$$\Delta H_{\text{i-s}} = 84 - \frac{4N_A z_i e_o \mu_W}{4\pi\epsilon_o(r_i + r_W)^2} - N_A \frac{(z_i e_o)^2}{8\pi\epsilon_o(r_i + 2r_W)}\left(1 - \frac{1}{\epsilon} - \frac{T}{\epsilon^2}\frac{\partial\epsilon}{\partial T}\right) \quad (11.7\text{-}2)$$

where r_W and μ_W represent the radius of the hydration sheath in water and the dipole moment of water, respectively.

Do these added terms increase the accuracy of the theoretical model in its predictions of the heats of hydration of individual ions? Figure 11.7-2 shows the calculated versus the empirical heats of solvation for the ions. Consideration of the structural aspects of ion–solvent interactions leads to a vast improvement in the agreement between model and experiment.

This model suggests that the major contributions to the energy of interaction between solvent and ions come from the electrostatic terms of the modified Born charging model and the dipole–ion interactions of the solvation sheaths. It can be shown in a more rigorous analysis that a few more modifications can lead to an even better model. While such an analysis is beyond the intended scope of this book, it is worth noting that when water is considered as a quadrupole (an assembly of four charges), as originally described by the Bjerrum model of the molecule (cf. Section 12.4), the energetics of the interactions in the primary solvation sheath are more accurately described. The most important concept to understand is that the nature of the ion–solvent interaction in aqueous solutions is highly dependent on the polar nature of the water molecules.

11.7.1. Ion–Solvent Interactions in Biological Systems

At this point, it is clear that the interaction of an ionic species with its solvent will depend greatly on the polar and structural behavior of the solvent and on the electrostatic nature of the ion. Just as the behavior of the ion will be changed as a kinetic entity by the association of solvated molecules, the properties of the solvent will also be affected. This can be extremely important when the case of biological solutions is considered. Biologically relevant solutions contain mixtures of small polar and non-polar molecules as well as macromolecules. A detailed treatment of the interactions of such molecules with water is forthcoming in the following chapters. Even without a detailed knowledge of the forces acting between nonionic molecules and an aqueous solvent, it is pertinent to ask at this point what the effect of adding ionic species to an aqueous solution of other molecules will be.

The most obvious effect of the addition of ions to a solution will be manifest on the solubility of a particular molecular species. Consider the case of a macromolecule that has a limited solubility in water. While the details have not yet been described, this molecule is in solution because of its interactions with the water. If a quantity of ions is now added to the solution, the ions will usually strongly associate with the water, forming the familiar hydration sheaths. The association of the water molecules with the ion effectively removes water from the solution. Thus, the effective concentration of the water (its activity) decreases with the addition of the ions. If the ion that is added has a solvation number of four molecules of water, then for every mole of ion added, the effective concentration of water will be decreased by four moles. For example, in a liter of water that contains 55.5 moles of water, the addition of one mole of ions will leave only 51.5 moles of water free. This is a significant decrease in activity. The macromolecule that was barely soluble in water before has now been stripped of some of the water molecules necessary to keep it in solution and will precipitate out. This is one basic reason why the time-honored "salting out" step is so useful in protein purifications.

The reader should at this point have a much greater appreciation for the effect of the structure of water on the interactions between ionized solutes and water. So far, the only interactions that have been considered are those between the fields of the ion and the dipoles or quadrupoles of water molecules. The idea that an ion might interact with other ions has been ignored to this point. However, few solutions are sufficiently dilute that such interactions can reasonably be ignored. Therefore, it is necessary to consider at this point the more typical case of a solution where an ion will interact not only with its solvent but with other ions as well.

Appendix 11.1. Derivation of the Work to Charge and Discharge a Rigid Sphere

The work necessary to charge a rigid sphere is found simply by calculating the work required to bring an amount of charge q from an infinite distance to reside on the sphere. This is done by bringing infinitesimally small quanties of charge dq to the sphere until the total charge is reached. The work is found by the product of the charge and the electrostatic potential:

$$dw = \psi_r \, dq \tag{A11.1-1}$$

Starting with a sphere of zero charge and adding infinitesimal amounts of charge until the charge $z_i e_o$ is reached is accomplished by integrating:

$$w = \int dw = \int_0^{z_i e_o} \psi_r \, dq \tag{A11.1-2}$$

The electrostatic potential is given by

$$\psi_r = \frac{q}{r} \tag{A11.1-3}$$

Substituting this into Equation (A11.1-2) gives

$$\int dw = \int_0^{z_i e_o} \frac{q}{r} \, dq = \left(\frac{q^2}{2r}\right)_0^{z_i e_o} \tag{A11.1-4}$$

$$w = \frac{(z_i e_o)^2}{2r} \tag{A11.1-5}$$

This is the work for charging a sphere in a vacuum and in the cgs system.

In SI units and in any dielectric medium, Equation (A11.1-5) can be written

$$w_c = \frac{(z_i e_o)^2}{8\pi\epsilon\epsilon_o r} \quad \text{(A11.1-6)}$$

The work of discharging is simply the opposite of the work of charging and is written as

$$w_{dis} = -\frac{(z_i e_o)^2}{8\pi\epsilon\epsilon_o r} \quad \text{(A11.1-7)}$$

Appendix 11.2. Derivation of $X_{ext} = 4\pi(q - q_{dipole})$ by Gauss's Law

Gauss's law states that the electric field that is perpendicular to a surface bounding a volume (X_{ext}) is equal to $4\pi q$, where q is the charge contained in the volume. The net electric field between two parallel plates in a capacitor is found by imagining a rectangular-shaped volume that encloses on one side the charge q from the plate of the capacitor and on the other side the countercharge resident on the ends of the dipoles (q_{dipole}). Therefore, the charge in the volume is exactly $q - q_{dipole}$. By Gauss's law, the electric field is therefore

$$X_{ext} = 4\pi(q - q_{dipole}) \quad \text{(A11.2-1)}$$

which is the result sought.

CHAPTER 12

Ion–Ion Interactions

12.1. Ion–Ion Interactions

If a potential electrolyte is added to an aqueous solution, it is conceivable that its dissociation may be so low that few ions are produced. It is also possible to make a solution from a true electrolyte so dilute that the number of ions in solution is very small. In both of these cases, the analysis completed in the previous chapter would be adequate to describe the interactions that cause the solution to deviate from ideality. These solutions would be so dilute that each and every ion in solution could in a sense look out past even its secondary hydration sheath and see only bulk water. These ions would thus seem to exist completely alone and isolated in solution.

In biological systems however, ions are not often found in dilute conditions. Biological systems contain a great number of ions in solution even when the proteins, carbohydrates, and colloidal particles are ignored. What is the effect of other ions on a reference ion that looks out and feels the presence of other ions? The question is answered by finding out what forces act on the reference ion. Once these forces are identified, quantification of these forces can be used to study how the behavior of the reference ion will change. The test of the model will be to see how effectively the activity of the ion can be predicted.

It is worth taking a moment to reflect on how the understanding of these various interactions in an ionic solution is of value to the biological scientist. The aim of these reflections is twofold. First, it is desirable to arrive at an intuitive sense of how molecules will behave in the complex systems that characterize the biological world. Second, it is important to have some quantitative formulations that will allow these complex systems to be predicted and understood. When the biochemist or biophysicist measures a chemical change or a force acting in a biological system, the behavior of the reference ion, molecule, or force will be affected by the interactions throughout the system. Since the reference object is subject to the complex array of forces that are now being derived, it would be impossible to understand the implications of the measured quantity if the parts of the system are not appreciated.

If the interactions between a single ionic species and all the other species in solution are to be accurately described, it will be necessary to quantify the forces of interaction. What is sought is a description of the free energy of ion–ion interactions, $\Delta G_{\text{i-i}}$, in a system where the initial state is one of ions in solution with no ion–ion interaction and the final state is one in which the ions in solution are interacting. The energy of this change in state is $\Delta G_{\text{i-i}}$. In most cases, and especially in cellular systems, the interest will be to describe the partial free energy change associated with a single ionic species with respect to the entire ionic assembly described by $\Delta G_{\text{i-i}}$. This is precisely the situation described previously (cf. Section 6.2), and the required partial free energy change is given by the chemical potential, μ. In a fashion similar to Equation (6.2-3), the quantity $\Delta G_{\text{i-i}}$ will be given by the sum of the chemical potentials, $\mu_{i\text{-I}}$, for each ion that interacts with the assembly of other ions:

$$G_{\text{i-i}} = \sum n_i \mu_{i\text{-I}} \qquad (12.1\text{-}1)$$

It is reasonable to assume (at least initially), as was done in the model of ion–solvent interactions, that the ion behaves as a charged rigid sphere and that the forces acting on it are essentially electrostatic in nature. Consequently, the chemical potential for a single ionic species will be related to the work of charging up a mole of the ions of interest while in the proximity of other ions. This is more than vaguely reminiscent of the energy of interaction found in the Born charging process and can be written

$$w_{\text{charging}} = \frac{(z_i e_o)^2}{8\pi\epsilon_o \epsilon r_i} \qquad (11.1\text{-}6)$$

In terms of the chemical potential, $\mu_{i\text{-I}}$, this can be written

$$\Delta\mu_{i\text{-I}} = N_A w_{\text{charging}} \qquad (12.1\text{-}2)$$

Combining these two equations and recognizing that $z_i e_o/4\pi\epsilon_o \epsilon r_i$ is actually ψ, the electrostatic potential of the ion, the following can be written:

$$\Delta\mu_{i\text{-I}} = N_A \frac{z_i e_o}{2}\psi \qquad (12.1\text{-}3)$$

The chemical potential change then of the interaction between the total ionic assembly and the ionic species of interest can be found by determining the electrostatic field at each individual ion that is a result of the other ions in solution. This field could be found if the spatial distribution of the ions in the solution were known relative to the reference ion. If such structural information were known, then the field could be calculated by the law of superposition, and the energy of interaction would result.

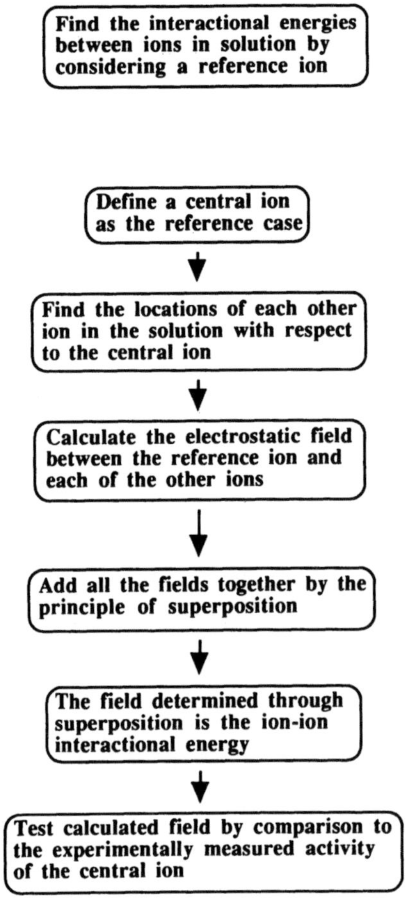

Find the interactional energies between ions in solution by considering a reference ion

Define a central ion as the reference case

↓

Find the locations of each other ion in the solution with respect to the central ion

↓

Calculate the electrostatic field between the reference ion and each of the other ions

↓

Add all the fields together by the principle of superposition

↓

The field determined through superposition is the ion-ion interactional energy

↓

Test calculated field by comparison to the experimentally measured activity of the central ion

Figure 12.1-1 General approach to solving the problem of ion–ion interaction forces.

Consequently, by constructing a model of the orientation of the ions that surround the reference ion and comparing the calculated energy of interaction with that found by experiment (by measuring activity), a test of the accuracy of the proposed structure will be possible. Figure 12.1-1 summarizes the approach.

The first step in solving this problem is to examine the distribution of charges that surround a reference ion as described by Debye and Hückel in 1923. In this analysis, a central or reference ion is considered and is given a specific charge. Like the Born model, the Debye–Hückel treatment assumes that everything other than this central ion is to be treated as a nonstructured continuum of charge residing in a dielectric continuum. For the purposes of the discussion, the dielectric constant is taken to be

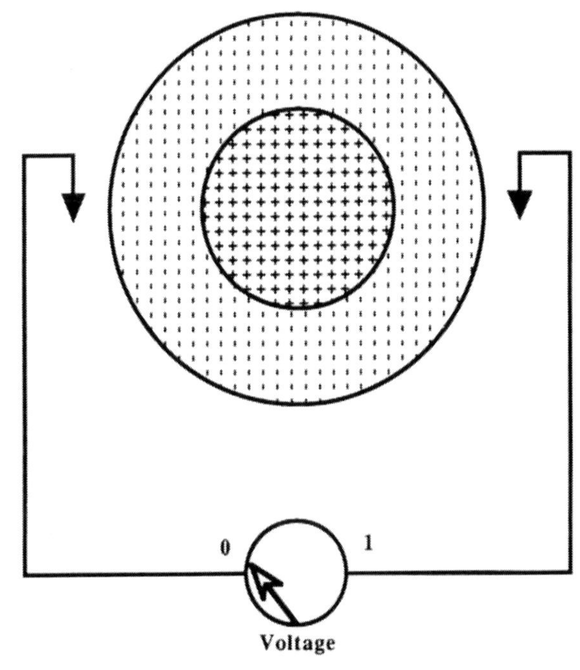

Figure 12.1-2 The principle of electroneutrality gives rise to the existence of clouds of countercharge to balance the electric field emanating from the central ion.

that of bulk water. Now, consider the system just proposed. The central discrete ion is charged, and therefore in the region close to the ion an electrical imbalance exists due to a charge separation. As pointed out previously, nature does not like regions of imbalance, and therefore an attempt will be made to neutralize this unbalanced charge distribution. The charge from the central ion will be neutralized by the continuum of charge that surrounds it. At equilibrium, the charge on the ion will be exactly countered by a countercharged atmosphere that will be arranged in some charge distribution around the ion as shown in Figure 12.1-2. Locally, there will be regions of excess charge density, but, taken as a whole, the solution will be electroneutral since each of the central ions will be surrounded by atmospheres of charge that are exactly equal in magnitude but opposite in sign to the charge on the central ion. This is an important principle and is called the **principle of electroneutrality**. Mathematically, the principle can be expressed as follows:

$$\sum z_i e_o X_i = 0 \tag{12.1-4}$$

where z_i is the number of elementary charges, e_o, carried by each mole fraction, X_i, of the species making up the entire solution.

Since the arrangement of the ions in the Debye–Hückel model is represented by the arrangement of a smeared charged atmosphere, describing the excess charge density around the central ion will provide information related to the spatial configuration of the actual ions in solution. In the earlier discussion of the dielectric constant, the balance between the directed electrostatic forces and the randomizing thermal forces was presented. As will shortly be seen, similar effects will need to be taken into account here.

It is important to make a historical note as to the origin of the idea of treating the interionic forces mathematically. Milner in 1912 first attempted to mathematically link the interionic forces acting in solution to the behavior of the solution. His treatment however was based on statistical mechanics; it was difficult to understand and therefore not easy to subject to experimental test. Consequently, the work went essentially unrecognized. The idea of the ionic atmosphere was proposed by Gouy in 1910. Gouy used the approach of smoothing out the ionic charges into a continuum and applying Poisson's equation. It was by building on these pioneering works that Debye and Hückel made their own highly significant contributions that led to an experimentally testable model.

The model of Debye and Hückel starts with several basic assumptions that ultimately lead to an equation that relates the activity coefficient of an ion in an electrolyte solution to the concentration of the other ions present. This is the relationship that, as described above, is sought to define the chemical potential change due to ion–ion interactions of the ion under scrutiny. The theoretical treatment by Debye and H:auuckel makes several principal assumptions:

1) A central reference ion of a specific charge can be represented as a point charge.
2) This central ion is surrounded by a cloud of smeared-out charge contributed by the participation of all of the other ions in solution.
3) The electrostatic potential field in the solution can be described by an equation that combines and linearizes the Poisson and Boltzmann equations.
4) No ion–ion interactions except the electrostatic interaction given by a $1/r^2$ dependence are to be considered (i.e., dispersion forces and ion–dipole forces are to be excluded).
5) The solvent simply provides a dielectric medium, and the ion–solvent interactions are to be ignored, so that the bulk permittivity of the solvent can be used.

In the **Debye–Hückel model**, because the central reference ion is surrounded by a countering cloud of charge, knowledge of the density of the cloud relative to the distance away from the central ion will give information about the arrangement of the ions that actually make up the cloud. Because the simplifying premise of the theory smears the individual ions

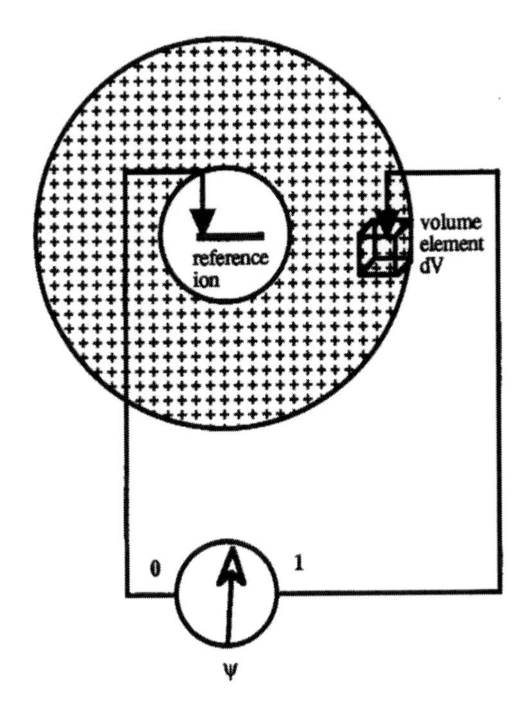

Figure 12.1-3 Examining the electrostatic field existing between a reference ion and a volume element provides information about the spatial position of the charge density in the ionic atmosphere.

into a continuum, the question of charge density in the cloud relative to the distance away from the central ion can be approached as a problem in electrostatics where the charge under consideration is a time-averaged smear. The question to be solved is, How does the charge density of the cloud change with respect to the distance away from the central ion? Because the excess charge density in the smeared cloud represents a separation of charge, the relationship between the two separated charges, that is, the central ion and a specific volume element of the cloud, can be discovered by examining the electrostatic potential that exists between the two charges (Figure 12.1-3).

The excess charge in a given volume, dV, of the cloud can be related to the electrostatic potential existing between the central ion and the small volume element under consideration. This relationship is given by **Poisson's equation**, which relates the charge distribution, ρ_r, to the electrostatic potential, ψ in a spherically symmetrical system such as the one under discussion. The equation is written as follows:

$$\frac{1}{r^2}\frac{d}{dr}r^2\left(\frac{d\psi}{dr}\right) = -\frac{\rho_r}{\epsilon_o\epsilon} \tag{12.1-5}$$

This equation relates the electrostatic potential to the excess charge density in the volume element as it varies with distance from the central reference ion. The dielectric constant in this equation is usually taken as that of bulk water.

The total charge in a given volume element is found by adding together all of the charges that reside in that volume (taking into account both the number of ions and the valence number for each species). This charge can be described by

$$\rho_r = \sum n_i z_i e_o \tag{12.1-6}$$

It follows therefore that the total electric charge available in a system will be described by the sum of all charged elements in solution.

Quantification of the charge in an electrolyte solution can be achieved through the use of a parameter called the **ionic strength**. The ionic strength is given by $\boldsymbol{I}$, where

$$I = \frac{1}{2} \sum n_i z_i^2 \tag{12.1-7}$$

and expressed in terms of concentration, I can be written

$$I = \frac{1}{2} \sum c_i z_i^2 \tag{12.1-8}$$

As an example, consider the ionic strength of a 1:1 electrolyte, 0.15 M NaCl solution:

$$\begin{aligned} I &= \frac{1}{2} \{[Na^+]\,(1)^2 + [Cl^-]\,(-1)^2\} \\ &= \frac{1}{2} [(0.15)(1) + (0.15)(1)] = 0.15 \end{aligned}$$

Taking as a second example a 1:2 electrolyte, I for a 0.15 M $CaCl_2$ solution would be

$$\begin{aligned} I &= \frac{1}{2} \{[Ca^{2+}](2)^2 + [Cl^-](-1)^2 + [Cl^-](-1)^2\} \\ &= \frac{1}{2} [0.15(4) + (2)(0.15)(1)] \\ &= \frac{1}{2} (0.6 + 0.3) = 0.45 \end{aligned}$$

It should be obvious that a relationship of ionic strength to concentration, $I = \boldsymbol{kc}$, exists, where the value of k is defined by the type of electrolyte in solution, M^xA^y. The values of $\boldsymbol{k}$ are given in Table 12.1-1. As will be seen shortly, the nonideal behavior of electrolyte solutions is related to the total number of ions (and given by the ionic strength), rather than to the actual chemical nature of the species making up the solution.

Table 12.1-1. Values of k, defining the relationship between ionic strength and concentration, depending on valence type of electrolyte.

	M^+	M^{2+}	M^{3+}	M^{4+}
A^-	1	3	6	10
A^{2-}	3	4	15	12
A^{3-}	6	15	9	42
A^{4-}	10	12	42	16

Returning now to the expression for charge excess, ρ_r, in a specific volume element, dV:

$$\rho_r = \sum n_i \, z_i e_o \tag{12.1-6}$$

It is necessary to be able to characterize each of the elements n_i, and to do this the Boltzmann distribution law is used. Earlier, the form of this law was given:

$$n_i = n_i^o \, e^{-U/kT} \tag{12.1-9}$$

U in the case described here represents the potential energy change associated with the change in distribution of n_i particles in the volume element as distinguished from the distribution of particles in the bulk given by n_i^o. U is a time average of all the forces that act on the particles and as such represents the sum of all forces that influence the distribution of the particles or ions. If U is positive, then the distribution of ions in the volume element is less than the bulk distribution, while a negative value for U indicates that the distribution of ions is increased relative to the bulk distribution. When U is zero, the distribution of ions in the volume element is identical to the bulk distribution. Because a central tenet of the Debye–Hückel theory is that the only interactional forces acting between the ions are electrostatic in nature and follow a $1/r^2$ relationship, the term U can be evaluated as

$$U = z_i e_o \psi_r \tag{12.1-10}$$

The Boltzmann relationship now becomes

$$n_i = n_i^o \, e^{-z_i e_o \psi_r/kT} \tag{12.1-11}$$

The distribution of the ions in the volume element as a function of the bulk distribution is now described, and Equation (12.1-6) can be modified to reflect this:

$$\rho_r = \sum n_i^o \, z_i e_o \, e^{z_i e_o \psi_r/kT} \tag{12.1-12}$$

This expression can be made linear if a simplifying condition is established. This is another of the assumptions on which Debye and Hückel built their model. If only cases are chosen where the electrostatic potential

ψ_r is so small that the term $z_i e_o \psi_r / kT \ll 1$, then the exponential in Equation (12.1-12) can be rewritten as a Taylor power series:

$$e^{-z_i e_o \psi_r / kT} = 1 - \frac{z_i e_o \psi_r}{kT} + \frac{1}{2}\left(\frac{z_i e_o \psi_r}{kT}\right)^2 \ldots \tag{12.1-13}$$

and all but the first two terms can be ignored. Equation (12.1-12) becomes

$$\rho_r = \sum n_i^o z_i e_o \left(1 - \frac{z_i e_o \psi_r}{kT}\right) \tag{12.1-14}$$

or

$$\rho_r = \sum n_i^o z_i e_o - \sum \frac{n_i^o z_i^2 e_o^2 \psi_r}{kT} \tag{12.1-15}$$

The first term is the same as Equation (12.1-4) which gave the total charge of the solution, and by electroneutrality must be equal to zero. Equation (12.1-15) therefore simplifies to

$$\rho_r = -\sum \frac{n_i^o z_i^2 e_o^2 \psi_r}{kT} \tag{12.1-16}$$

This is the **linearized Boltzmann equation.**

By combining the linearized Boltzmann equation and the Poisson equation, each relating the charge density, ρ_r, in the volume element, dV, to the distance, r, from the reference ion, the **linearized Poisson–Boltzmann equation** is obtained:

$$\frac{1}{r^2}\frac{d}{dr} r^2\left(\frac{d\psi}{dr}\right) = \frac{1}{\epsilon_o \epsilon kT} \sum n_i^o z_i^2 e_o^2 \psi_r \tag{12.1-17}$$

If all of the right hand terms are collected into a single variable, κ^2, the expression may be rewritten

$$\frac{1}{r^2}\frac{d}{dr} r^2\left(\frac{d\psi}{dr}\right) = \kappa^2 \psi_r \tag{12.1-18}$$

The variable κ is important and will be discussed shortly.

Equation (12.1-18) can be solved by considering the boundary conditions that derive from the consideration of the ion as a material point charge with a field that extends to infinity. Consequently, the problem is integrated from $r = 0$ to $r = \infty$ and the result is the following:

$$\psi_r = \frac{z_i e_o}{4\pi\epsilon_o \epsilon r} \tag{12.1-19}$$

This equation can be approximated and expanded to describe a contribution to the electrostatic field by the central ion and the cloud that surrounds it. The ion's contribution is $z_i e_o/4\pi\epsilon_o\epsilon r$ while the contribution of the cloud is $z_i e_o/4\pi\epsilon_o\epsilon L_D$, where L_D represents a distance from the central ion and is equal to κ^{-1} as described in the following paragraph. The field contributed by the cloud of charge must counter that contributed by the ion, and the expansion of Equation (12.1-19) confirms this:

$$\psi_r = \frac{z_i e_o}{4\pi\epsilon_o\epsilon r} - \frac{z_i e_o}{4\pi\epsilon_o\epsilon L_D} \tag{12.1-20}$$

The total field given by ψ_r then is comprised of contributions from the ion and from the charge cloud:

$$\psi_r = \psi_{\text{ion}} + \psi_{\text{cloud}} \tag{12.1-21}$$

It follows that

$$\psi_{\text{cloud}} = -\frac{z_i e_o}{4\pi\epsilon_o\epsilon L_D} = -\frac{z_i e_o}{4\pi\epsilon_o\epsilon\kappa^{-1}} \tag{12.1-22}$$

A brief survey of the problem so far will be valuable. In an effort to determine the alteration in the chemical potential of an ionic species in solution due to the interactions between ions, a model first proposed by Debye and Hückel has been presented. This model is based on the assumption that the most significant interactional forces can be described in terms of electrostatics. A central reference ion is chosen with a charge $z_i e_o$ and is considered as a point charge. Since it is a point charge, it generates a potential field that extends into space in a spherically symmetrical manner. The remaining ions in the solution constitute an electrical charge that surrounds the central ion and will be arranged so that another electrostatic field is generated. This field will act to neutralize the potential field created by the central ion. In the Debye–Hückel model, the surrounding ionic charge is treated as a continuum of charge much like that described previously in the discussion of the Born model. By treating the question of alteration in chemical potential or activity in this fashion, the problem becomes one in electrostatics. If the interactional forces that are described by the electrostatic potential fields between the reference ion and the surrounding charge cloud can be defined, then the change in chemical potential can subsequently be found. The contributions to the electrostatic field can be found for the central ion since the field is determined simply by the charge of the ion, $z_i e_o$. In the case of the field contributed by the countering charge cloud, the field could be calculated after the charge density in a specific volume element under study, dV, was found. Through the use of the Boltzmann equation, the distribution of charge in the volume element was defined, and then a linear relationhip was derived by making an assumption that thermal forces were much more significant than electrostatic forces in the distribution of ions. If this were not the case, the ions making up the cloud would condense around the central ion and form a crystal. At this point, Equation (12.1-20) relates the contributions of the central ion and the surrounding cloud of charge to the electrostatic potential, ψ_r, at a particular distance from the central ion. A very important feature of the derivation to this point is the result that the field contributed by the charge cloud is related to the parameter, L_D, which has units of length. The physical interpretation of this mathematical expression is that the ionic atmosphere can be replaced by a charge at the distance L_D from the central ion. L_D is equal to κ^{-1} and is often called the **Debye length** or the **effective radius** of the charge atmosphere surrounding the central ion.

The effective radius is obviously an important physical aspect of an ion in relationship to other ions in solution. Earlier in the derivation, the term κ was created by grouping a series of constants together. What are the parameters of κ that have an effect on the thickness of the ionic atmosphere? Referring to Equation (12.1-17), κ^2 can be written as follows:

$$\kappa^2 = \frac{1}{\epsilon_o \epsilon kT} \sum_i n_i^o z_i^2 e_o^2 \tag{12.1-23}$$

By rearrangement, κ^2 can also be written as follows:

$$\kappa^2 = \frac{e_o^2 \sum_i n_i^o z_i^2}{\epsilon_o \epsilon kT} \tag{12.1-24}$$

This relationship can also be expressed on a molar basis:

$$\kappa^2 = \frac{e_o^2 N_A}{\epsilon_o \epsilon kT} \sum_i n_i^o z_i^2 \tag{12.1-25}$$

The last term in Equation (12.1-25) should be familiar as closely related to the definition of ionic strength:

$$I = \frac{1}{2} \sum n_i \, z_i^2 \tag{12.1-7}$$

Therefore, by substitution, κ^2 can be expressed in terms of ionic strength:

$$\kappa^2 = \frac{2 e_o^2 N_A I}{\epsilon_o \epsilon kT} \tag{12.1-26}$$

κ^{-1} or L_D can be written as follows:

$$\kappa^{-1} = L_D = \left(\frac{\epsilon_o \epsilon kT}{2 e_o^2 N_A I} \right)^{1/2} \tag{12.1-27}$$

This means that the effective radius of the ionic atmosphere is inversely related to the ionic strength of the solution. The radii of ionic atmospheres for electrolytes of several valence types at several concentrations are given

Table 12.1-2. Radii of ionic atmosphere ($L_D = \kappa^{-1}$) for electrolytes of different valence types at various concentrations.

	Valence type				
Molarity	1:1	1:2	1:3	2:2	2:3
10^{-5}	96.1	55.5	39.2	48.00	24.80
10^{-4}	30.40	17.50	12.40	15.20	7.80
10^{-3}	9.61	5.55	3.92	4.80	2.48
10^{-2}	3.04	1.75	1.24	1.52	0.78
10^{-1}	0.96	0.56	0.39	0.48	0.25
1	0.30	0.18	0.12	0.15	0.08

[a]298 K; all values in nm.

in Table 12.1-2. The physical ramifications of this relationship will be examined shortly.

The link between the ionic atmosphere and ψ_r and the effect on the chemical potential of the ion as expressed by the activity coefficient is developed by employing a charging process similar to that described in the formulation of the Born model. To find the potential energy of the ion–ion interaction, the ion is first considered to be discharged and then is brought to its charge, $z_i e_o$. The work associated with charging the ion is the potential energy and gives the alteration in chemical potential for the system. Earlier, this relationship between the change in chemical potential and the charging process was given as

$$\Delta\mu_{i\text{-I}} = N_A \frac{z_i e_o}{2} \psi \tag{12.1-3}$$

ψ is now well described (Equation 12.1-22), and this equation can be rewritten

$$\Delta\mu_{i\text{-I}} = -\frac{N_A (z_i e_o)^2}{8\pi\epsilon_o \epsilon \kappa^{-1}} \tag{12.1-28}$$

Previously, the activity coefficient was introduced and defined in terms of a difference between the ideal and real chemical potentials:

$$\begin{aligned} \Delta U_{(\text{real-ideal})} &= (\mu_o + RT \ln X_i + RT \ln \lambda_i) - (\mu_o + RT \ln X_i) \\ &= RT \ln \lambda_i \end{aligned} \tag{12.1-29}$$

An ideal system is defined as being composed of noninteracting particles, while for a real system the interactions between particles must be taken into account. In the Debye–Hückel model, the interactions are the ion–ion interactions, and consequently $\Delta U_{(\text{real-ideal})}$ is $\Delta U_{\text{i-I}}$. Combining Equation (12.1-28) with Equation (12.1-29) therefore gives the result:

$$-\frac{N_A (z_i e_o)^2}{8\pi\epsilon_o \epsilon \kappa^{-1}} = RT \ln \lambda_i \tag{12.1-30}$$

At this stage, a theoretical relationship has been derived, based on a particular model, that will predict the activity coefficient for a single ionic species in solution. Originally, the concept of the activity coefficient was introduced as an empirical device to allow continued use of the derived ideal equations for thermodynamic systems, and so it is grounded in experiment. The benefit in deriving Equation (12.1-30) lies in the fact that if the theory accurately predicts the activity coefficients found experimentally, it is validation of the structural understanding implicit in the theoretical model. Therefore, before going further, it is crucial to know just how well this model and experiment agree.

12.2. Testing the Debye–Hückel Model

As is so often the case, the simplistic elegance of a model is not so easily tested in the laboratory. A similar problem to that described in the experimental validation of the Born model exists here. To measure the interactions, the solution must remain uncharged and electroneutral. If a single ionic species is added, this condition cannot be met and the measurement will include the confounding interaction of the reference ion with an electrified solution. For this reason, activity coefficients cannot be experimentally measured for single ions, but instead the activity coefficient of a net electrically neutral electrolyte is used. This consideration leads to the **mean ionic activity coefficient**. If a mole of electrolyte MA is added into solution, the chemical potential for the new solution is described by the sum of each ion's chemical potential, i.e.:

$$\mu_{M^+} + \mu_{A^-} = (\mu^0_{M^+} + \mu^0_{A^-}) + RT \ln (X_{M^+}X_{A^-}) + RT \ln (\lambda_{M^+}\lambda_{A^-}) \quad (12.2\text{-}1)$$

However, this equation actually describes a 2 M solution, since one mole of each ion has been added. In order to determine the free energy contribution from one mole of both ions comprising the electrolyte, Equation (12.2-1) must be divided by two:

$$\frac{\mu_{M^+} + \mu_{A^-}}{2} = \frac{\mu^0{}_{M^+} + \mu^0{}_{A^-}}{2} + RT \ln (X_{M^+}X_{A^-})^{1/2} + RT \ln (\lambda_{M^+}\lambda_{A^-})^{1/2} \quad (12.2\text{-}2)$$

This equation defines a mean quantity derived from the average contributions of each species to the system as a whole. Each of these average quantities is measurable. These avaerage quantities are called **the mean chemical potential, $\mu_{\pm}$, the mean standard chemical potential, $\mu^o{}_{\pm}$, the mean mole fraction, $X_{\pm}$,** and the **mean ionic activity coefficient,** $\lambda_{\pm}$. The general formula relating the measurable mean ionic activity coefficient to the activity coefficient for each individual ion is

$$\lambda_{\pm} = (\lambda^{\nu+}\lambda^{\nu-})^{1/\nu} \quad (12.2\text{-}3)$$

where

$$\nu = \nu^+ + \nu^- \quad (12.2\text{-}4)$$

Table 12.2-1. Values of A for water at various temperatures.

Temperature (K)	A ($dm^{3/2}\ mol^{-1/2}$)
273	0.4918
293	0.5070
298	0.5115
303	0.5161
313	0.5262
323	0.5373

Using this relationship and the Debye–Hückel formulations for λ^+ and λ^-, an equation is found that relates the theoretical calculation of the activity coefficient to the experimentally available mean ionic activity coefficient:

$$\log \lambda_{\pm} = -A(z_+z_-)\, I^{1/2} \qquad (12.2\text{-}5)$$

For water, the constant A has values as shown in Table 12.2-1 with units of $dm^{3/2}\ mol^{-1/2}$. In SI units, the value of A for water at 298 K is 1.6104 $\times\ 10^{-2}\ m^{3/2}\ mol^{-1/2}$. Equation (12.2-5) is called the **Debye–Hückel limiting law** and is the link between experimental values and the model developed so far.

When the comparison between theory and experiment is made (Figure 12.2-1), the limiting law is found to be quite accurate in solutions of 1:1

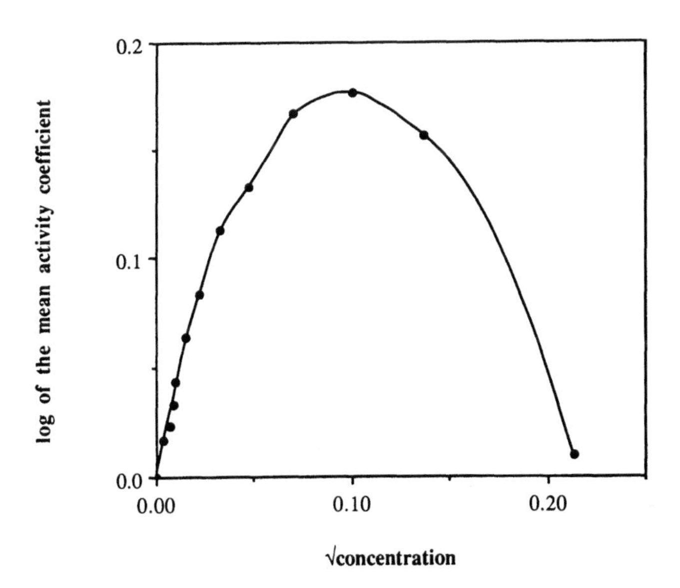

Figure 12.2-1 Plot of the mean activity coefficient for NaCl at various concentrations.

electrolytes of concentrations no greater than 0.01 *N*. However, for more concentrated solutions the theoretical and experimental values diverge significantly and even in moderately concentrated solutions of only 1 *N*, the mean activity coefficient will begin to rise. Additional problems exist; variations in the mean activivty coefficient are found with electrolytes of different valence types (i.e., 1:1, 1:2 electrolytes), as well as with different electrolytes of the same valence type.

The limits of the Debye–Hückel limiting law are significant for the biological worker. The ionic strengths of biological fluids are considerably higher than the useful limit of Equation (12.2-5). It is clear that the ionic strength of solutions in vitro plays a significant role in cellular and biochemical behavior. It is relevant therefore to attempt to gain a greater insight into the behavior of more concentrated solutions. The first step will be to examine the assumptions on which the limiting law is based and to see if modifications of these assumptions will lead to greater agreement between theory and experiment.

12.3. A More Rigorous Treatment of the Debye–Hückel Model

One of the basic tenets of the Debye–Hückel model was the treatment of the central ion as a point charge. Obviously, this is a drastic oversimplification, since ions are finite in size and must occupy a certain space. Could this assumption account for some of the problems with the limiting law? Consider that the limiting law linearly relates the mean ionic activity coefficient to the square root of the ionic strength. Earlier (Equation 12.1-27), it was shown that the radius of the ionic atmosphere is inversely related to the concentration through the constant κ. The ionic atmosphere is relatively large for a dilute solution and becomes progressively more compact as the concentration increases. Table 12.1-2 shows that the radius of the ionic atmosphere for a 1:1 electrolyte such as NaCl is 96 nm at a concentration of 10^{-5} *N*, yet at a concentration of 10^{-1} *N*, it has narrowed 100-fold to 0.96 nm. Since an ion is approximately 0.2 nm in diameter, the cloud at a concentration of 1 *N* will be of a similar dimension to the ion. This is a problem since the point charge approximation depends on the idea that the radius of the ionic atmosphere is much greater than the ionic radius, a simplification that becomes increasingly inaccurate as the dimension of the ionic cloud diminishes. The Poisson formulation also is based on the premise that discrete ions can be treated as a smeared charge, an assumption that becomes unsustainable when the ions are concentrated into a much smaller space. At high concentration, the ions respond to other ions as discrete charges, and the Poisson equation becomes less valid.

The point charge assumption was made when the limits were chosen in solving Equation (12.1-19), and it is here that the finite size of the ion will be introduced. The lower limit of integration will no longer be zero, since the central ion occupies this point. The lower level is more logically chosen to start at the edge of the finite-sized ion, a specific distance, a, from the center point of the ion. The upper limit can still be considered to be at infinity, and the result of the new equation is

$$\psi_r = \frac{z_i e_o}{4\pi\epsilon_o\epsilon} \frac{e^{\kappa a}}{(1 + \kappa a)} \frac{e^{-\kappa r}}{r} \tag{12.3-1}$$

This defines the potential ψ_r at the distance r from a central ion of finite size a. A procedure similar to that used to obtain Equation (12.2-5) may now be applied to yield the following formula that relates the mean ionic activity coefficient to the finite-central-ion model:

$$\lambda_{\pm} = -\frac{A(z_+z_-)I^{1/2}}{1 + BaI^{1/2}} \tag{12.3-2}$$

The constant A is the same as that in the limiting law (Equation 12.2-5), and B can be written as Equation (12.3-3) and has the value of 3.291 $\times$ 10^9 m^{-1} $mol^{-1/2}$ $kg^{1/2}$ for water at 298 K:

$$B = \left(\frac{2e^2N_A}{\epsilon_o\epsilon kT}\right)^{1/2} \tag{12.3-3}$$

As the solution becomes increasingly dilute and $I \rightarrow 0$, the second term in the denominator of Equation (12.3-2) approaches zero, and Equation (12.3-2) reduces to Equation (12.2-5). The physical interpretation of this is that at very dilute concentrations the radius of the ionic atmosphere is so much larger than the finite radius, a, of the central ion that the ion can be effectively treated as a point charge. However, as the concentration increases, the finite size of the ion must be taken into account.

In fact, the value of a, the effective radius of the central ion, is difficult to know exactly but is usually considered to vary between 0.3 and 0.5 nm. These values are found from experiment and represent distances of closest approach between two ions. When values in this range are used, Equation (12.3-2) shows a much greater accuracy in predicting the mean activity coefficient up to a concentration of 0.1 N. Furthermore, because the model of the ion of finite size takes into account the actual identity of the ions, at least as reflected in their effective radii and not just their valence types, this model can account for the differences in observed activity coefficients for electrolytes of the same valence type such as NaCl and KCl.

The Debye–Hückel theory as quantified in Equation (12.3-2) is truly a remarkable success, and the model it describes is an important one both conceptually and practically. However, the disparity between the theoretical predictions and experimental values for mean activity coefficients becomes a serious one in the range of ionic strengths of many biologically relevant solutions. Are there further modifications that can be considered to explain the experimental values? The logical next step is to examine the assumptions on which the model was originally constructed. Substitution of increased detail for the simplifications, as in the case of the addition of finite ion size, will lead to a more sophisticated model. In the case of this model, the original assumptions were:

1) Point charge representation of the central ion.
2) Availability of all ions in solution to participate freely with the central reference ion.
3) Valid use of the linearized Poisson–Boltzmann equation.
4) Only ion–ion interactions with a $1/r^2$ dependence considered.
5) No ion–solvent interactions.

So far, it has been shown that the first and third assumptions may be oversimplifications in systems of practical biological relevance. The addition of the finite radius, a, improved the agreement between theory and experiment. Will consideration of the other approximations provide greater understanding of the shape of the curve found experimentally as shown in Figure 12.2-1?

12.4. Consideration of Other Interactions

In Chapter 11, the interactions between ions and their aqueous solvent were considered in some detail. These interactions have been completely ignored up to this point. What effect might the interaction between the ions and water have on the mean activity coefficient? One of the most important aspects of the ion–solvent interactions was the formation of hydration sheaths. Molecules of water that are tightly bound to an ion in the hydration sheath will no longer be free to interact with new electrolyte added. By an effect analogous to the salting-out effects discussed earlier, as an increasing concentration of electrolyte is added, a decrease in the effective concentration of the water occurs. For a 1 M solution of LiCl, whose primary hydration number can be considered as 7, the number of moles of water immobilized by the addition of the electrolyte will be 7. This represents a 12.6% drop in the effective solvent concentration. A 3 M LiCl solution, commonly used in the extraction of RNA from cells, effectively removes 21 moles of solvent or 38% of the water available. As the effective concentration of the solvent falls, the effective concentration of the electrolyte increases. Since it is the activity coefficient that relates actual concentration to the effective concentration, the activity coefficient of the water will decrease, and the activity coefficient of the LiCl will increase. This effect accounts partly for the increase seen in the mean activity coefficient at increasing concentrations.

12.4.1. Bjerrum and Ion Pairs

The model as it now stands has been modified to take into consideration that (1) the ionic radii are not point charges, and (2) there are significant ion–solvent interactions. Another basic assumption is that all of the ions in solution are free to contribute to the ionic atmosphere. The motions and positions of the ions are considered random, secondary to thermal forces, except as restricted by the coulombic forces that act between the central ion and the localized ionic atmosphere. Even though it is assumed that the thermal forces are generally much greater than the coulombic forces (Equation 12.1-13), there is a probability that an ion acting as a member of the ionic atmosphere may get close enough to the central ion that it will become locked in a coulombic embrace and no longer will either ion be independent of the other. When such an event occurs, Bjerrum suggested that an **ion pair** would be formed. When an ion pair forms, the number of free ions in solution decreases. The electric charge of an ion pair is less than that of a free ion; in fact, the charge of an ion pair is frequently zero. Any property of an electrolytic solution that depends on the number of free charged particles (i.e., the ionic strength) will be affected by ion pairing. The probability that ion pairing will occur depends on whether the electrostatic fields of two mutually attractive ions will overcome thermal randomizing forces. A minimum distance, q, may be defined at which the mutual attraction of the oppositely charged ions will be comparable to their thermal energy:

$$q = \frac{|z_+|\,|z_-|\,e_o^2}{8\pi\epsilon kT} \tag{12.4-1}$$

For ion pair formation to occur, the physical size of the ions must be taken into account. Obviously, if the ions are too large to approach one another within the distance q, ion pairing will be impossible. If the sum of the effective radii of the ions, a, is less then q, then ion pairing can occur, i.e.:

If $a < q$ then ion pairing occurs,
if $a > q$ then no ion pairing.

For electrolytes in water at 298 K, q can be written

$$q = 0.336\,|z_+|\,|z_-|\ \text{nm} \tag{12.4-2}$$

For 1:1 electrolytes in water, virtually no ion pairing occurs, since a is usually between 0.3 and 0.5 nm and these electrolytes are completely dissociated. However, ions of higher valence, such as Ca^{2+} and Mg^{2+}, may form ion pairs in aqueous solution.

12.5. Perspective

At this point, the reader should have an appreciation for some of the relationships that contribute to the behavior of ions in solution. The principles and approaches to the study of the behavior of electrolyte solutions that have been presented are useful to the biological scientist for two reasons. First, ions play a crucial role at every level of biological systems, from the conformation and reactivity of enzymatically active sites to the signaling functions of cells and organs. Second, the forces acting on ions and their formulation are more straightforward than the more complex interrelationship of forces acting on a complex structure such as a macromolecule. As will be seen in the next chapter, macromolecules have many forces that contribute to their properties. They have a large number of functional groups that are ionized in solution, and hence the macromolecule is a polyelectrolyte. Polyelectrolytes have small ions called **gegenions** associated with them to ensure electroneutrality. As might be expected, the behavior of a solution of polyelectrolytes is different from that of a system composed of low molecular weight electrolytes. The tools developed in this discussion of ion–ion and ion–solvent interaction will be invaluable as a starting point, since the biological systems that will be described will now grow increasingly complex.

CHAPTER 13

Molecules in Solution

13.1. Solutions of Inorganic Ions

The focus will now shift to the examination of the structural aspects of aqueous solutions, considering polar and nonpolar solutes, and as well solutes that are amphiphilic. This change in focus will allow models to be constructed that more closely approximate the behavior of biological macromolecules.

Reviewing the structural picture developed in Chapter 10, small ions disrupt the organized structure of liquid water (Figure 13.1-1). The electrostatic interaction between the ion and the water dipole forces an orientation of the water molecules with respect to the ion. The hydrogen bonds between water molecules are weakened and then broken, as they deviate from their normal 180° angle during their reorientation. In the final fully hydrated structure, each ion is surrounded by water molecules oriented such that the partial opposite charge faces the ion. The number of water molecules involved in the primary hydration sheath depends upon the ion's size and charge.

The enthalpy associated with the hydration of these ions is always negative ($\Delta H < 0$), with multivalent ions having a much more negative

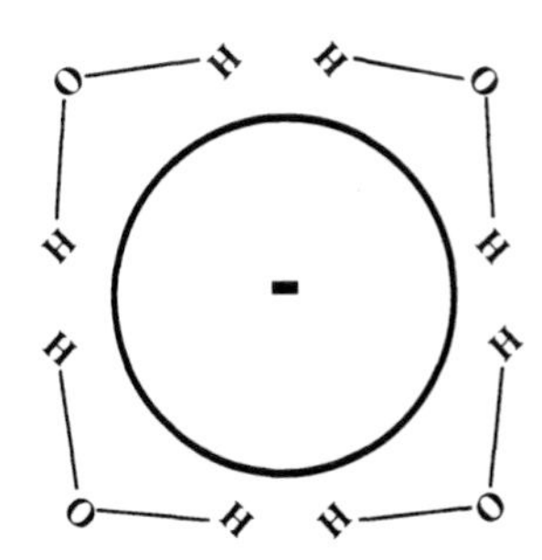

Figure 13.1-1 Schematic of the arrangement of water molecules around an ion.

Table 13.1-1. Enthalpies of hydration for single ions.

Ion	$\Delta H_{\text{hydration}}$ (kJ mol^{-1})
H^+	−1071
Li^+	−548
Na^+	−485
K^+	−385
Ca^{2+}	−1715
Al^{3+}	−4807

ΔH than monovalent ones (Table 13.1-1). The enthalpy of solvation is sizable and is attributable to a combination of interactions in the primary and secondary hydration sheaths, as well as the hydrogen bond energy lost when the bulk water structure is perturbed. Additionally, there will be an entropy decrease, its magnitude depending on the number of H_2O molecules involved. This decrease in entropy is a consequence of the removal of H_2O molecules from the bulk solvent, where the entropy is high, to the lower entropic environment of the hydration sheaths. The entropy falls because the number of possible microstates in bulk water is much higher than the number of relatively fixed locations around an ion. The enthalpy and the entropy of aqueous solvation of ions depend partly on the identity of the ion and not solely on its charge.

Removal of the water molecules from the hydration sheaths around inorganic ions is a process requiring a significant amount of energy. This accounts for the co-transport of H_2O across biological membranes with a number of charged species. It also accounts for the specificity of certain compounds which act as transporters of ions across the membrane, such as the ionophore **valinomycin**. Note that this usage of the term ionophore is distinct from the definition given in Chapter 10. In biological studies, the term ionophore is most commonly used to refer to compounds such as valinomycin, ionomycin, and nigericin which act to allow transport of certain ionic species across membranes. As shown in Figure 13.1-2, valinomycin is a cyclic **depsipeptide** which folds so that all of its carbonyl groups point toward the center of a sphere. Its shape is like the seam on a tennis ball, having the depsipeptide as a backbone and the nonpolar side chains pointing toward the surface of the sphere. Its exterior is hydrophobic and lipophilic, and its interior is quite hydrophilic. Lipid membranes are freely permeable to valinomycin, whether its interior space contains a molecule of H_2O or an ion. The space within valinomycin is too small to accommodate any hydrated ion, and the compound can therefore act as a transmembrane transporter only if the energy required to strip the solvating water from the ion is compensated by the energy of binding to the carrier. Only for K^+ is such an interaction energetically

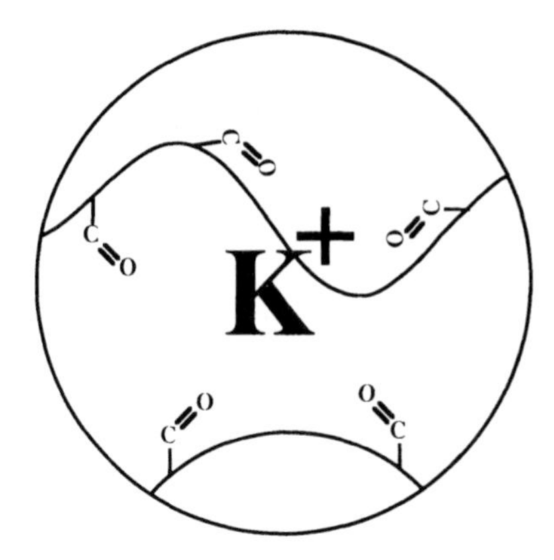

Figure 13.1-2 The structure of valinomycin. The peptide backbone (L-Lac-L-Val-D-hydroxyisovalerate-D-Val)$_3$ is cyclic. The carbonyls all point into the center of the sphere.

favorable. Smaller ions have a greater energy of solvation and a larger separation distance and therefore have a weaker interaction with the oxygen atom of the carbonyl groups. Ions larger than K^+ cannot be accommodated in the limited intraspherical space. Therefore, valinomycin is an ionophore which is virtually specific for K^+ transport across membranes and is indeed the only known ionophore which exhibits such complete selectivity toward one ionic species.

The existence of an ionic atmosphere in which normal hydrogen bonds do not exist accounts for the fact that the heat capacity of a dilute solution of an inorganic electrolyte (such as KBr) is very much smaller than that of water and decreases as the concentration of electrolyte increases (Figure 13.1-3). Furthermore, because the hydrogen bonds do not exist near the ion, the hexagonal structure with its large spaces is disturbed. This leads to a smaller volume of a molal solution of an inorganic salt when compared to the volume of pure water alone (i.e., $\Delta V < 0$).

The discussion here has centered on the solvation of small inorganic ions, and on the effect of this solvation sheath on thermodynamic parameters such as heat capacity, enthalpy, and entropy of solution. In spite of the increased level of organization of the H_2O molecules nearest to an inorganic ion, the overall effect of solvation of these ions in water is to lower the energy requirement for any changes in the thermodynamic state of the system, (e.g., heat capacity and heat of vaporization) in which disruption or formation of hydrogen bonding is involved.

13.2. Solutions of Small Nonpolar Molecules

The hydration of small nonpolar molecules is qualitatively quite different from the electrostatically dominated interaction described for polar molecules. In nonpolar solvation, there is "hydrophobic hydration" or struc-

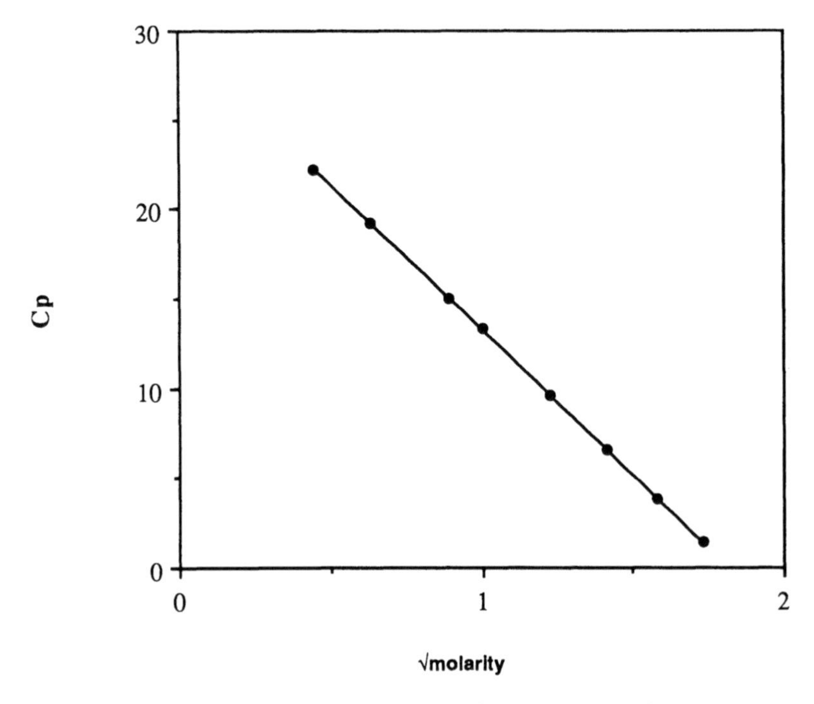

Figure 13.1-3 Plot of the heat capacity (C_p) for a solution of KBr in water at various concentrations.

turing of the solvent around the solute molecule. This is an entropically unfavored step, and therefore $\Delta S_{\text{solvation}}$ for nonpolar molecules is always negative. When an increasing concentration of nonpolar molecules is introduced into the aqueous system, there is a strong gain in entropy as a result of these molecules associating with one another and as a result of the release of the water molecules involved in the entropically unfavored hydration of these molecules. This association is universally called a **hydrophobic interaction**.

When a nonpolar solute is small, that is, similar in size to H_2O, the thermodynamic properties (as exemplified in Table 13.2-1) are remarkably independent of the solute, of its nature, or of its chemical reactivity. This is clearly a major difference from the solute nature-dependent heat of solvation of ions just discussed. The heats of solvation of small nonpolar molecules reflect an interaction energy common to all the solutes. This energy is associated with the insertion of these small compounds into the partially ordered structures of liquid H_2O, which become more stable as a result of the insertions.

Initially, it was thought that each of the entities listed in Table 13.2-1 would fit into the space within the center of the hexagonal arrays described earlier (cf. Chapter 9), much as other molecules of H_2O fit into the space

Table 13.2-1. Enthalpies of hydration for selected small nonpolar molecules.

Compound	$\Delta H_{solvation}$ (kJ mol^{-1})
Ar	−69.5
Kr	−58.2
H_2S	−69.0
PH_3	−62.8
SO_2	−69.5
CH_4	−60.7
C_2H_2	−62.8
C_2H_4	−62.8
C_2H_6	−62.8
CH_3Cl	−62.8
CH_3SH	−69.5

in liquid water. The diameter of these inserted nonpolar substances would then have to be under approximately 0.54 nm. However, it is now known that the stabilized H_2O structure around the nonpolar solute changes from a hexagon to a pentagon with slightly bent and strained hydrogen bonds. Each of the water molecules in the pentagonal array found in the vicinity of nonpolar molecules is still tetrahedral, that is, still has a coordination number of 4, but the orientation of the individual H_2O molecules is different and the tetrahedral angle is deformed, with the hydrogen bonds no longer being linear (Figure 13.2-1).

The pentagonal structures have been called "icebergs" by Pople and are of definite size and spacing. Their stabilization requires the inserted nonpolar molecules. Removal of the solute or addition of an ionic species which effectively disrupts the solvent structure causes reversal of the solubilization and precipitation of the solute. The procedure whereby ionic species are added to a solution containing a solvated nonpolar species in order to remove it from solution was discussed in Chapter 11 and is called "salting out." It is frequently used in biochemical separation techniques where the solute is a macromolecule (cf. later in this chapter).

The stabilization of H_2O structure by a nonpolar solute leads to a higher heat capacity and higher melting temperatures for solutions of these compared to bulk water. Furthermore, the pentagonal structures shown in Figure 13.2-2 have very defined spaces within the pentahedral array. Thus, a nonpolar residue which has been "solubilized" will have a predictable number of H_2O molecules associated with it, giving it the regularity of a crystal and the stoichiometry of a hydrate. The smallest hydrate has eight possible insertion sites for small nonpolar solutes (two sites have 0.52-nm diameters, and six sites are 0.59 nm in diameter).

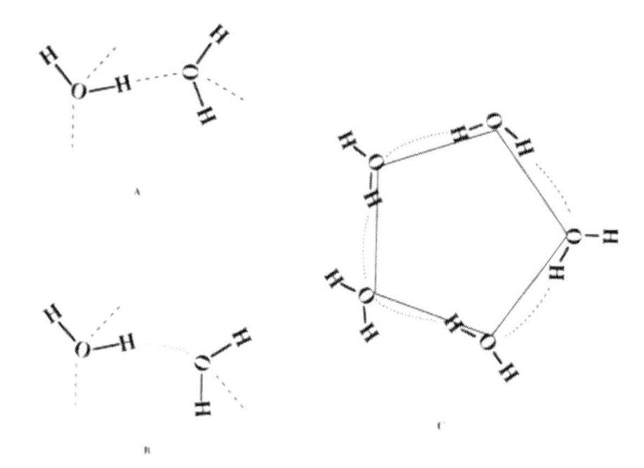

Figure 13.2-1 Comparison of hydrogen-bonded water arrays: (A) hydrogen bonding in bulk water; (B) water molecules in a nonpolar array with deformation of the previously linear hydrogen bonds; (C) the resultant pentagonal structure in the presence of a nonpolar molecule.

This array can include 46 molecules of H_2O. If the solute is larger and cannot fit into this array, there are a whole series of regular structures called **clathrates** with a defined number of solute and solvent sites. The latter also have a regular pentagonal array, which corresponds to a stable hydrate in which there is a given stoichiometry between specific nonpolar solute molecules and H_2O molecules. The actual existence of such clath-

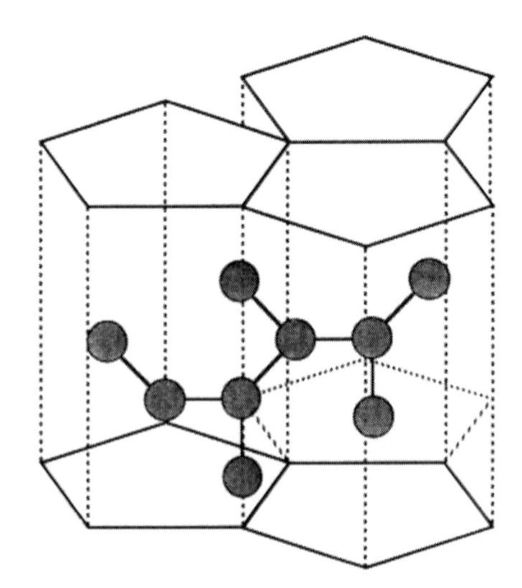

Figure 13.2-2 Structure of a pentagonally organized water cavity surrounding a large solute cavity.

rates, as distinguished from a nonstoichiometric highly ordered structure around a large nonpolar solute, is still the subject of some debate and is discussed in detail in the series edited by Franks (see Appendix I). For the present text, this is not a vital point; it is merely important that the reader understand that in the presence of nonpolar solutes, water forms a somewhat different and inherently more strained regular structure in which the solute is accommodated. The solute molecules are in close proximity to the solvent, approaching their van der Waals radii, but the solute develops no enthalpic interactions with the water molecules. As will be seen in Chapter 14, H_2O plays a major role in changes in conformation of such molecules.

13.3. Solutions of Organic Ions

The effects of the disruption of the normal hydrogen-bonded structures in bulk water by small inorganic ions have been emphasized. The situation is different for nonpolar organic ions. These include the small aliphatic or aromatic quaternary ammonium salts and the large detergents or charged lipids. Since small and large nonpolar ions behave differently, they will be considered separately.

13.3.1. Solutions of Small Organic Ions

The small organic ions carry their single charge on an otherwise hydrophobic moiety. With respect to thermodynamic parameters, organic ions are distinct from inorganic ions because the disruption of hydrogen bonding in the aqueous solvent occurs differently. The association between an inorganic ion and the irrotational water layer is a stronger interaction than an ordinary hydrogen bond, but a small organic ion is essentially incapable of such bonding. A small organic ion tends to organize its aqueous sheath as if it were an uncharged nonpolar solute. Therefore, the hydration sheath around the exterior of such an organic ion will tend to have a strongly hydrogen-bonded pentagonal structure. The positive concentration dependence of the heat capacity of solutions of quaternary ammonium salts (Figure 13.3-1), for example, indicates that H_2O is more ordered around these tetraalkylammonium ions than around inorganic ions. The complex nature of the heat capacity curve for the solutions of organic ions has been difficult to explain; it implies that at low concentrations of solute the nature of the interactions is mainly hydrophobic. At higher concentrations, these solutions begin to resemble inorganic ionic ones, and C_p, having gone through a maximum, begins to decrease. It is not yet possible to predict the values of ΔH and ΔS of solvation for these nonpolar ions, although it is clear that C_p for nonpolar ions is highly temperature, as well as solute concentration, dependent (Figure 13.3-1).

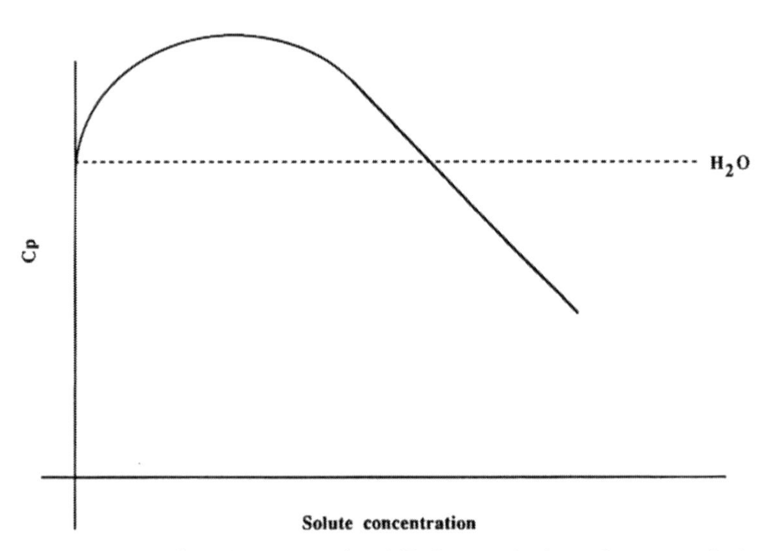

Figure 13.3-1 Plot of the heat capacity (C_p) for a solution of an organic ion in water as a function of concentration.

In the region in which the hydrophobic nature of the solute predominates, the increased order imposed upon the molecules of the aqueous sheath means that more, not less, energy is required to heat the sheath water.

While the aliphatic or aromatic group-substituted quaternary ammonium ions behave more nearly like nonpolar solutes with respect to the structure of the aqueous hydration sheath, solutions of these ions will exhibit conductance properties dependent on the hydrodynamic properties and the presence of a net charge on the solvated entity, as shown in Table 13.3-1.

13.3.2. Solutions of Large Organic Ions

Large organic ions that are comprised of a single charged entity at one end of a long aliphatic chain are amphiphilic molecules called soaps or

Table 13.3-1. Limiting ionic conductivity in aqueous solutions at 298 K.

Ion	Conductivity (Ω^{-1} m^2 × 10^{-4})	Ion	Conductivity (Ω^{-1} m^2 × 10^{-4})
Li^+	38.69	$(CH_3)_4N^+$	44.42
Na^+	50.11	$(C_2H_5)_4N^+$	32.22
K^+	73.50	$(C_3H_7)_4N^+$	23.22
Rb^+	77.20	$(C_4H_9)_4N^+$	19.31
Cs^+	77.29		

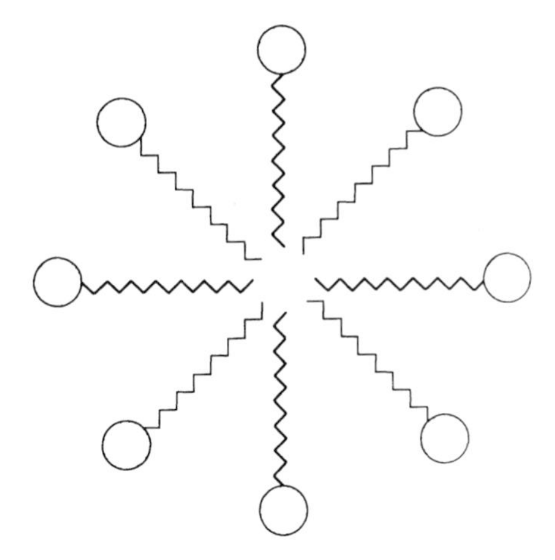

Figure 13.3-2 Arrangement of large organic ions into micelles. The polar head groups form the outer hydrophilic surface that protects the hydrophobic aliphatic chains from having to interact with water.

detergents. Some lipids also fall into this category. These molecules are distinct because their aliphatic portions will not interact with H_2O but will tend to form hydrophobic interactions with each other. Due to entropic driving forces, the aliphatic portions exclude H_2O and form van der Waals bonds, while the charged ends will form hydrogen bonds and electrostatic interactions with the aqueous milieu. The effect of these very different forces on the structure of the solute, as well as its interaction with solvent, is strongly concentration dependent.

Except in very dilute solutions, these forces will cause amphiphilic molecules to associate with each other into structures called **micelles** (Figure 13.3-2). In micelles, the hydrophobic portions of the molecules are separated from the aqueous solvent by a "self-generating" hydrophobic milieu, while the polar groups face the solvent and hydrogen bond with the water molecules. If the aliphatic chains are short, the hydrocarbon region is not structured at room temperature, and the interior of the micelle, the hydrocarbon region, can be a pure liquid phase. No hole can exist at the center of a micelle, which is a hydrophobic region. Therefore, hydrophilic entities cannot be accommodated, but a small hydrophobic entity could be included and carried within the micelle core. The predicted size and shape of micelles depends strongly on the length and degree of unsaturation of the alkyl chains. For longer alkyl chains and for greater degrees of asymmetry, the resulting micelle may no longer be spherical but can instead be ellipsoidal or resemble a flat disk. When the disk becomes large enough and flat enough, the amphiphiles essentially form a bilayer, a structure which will be considered in detail in Chapter 15.

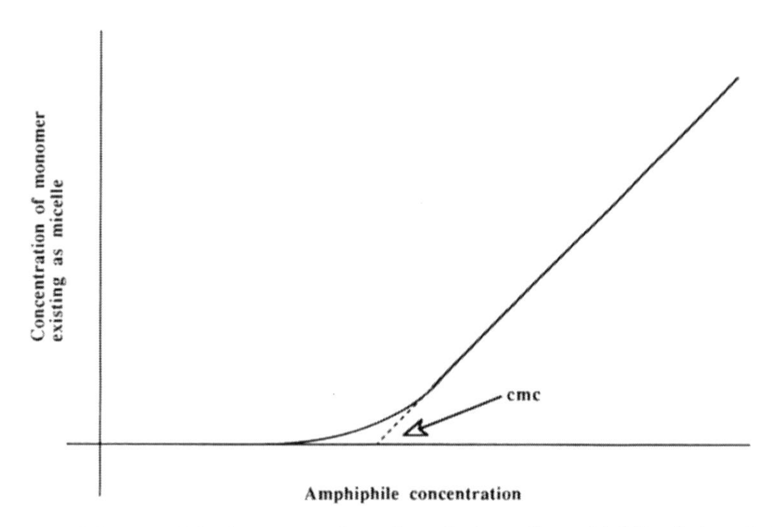

Figure 13.3-3 Plot of the incorporation of a solution of amphiphiles into micelles as the concentration of the monomer increases.

The concentration at which the change from a solubilized monomeric form to a micelle occurs is called the **critical micelle concentration**. It tends to be a convenient point of reference for the formation of the micellar phase. The actual critical micelle concentration depends strongly on the nature and length of the alkyl portion. The critical micelle concentration will also depend on the charge and the nature of the polar end. The critical micelle concentration is usually determined graphically for a given amphiphile under fixed conditions by varying the concentration alone (Figure 13.3-3). The ionic strength and the temperature also are important, since these all affect the electrostatic interactions between the polar or "head" groups. The importance of the combined ionic and hydrophobic behavior of large molecules in biological systems cannot be overestimated. In the following chapters, more complex molecules (proteins and lipids) that are essential building blocks of membranes and cellular systems will be considered.

CHAPTER 14

Macromolecules in Solution

14.1. Solutions of Macromolecules

The solvent–solute structure in aqueous solutions of small polar and nonpolar molecules has been described. The view presented of the disturbance of the normal hexagonal structure of liquid water by these small solutes can serve as a model for solutions of macromolecules, which, to a reasonable approximation, may be treated as a chain of small solute molecules attached to each other. Such molecules are **polymers.** Polymers of a single component or monomer are called **homopolymers** (A_n). **Copolymers** are comprised of two or more monomers. Copolymers can be random (AAABBABBBA . . .) or ordered (ABABABAB or ABCABCABC). Polymers of these types are mainstays of the chemical industry, and their solubility or insolubility in aqueous environments can now be predicted with a high degree of accuracy, allowing proper formulation of the polymerizing mixture.

This text deals with biological systems where the polymers of interest are not the synthetic plastics and fabrics, but biological polymers such as proteins (polyamino acids), polynucleotides, glycopolymers (polysugars), or mixed biological polymers such as proteoglycans. When a portion of such a polymer is comprised of some charged residues, a **polyelectrolyte** results. Few naturally occurring proteinaceous biological polymers consist of a single component (e.g., AAAAAAA), although cellulose and glycogen are polymers of the single hexose glucose. Thus, most polymers of biological origin are heteropolymers whose solvation must be approximated from simpler model systems.

14.1.1. Nonpolar Polypeptides in Solution

A model system of a macromolecule based on the properties of the constituent smaller residues can be built using thermodynamic information such as that listed in Table 14.1-1. The side chains of the nonpolar amino acids can be considered to behave as if they were small nonpolar solutes.

Table 14.1-1. Thermodynamic parameters of transfer from hydrocarbon to aqueous environment for selected amino acids.

Amino acid	ΔG_t (kJ mol^{-1})	ΔH_t (kJ mol^{-1})	ΔS_t (J mol^{-1} deg^{-1})
Alanine	5.5	−6.30	−39.5
Valine	8.0	−9.2	−57.5
Leucine	8.0	−10.1	−60.1
Cysteine	8.0	−10.1	−60.9
Methionine	8.4	−11.7	−67.2
Proline	8.4	−9.2	−58.8
Phenylalanine	1.3[a]	−11.7	−42.4
	7.6[b]	−4.2	−39.9

[a]Transfer from aliphatic environment.
[b]Transfer from aromatic environment.

Just as small molecules form stable hydrates in dilute aqueous solutions, the nonpolar polymerized amino acids can also form hydrates, which are much larger. The water molecules forming these hydrates surround the entire polymer, and the resultant structures are named **hydrotactoids**. Water in these hydrotactoids has a very different structure from water surrounding the small nonpolar molecules. The small nonpolar entities exist within the ordered H_2O structures as separate entities and act as isolated molecules. In contrast, the residues of nonpolar polypeptides have limited freedom of motion because the covalent peptide linkage imparts a structure to the polypeptide chain and therefore forces a non-random relative orientation on the nonpolar side chains. This forces the surrounding water molecules into even more rigidly held structures.

As already described, solvation of the small nonpolar side chains involves the rearrangement of the surrounding water in order to form pentagonal structures. The enthalpy change for solvation of the isolated nonpolar molecules is negative (exothermic). Incorporating these clathrate structures into the hydrotactoid still results in a negative enthalpy of solvation for the side chain, but to a smaller degree than for the single molecules. The entropy change associated with the polymer solvation is highly negative, since a high degree of order is imposed upon the water molecules immediately adjacent to the polymer. The degrees of freedom for these water molecules are severely limited, since the number of possible microstates available becomes greatly reduced compared to that in the bulk water. The overall solvation of the polymer would surely be thermodynamically unfavorable ($\Delta G > 0$) if only the side chains were considered in the model, but most proteins, even highly hydrophobic ones such as tropoelastin, are soluble in aqueous solution. Perhaps consideration of the constraints due to the peptide bond structure also needs to be incorporated into the model. Therefore, describing the relationship

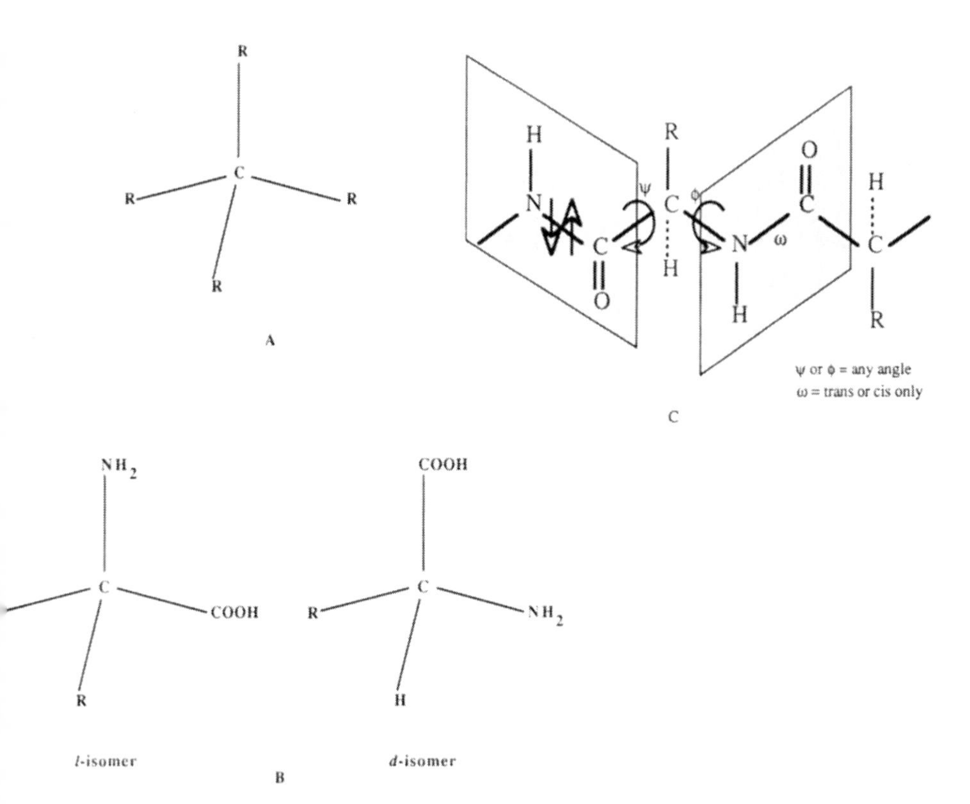

Figure 14.1-1 The steric constraints on a polypeptide chain result from (A) the tetrahedral bonding of carbon; (B) the asymmetry of the α carbon in virtually all amino acid residues; and (C) the planar nature of the peptide bond.

between the macromolecule and its aqueous environment depends on extending the model of nonionic hydration for small independent molecules to the more restricted configuration of a macromolecule and its hydrotactoid. The ultimate configuration depends partly on interactions between the nonpolar residues, the sterically restricted rotational freedom of the polypeptide chain, and the strucutre of the water of solvation.

The steric constraints of a polypeptide chain are severe as originally demonstrated by Pauling. These constraints arise from:

1) the tetrahedral nature of the carbon atom, which restricts its bonds to specific directions and angles (Figure 14.1-1A),
2) the asymmetry of the α carbon atom of all amino acids except glycine (Figure 14.1-1B), and
3) the existence of hybrid orbitals in the peptide bond attributable to keto–enol isomerization, which imparts a double bond character restricting rotation about that bond and making the peptide bond a planar entity (Figure 14.1-1C).

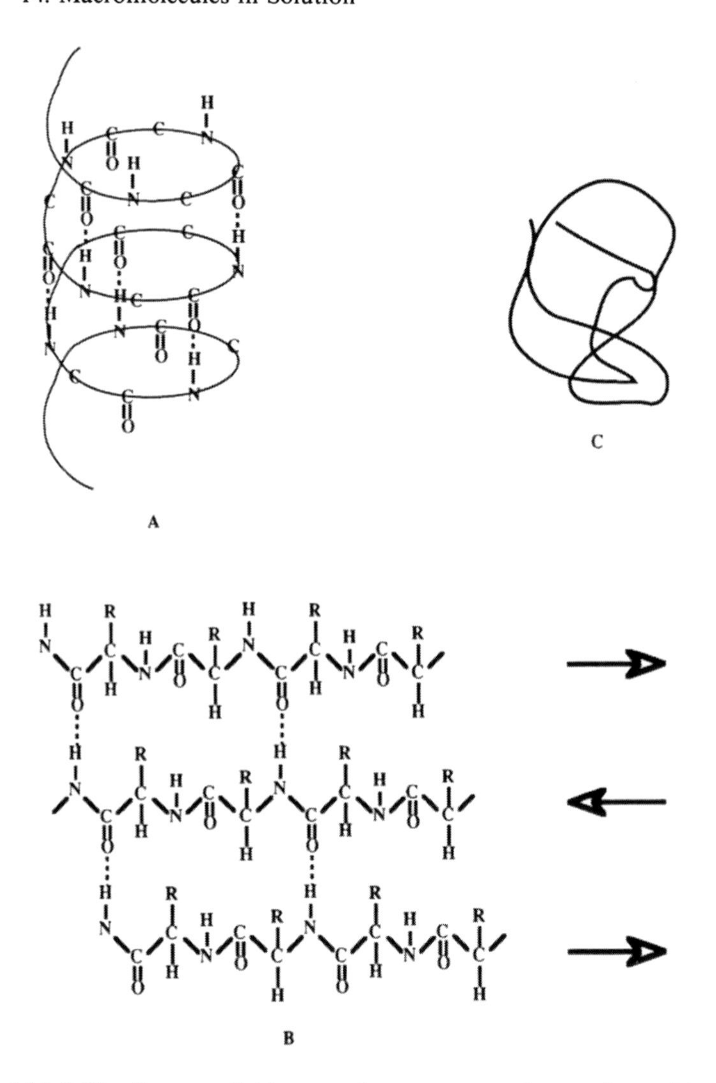

Figure 14.1-2 The three available secondary structures for a polypeptide chain: (A) helix; (B) pleated sheet; (C) random chain.

Normally, the planar peptide bond is a *trans* bond in polyamino acids, with such a high energy of activation that rotation into the *cis* form does not occur except in imino acid-containing polypeptides.

Under these restrictions, only three structures, or some mixture of these conformations, are available to a poly-α-*L*-amino acid. Both the strongly intrachain hydrogen-bonded helix (Figure 14.1-2A) and the extended pleated sheet (Figures 14.1-2B) have a high degree of periodicity. The third structure exhibits no periodicity and is called a random chain (Figure 14.1-2C).

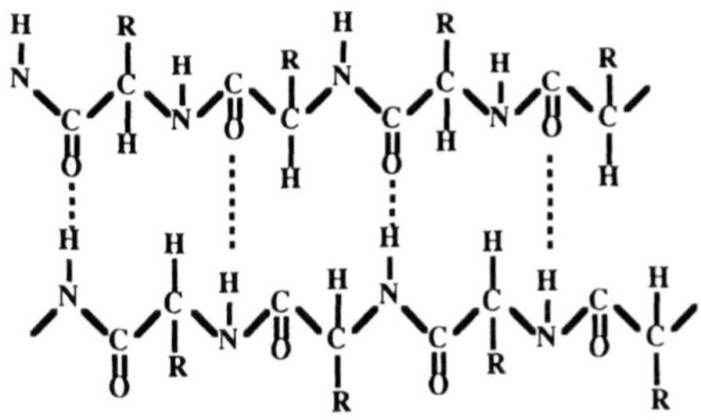

Figure 14.1-3 A stable antiparallel pleated sheet prevents steric collision when R groups are large. The R groups will be located above and below the pleated sheet with hydrogen bonding occurring within the sheet.

A number of helical arrays, not just the more famous α helix, is possible when external stabilization is present. The net effect of any of these ordered or periodic structures is to impose a higher degree of proximity on the nonpolar R groups. In fact, these nonpolar polymers generally make very compact helices in dilute aqueous solutions when there is no steric interference between the side chain R groups. A stable antiparallel pleated sheet structure is formed when the side chains sterically interfere with each other (Figure 14.1-3).

These structures are the stable ones calculated on the basis of the polypeptide "backbone" chain. The more extended the polymer, the greater is the number of H_2O molecules whose state is perturbed. The thermodynamic result of this perturbation depends upon the relative extent of interaction of the polymer with itself versus that with the aqueous solvent. If there is no steric hindrance, the polypeptide will fold so as to maximize the intrapolymer interactions and to expose the least surface to its aqueous surroundings. For example, a homopolymer of the nonpolar amino acid poly-*l*-alanine folds to yield an α helix. In such a conformation, there is a maximum number of internal hydrogen bonds, because every peptide bond NH and C=O is hydrogen bonded. Because of the pitch of the helix, there is a rise of 3.6 amino acid residues for each full turn of helix. Each hydrogen bond is therefore made to the fourth peptide bond along the chain, so that all the hydrogen bonds align themselves parallel to the helix's axis and the peptide backbone is fully hydrogen bonded. The exterior of this helix, which consists of the exposed side chains, is nonpolar. The helix is surrounded by H_2O molecules in their clathrate pentagonal structures forming the hydrotactoid but in a much more favored arrangement than that predicted for a long chain of unassociated nonpolar residues.

This behavior is observed in very dilute solutions of nonpolar homopolymers. However, in more concentrated solutions, the hydrotactoids will be close enough in proximity that removal of the H_2O molecules

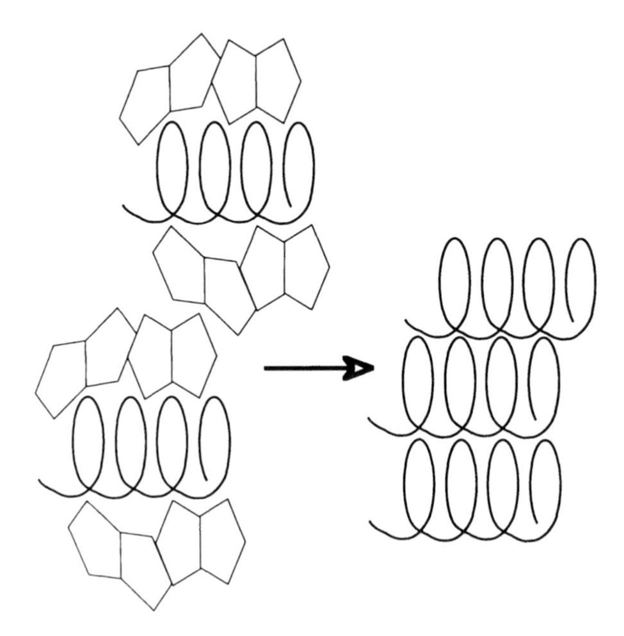

Figure 14.1-4 At a high enough concentration of nonpolar polypeptide, the chains approach closely enough that the hydrotactoid water is removed to the entropically favored bulk water environment. Aggregation and precipitation will then occur.

to the bulk water becomes possible. Under this condition, the overall entropy is increased. The polymer molecules associating so closely with each other will then no longer be adequately solvated and will aggregate and often precipitate (Figure 14.1-4).

14.1.2. Polar Polypeptides in Solution

In contrast to the simple model of a nonpolar homopolymer (e.g., polyalanine) which is stabilized by intrachain hydrogen bonding, a polar homopolymer (e.g., polyserine) will be stabilized by hydrogen bonding with the aqueous milieu. If the R groups cannot ionize but are able to hydrogen bond with H_2O (as is the case for, e.g., polyserine), the polymer can take a helical conformation, and the external R groups will hydrogen bond to the aqueous solvent. An apparently normal hexagonally arrayed H_2O sheath can surround the molecule (Figure 14.1-5).

Finally, a homopolymer of an ionizable amino acid, e.g., poly-*L*-glutamic acid, can be considered (Figure 14.1-6). The net charge on the polymer (the sum of the contributions from each side chain) will depend upon the degree of ionization of the side chains, which in turn depends upon the pH. If the polymer is unionized and consequently uncharged,

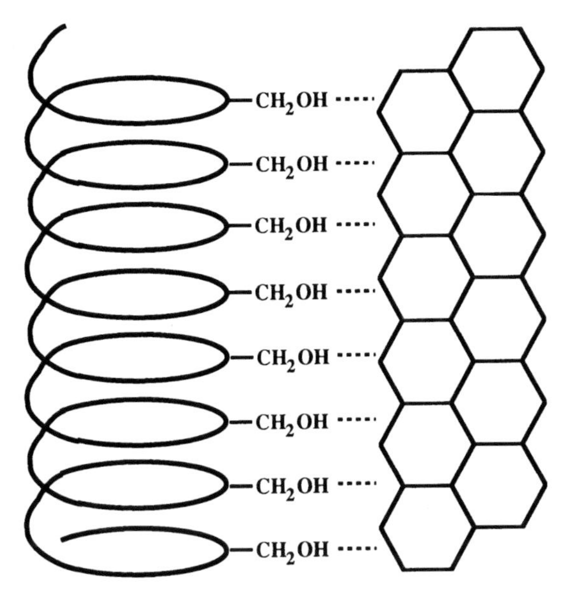

Figure 14.1-5 A nonionized polar polypeptide may form hydrogen bonds with the water structure surrounding it, with normal hexagonally arrayed water as the result.

the conformation can be helical and can under some circumstances (e.g., poly-*L*-glutamic acid at pH 3.0) be stabilized further by hydrogen bonds between the side chains analogously to polyserine. On the other hand, if the side chains of the polymer are ionized (e.g., poly-*L*-glutamic acid at pH 7.0), the identically charged side chains will repel each other and a helical structure cannot exist (Figure 14.1-6). Even a pleated sheet may not be stable. Instead, the homopolymer will assume the conformation in which the like charges are at the greatest possible distance from each other, forming a random chain configuration. The water structure around this charged polymer will be altered. All hydrogen bonds between the molecules immediately adjacent to the polymer will be broken, and the hydration sheath will attempt to achieve the structure already discussed for small ions (cf. Chapter 13).

Biological polymers are rarely as uniform as the polymers discussed so far. Instead of a single repeating amino acid side chain, natural proteins have some hydrophobic, some polar nonionized hydrophilic, and some charged residues. The protein will fold in such a manner that a minimum of energy will be required for its stabilization under the specific conditions of solvation (i.e., minimal ΔG). The actual sequence of these amino acid residues in any polypeptide chain as it is synthesized (i.e., before post-translational modification) dictates the conformation of the polypeptide. The folded conformation will have the maximal extent of intrachain

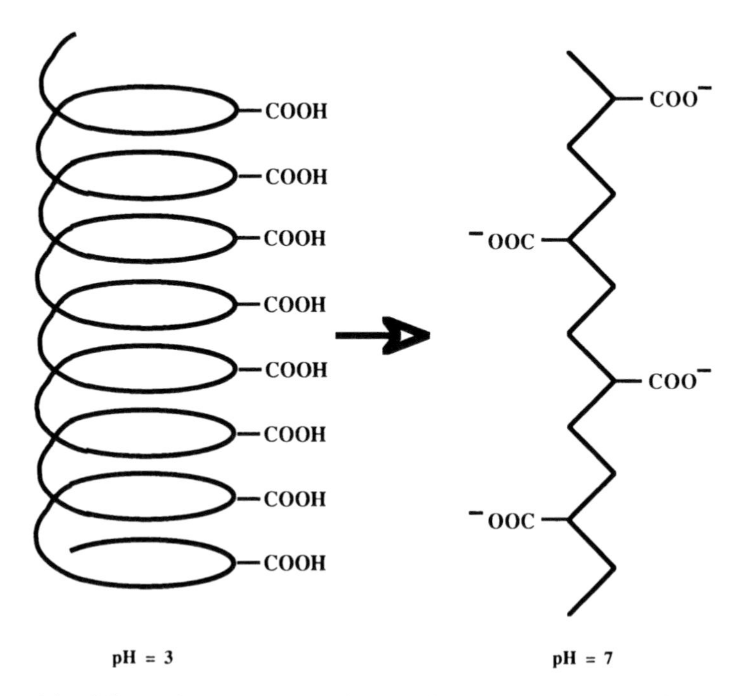

Figure 14.1-6 Secondary structures that result as a consequence of the state of ionization of a polyelectrolyte such as polyglutamic acid.

interactions that the given sequence of amino acids can accommodate. The forces operative in such folding are summarized in Figure 14.1-7, as are their approximate strengths within the folded macromolecule. The forces of interest include coulombic interactions between oppositely charged ions, ion–dipole interactions, hydrogen bonds, and hydrophobic interactions (such as van der Waals and London dispersion forces) when the entities are in virtual contact. It is worth emphasizing that hydrophobic interactions are highly entropy-driven. Because the entropy of hydrogen-bonded bulk water is so high, H_2O molecules tend to be excluded from hydrophobic regions where they would be trapped. Molecules of water in a hydrophobic pocket have limited microstates available and do not hydrogen bond. Therefore, removal of these restricted H_2O molecules from a hydrophobic pocket leads to a great increase in entropy. Many of the noncovalent interactions that lead to folding behavior are electrostatically derived. The dielectric constant of the environment must be explicitly considered in understanding these interactions. The lower the dielectric constant, the stronger the interaction will be. The driving forces in folding a macromolecule in an aqueous environment ($\epsilon = 80$) will generally be toward making the region where folding occurs as anhydrous as possible, thereby decreasing the denominator in each of the

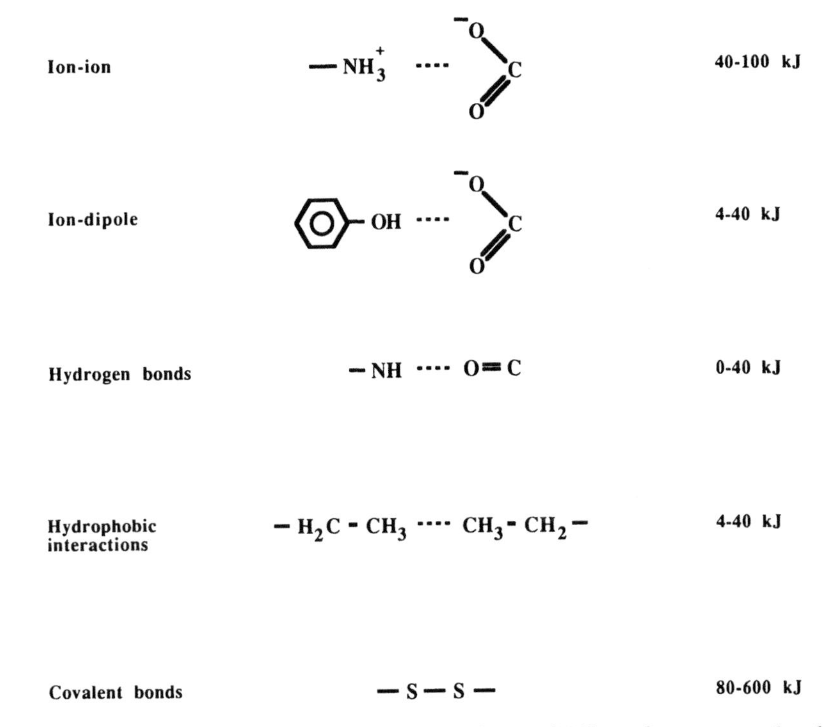

Figure 14.1-7 Types of interactions involved in the folding of a macromolecule.

force equations. X-ray investigations of pure protein crystals show very little trapped H_2O, generally less than three molecules per molecule of protein of molecular weight 14,000–100,000. Thus, the driving force for the folding of macromolecules in aqueous solutions is strongly in a hydrophobic direction and seeks to minimize the number of water molecules located away from the hexagonally arrayed lattices of bulk water. Once the water molecules have been driven out of the region and the couloumbic, ion–dipole, and hydrogen bonding associations are established, the stabilization energy for the folded structure will be so high that the tendency of the macromolecule to retain this conformation will be very strong. Therefore, conformational transitions which allow the exposure of these regions of interactions to the aqueous milieu will not be favored. A description of the dependence of such transitions on the actual equilibrium conformation will be discussed later.

Because conformational stabilization is achieved by the various interactions described above, there will be a strong dependence on the overall environment. In a homogeneous aqueous solution, the electrostatic and ionic interactions can be affected by the physical conditions at the time of structural analysis. For example, the electrostatic interactions between charged residues depend on the degree of ionization (i.e., on the pH), the ionic cloud of countercharge, and the local dielectric environment. In an aqueous system, ΔG is negative for an interaction between ionized residues of opposite charges, while ΔH and ΔS are both positive. Thus, the enthalpy change is unfavorable (energy must be expended to remove solvating H_2O molecules), but the entropy, and therefore the free energy changes, are highly favorable because the excluded H_2O molecules become part of the strongly hydrogen-bonded bulk H_2O. In the nonpolar interior of a macromolecule in its native conformation, these electrostatic or ionic interactions are strengthened by the low dielectric constant. In an aqueous environment, hydrogen bonds between residues or groups in the solute and the aqueous milieu will have a bond strength which approximately equals that of hydrogen bonds between H_2O molecules. Bonds between such groups do, however, become very much stronger in a nonpolar milieu. Finally, the van der Waals attractions that form the basis of the force of hydrophobic interactions are substantial. Once water has been excluded, it is highly unfavorable for the water to surmount the energy barrier to return to the nonpolar interior.

Even entities which possess quaternary structure, that is, which consist of several noncovalently bound subunits, remain strongly associated unless there is addition of some perturbing agent, such as a salt, a competing denaturant, a detergent, or an hydrogen-bonding entity, or unless the dilution is extreme. Only in very dilute solutions does a multimeric protein like hemoglobin dissociate into its monomeric subunits, while at physiologic concentrations these subunits are held together in the native multimer by strong hydrophobic interactions between the α and β subunits. In addition, electrostatic interactions (salt bridges) exist in this hydrophobic environment between the subunits, being stronger in the deoxygenated than in the oxygenated hemoglobin. Extraneous components or conditions (for example, the concentrations of ions or changes in temperature) can disturb the macromolecule's aqueous sheath and perturb the stable conformation of the biopolymer. These perturbing agents elicit changes in the macromolecular conformation of the solute and in its thermodynamic state. These transitions of state will be the topic of the next section.

The aqueous sheath ("irrotational water") is so highly ordered and conserved that the crystal structure of a protein formed in an environment where free water is not readily available (e.g., a high molar salt solution) is almost structurally indistinguishable from the structure of the protein in dilute aqueous solutions. This indicates that the formation of the aqueous sheath around a protein is probably highly favored and not simply a matter of the experimental technique involved in studying the protein structure. Since the amount or abundance of free bulk water in a cell is not clearly known, but is probably quite limited, this finding suggests that the structure of proteins produced in a cell is probably the same as the structure found by analysis from experiments done in dilute solutions.

The thermodynamic evaluation of solutions of macromolecules, especially of polyelectrolytes, is exceedingly complex even in simple dilute aqueous solutions. While there have been a number of theories attempting to explain the state of these solutions and to predict the nature of the aqueous medium which is their largest component, most of the theories are limited in their success. Analyses are therefore performed experimentally, and the changes in thermodynamic state parameters, especially ΔS, are more readily determined than the absolute state constant of entropy.

14.2. Transitions of State

As stated above, biopolymers are always in a hydrated state, even in the crystalline form. The crystal structure of a protein can be shown to be identical to the conformation in solution under comparable conditions. Although the polymer may no longer contain the entire primary amino acid sequence present at synthesis, the conformation in which biopolymers are isolated has been defined as the **native** conformation. Although it is clearly a function of the conditions of isolation, including pH, ionic strength, and specific ions present, these parameters are frequently unspecified when the native conformation is described. As discussed previously (cf. Section 14.1), the forces operative in stabilizing this native conformation include ionic, ion–dipole, and van der Waals forces, expressed as ionic and hydrogen bonds and hydrophobic interactions. All take place in as water-free an environment as the macromolecule can achieve, so that the interior dielectric constant can be kept low. Crystallographic evidence points to a very small number of internal "trapped" H_2O molecules in most proteins. Thus, while the hydration sheath is known to exist, its extent and properties are far from clear, although it is apparent that solvation plays a critical role in the stabilization of polyelectrolyte conformations. The greater the extent of intrapolymer interactions, the more native the conformation will be, while predominant polyelectrolyte–solvent interactions would favor an extended random conformation. Changes in the native state of a biopolymer yield a sizable amount of thermodynamic information on the energy and entropy changes involved in the transition, as well as an overview of the type, strength, and interdependence or **cooperativity** of the stabilizing interactions in each state.

In describing the conformational stages that a macromolecule goes through as it changes from a native to a denatured state, the simplest model for the change in state which can be achieved by a change in temperature, pH, ionic composition, or solvent can be represented as a

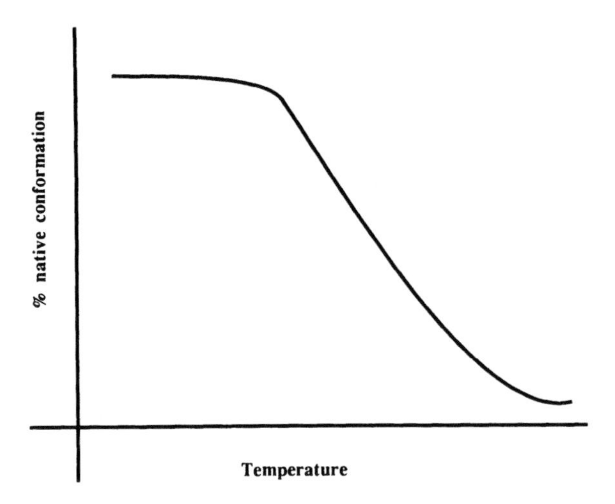

Figure 14.2-1 Sample denaturation curve.

series of steps from the native state, n, through states y_1, y_2, y_3, . . ., to the denatured state, d:

$$n \rightarrow y_1 \rightarrow y_2 \rightarrow y_3 \rightarrow\rightarrow\rightarrow d \quad (14.2\text{-}1)$$

At any point during the transition the total system can be represented as the sum of the individual component states, that is, as the sum of the fractional contributions from each state. This is written in terms of the mole fractions as the sum of the mole fraction X_x times the state y_x, where y_n is the native state, y_d is the denatured state, and y_i are the intermediate states. This can be written

$$1 = X_n y_n + \sum_i X_i y_i + X_d y_d \quad (14.2\text{-}2)$$

For each step in the denaturation, there is a different equilibrium constant $K = X_i/X_{i-1}$, which is the ratio of the mole fraction of state i divided by the mole fraction of the previous state, i - 1. The denaturation curve for such a system might look like Figure 14.2-1. The abscissa represents the magnitude of any given perturbation which brings about the change in state, and the ordinate shows the ratio of the native to the denatured state for a macromolecule.

Most actual denaturation curves follow this pattern, which is asymmetric about its midpoint halfway between the native and the denatured states. This type of curve indicates that there are a number of intermediate quasi-stable, although not necessarily isolable, states between the native and the denatured end points. Mathematical treatment of such asymmetric curves, and the extraction from them of the free energy and the entropy change associated with each change in state, is extremely complex

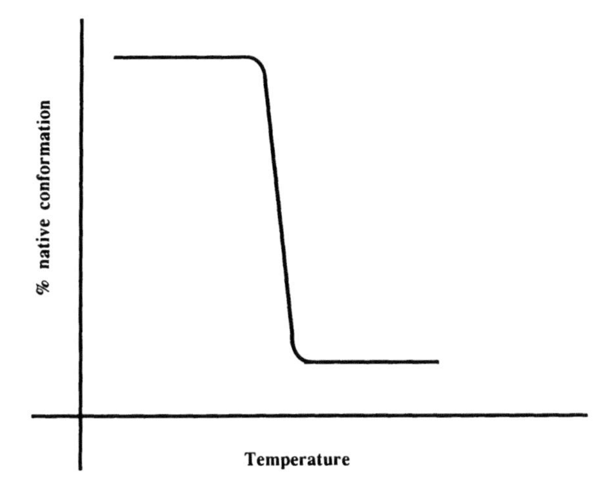

Figure 14.2-2 Denaturation curve for the special two-state case.

and often impossible unless the intermediate states themselves can be characterized. Therefore, the discussion will be limited to the special case in which only two stable states, the native, *n*, and the denatured, *d*, can exist. In this special case, at every intermediate point between 100% native and 100% denatured, a simple mixture of X_n and X_d in various proportions exists and accounts for all the material:

$$X_n + X_d = 1 \tag{14.2-3}$$

X_n is the mole fraction that is native and X_d is the mole fraction in the denatured state. Then, since

$$y_n \rightleftharpoons y_d \tag{14.2-4}$$

and, at any given quantity of the perturbing agent, the total, y_t, is

$$y_t = X_n n + (1 - X_n)d \tag{14.2-5}$$

where X_n is the fraction remaining in the native state, then at any chosen value of the parameter causing the change in state (e.g., T), the ratio of native to denatured states, f_T, can be represented as:

$$f_T = \frac{X_n}{1 - X_n} \tag{14.2-6}$$

There is just one equilibrium constant:

$$K = \frac{(X_n)_{\text{eq}}}{(1 - X_n)_{\text{eq}}} \tag{14.2-7}$$

Such a denaturation can be represented by Figure 14.2-2.

In this two-state-case system, the curve must be fully symmetric about the midpoint at $f = 1.0$; this is the easiest way to recognize a system in which only the native and the denatured states can exist. The curves can indicate not only the presence of two states, but also the **cooperativity** inherent in the change from one to the other. Cooperativity is a measure of the increased likelihood that once a given residue in the polymer has undergone a conformational change associated with denaturation, the next residue along the chain will also undergo such a change. Graphically, cooperativity is represented by the slope of the denaturation curve at its midpoint. A parameter, **s**, was defined by Zimm as the probability that once a residue is in a given conformation (e.g., denatured) the next one will also be in the same conformation. Therefore, s is a measure of the interdependence of the conformations of adjacent residues. If these conformations are independent, the probability s will be ½; if totally interdependent, s will be 1. A second parameter, σ, is a measure of the ease with which a residue exhibits a conformation different from that of its predecessor, that is, of the ease of initiating a change in conformation. Therefore, σ will be 1 when there is no restriction on such an initiation, and it will be 0 when there is no chance of an initiation, that is, when the entire molecule *must* be in the same conformation.

The equilibrium constant for the nucleation or start of the conformational change can be defined as σs:

$$\sigma s = \frac{\ldots \mathrm{rrrhrrrr}}{\ldots \mathrm{rrrrrrrr}} \tag{14.2-8}$$

where r is a polypeptide residue in the random configuration and h is one in the helical configuration. This could also be written as

$$\sigma s = \frac{\text{Single change in conformation}}{\text{No change}}$$

The constant s then represents the likelihood that once a region is started in a given conformation, it will continue in that conformation (a "zipper" effect), exhibiting cooperativity or interdependence of components:

$$s = \frac{\ldots \mathrm{rrrhhhhhrrrr} \ldots}{\ldots \mathrm{rrrhhhhrrrrr} \ldots} \tag{14.2-9}$$

That is,

$$s = \frac{\text{Propagated}}{\text{Nonpropagated}}$$

If s is high, there is extensive interdependence. The likelihood that the next residue will also undergo denaturation will be greater, the higher is the value of s. The smaller σ, the fewer is the possible number of locations where initiation of denaturation will start. This type of cooperativity

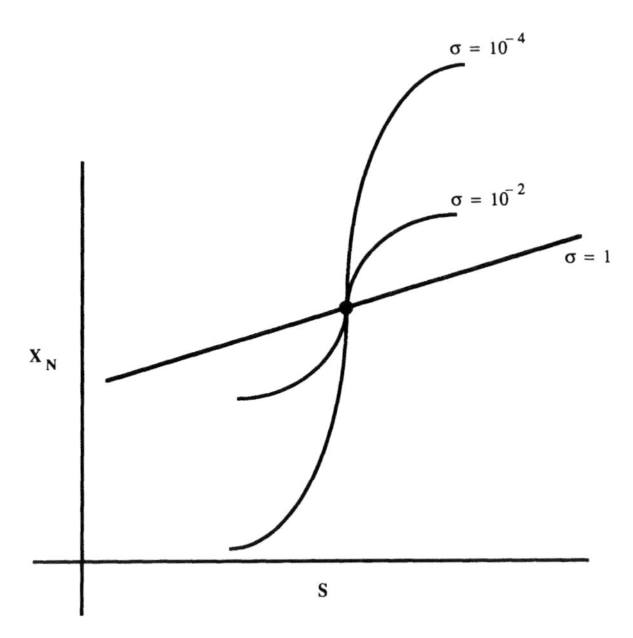

Figure 14.2-3 Family of curves showing the relationship between s and σ.

accounts for the steepness of most biopolymer or protein denaturations when the overall conformation is observed (Figure 14.2-3).

As indicated in Figure 14.2-4, when a protein such as ribonuclease is denatured by heat, the appearance of the denaturation curve and the cooperativity of the process depend upon the conditions employed and the parameter of denaturation which is chosen to be measured. If the parameter reflecting the extent of native structure is characteristic of the entire molecule, which intrinsically means that it is less sensitive to a change in conformation involving only a few (or a small percentage) of the residues, the apparent cooperativity will be great. Ribonuclease denaturation will exhibit strong cooperativity when it is measured by a circular dichroism change in the peptide bond absorbance region around 200 nm (Figure 14.2-4). Under the same conditions, if the parameter observed is characteristic of only a few residues, for example, for ribonuclease the circular dichroism in the region of aromatic side chain absorbance, then the denaturation curve is no longer either symmetric or highly cooperative (Figure 14.2-4).

In an aqueous system, a biopolymer is surrounded by a hydration shell, but in the native state it contains very few internal H_2O molecules. A change in conformation by a single residue in such a biopolymer may permit the insertion of H_2O into what had been a water-free environment. Since the internal interactions stabilizing the native conformation are all weakened in the presence of H_2O, the next residue will also become

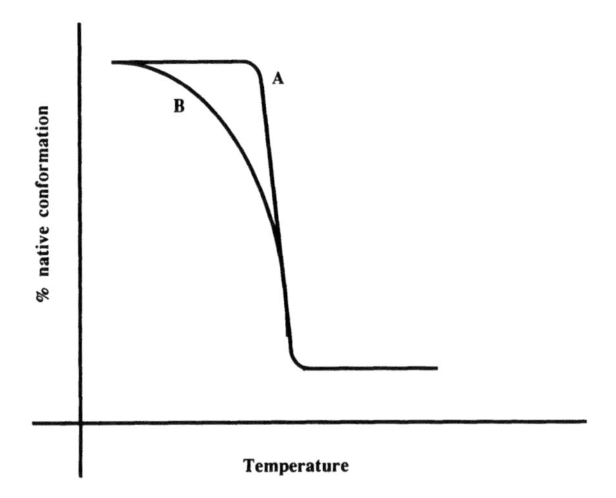

Figure 14.2-4 Thermal transition curve obtained using circular dichroism measurements that shows the differential effect of measuring a parameter representative of all the peptide bonds (curve A) versus one that examines only a small proportion (1/6) of a smaller population (6/123) that is sensitive to a conformational change (curve B).

denatured, then the next, and so on. Therefore, denaturation of a biopolymer in an aqueous medium is likely to be accompanied by a high value of *s* and is highly cooperative. In the example of a biopolymer such as DNA, the steepness of the denaturation curve, *s*, will be a measure of the homogeneity of the DNA and of the existence of well-hybridized double-helical regions, rather than a mixture of partially single-stranded or poorly hybridized species. Poorly hybridized species exhibit a higher value of s, since the base pairing will be partially imperfect in poorly hybridized specimens. These facts form the basis for the pragmatic curves by which DNA is often characterized.

Thermodynamic values can be determined for any denaturation, those of greatest interest being ΔG, ΔH, and ΔS for thermally induced changes of state. If the process can be approximated by a two-state model with the fraction of molecules in the native state being represented by X_n, and that in the denatured state by X_d, then

$$X_n + X_d = 1 \qquad (14.2\text{-}3)$$

Then for ΔG for a state that is not the standard state:

$$\Delta G = \Delta G^{\circ} + RT \ln \frac{X_d}{X_n} = -RT \left(\ln \frac{X_d}{X_n} \right)^{\circ} + RT \ln \frac{X_d}{X_n} \qquad (14.2\text{-}10)$$

A term K_{app} can be defined:

$$K_{app} = \left(\frac{X_d}{X_n}\right)^{o} \left(\frac{X_n}{X_d}\right) \tag{14.2-11}$$

which gives the result

$$\Delta G = -RT \ln \left(\frac{X_d}{X_n}\right)^{o} \left(\frac{X_n}{X_d}\right) = -RT \ln K_{app} \tag{14.2-12}$$

But $\Delta G = \Delta H - T\Delta S$, and therefore

$$\Delta \frac{H}{T} = -R \ln K_{app} + \Delta S \tag{14.2-13}$$

Equation (14.2-13) is the **van't Hoff equation.** If it can be assumed that ΔH and ΔS are independent of T, a plot of ln K_{app} versus $1/T$ (in degrees Kelvin) should yield a straight line whose slope is $-\Delta H/R$. If the assumption that only two states (native and denatured) exist is valid, the change in enthalpy at the midpoint of the transition, T_m, when $K_{app} = 1$ can be calculated from the slope of the denaturation curve at that inflection point. The enthalpy change obtained by application of the van't Hoff equation is called $\Delta H_{\text{van't Hoff}}$. The change in enthalpy derived in this way is valid only for the particular conditions of pH, salt concentrations, etc., under which the transition was measured but is nevertheless a useful parameter in the comparison of the denaturation of polyelectrolytes. In general, the most direct way to obtain ΔH is by calorimetry, but such measurements on macromolecules have proven difficult to perform.

If the assumption of a temperature-independent ΔH is not valid, the relationship between ΔH, ΔS, and K_{app} can nevertheless be used to determine ΔS for the denaturation at various temperatures. It will generally be positive, since the polymer chain has more available configurations than a more tightly held periodic structure. Although the random form may require a less favorable solvation sheath structure, thereby reducing the overall system entropy, unfavorably solvated random chains will lead to aggregation of the denatured species in an attempt to reduce the aqueous sheath. The overall ΔS for denaturation therefore remains positive.

This chapter has concentrated on thermally induced changes of state, because the thermodynamic parameters ΔH, ΔG, and ΔS are more readily understood in this context. It should however be noted that denaturation or change of phase can be brought about by any of a number of perturbations other than heat, and the same kind of thermodynamic analysis can be applied. Using ribonuclease as an example of a typical system, the curve for the change in state from the native to the denatured state as a function of added urea concentration can be plotted. Urea acts as a denaturant, because it forms hydrogen bonds to H_2O in the aqueous solvation sheath surrounding a solute, as well as to other polar groups which may have been involved in hydrogen bonds (e.g., alcohol-, carbonyl-, or amine-containing side chains of proteins). Denaturation by addition of urea yields a curve which can be analyzed in exactly the same way as thermal denaturations in terms of deducing ΔH, ΔG, and ΔS of the denaturation as a function of urea concentration rather than as a function of temperature. The higher the concentration of urea at the midpoint of the curve, the stronger the interactions within the native conformation of the solute will be. The steeper the denaturation curve, the more cooperative the denaturation will be once denaturant has begun to compete successfully for hydrogen bonds outside or within the protein. If the curve is symmetric about its midpoint at 50% native conformation, the system has only two possible conformations, a native and a denatured one, and only the approximations valid for a two-state system may be applied.

This chapter has presented some of the thermodynamic and conformational information necessary to analyze and predict the behavior of proteins and biopolymers and their changes of state. The dependence of these states on the aqueous solvent which surrounds these solutes even in the crystalline state has been emphasized. Throughout this section, the behavior of solutes of various sizes and classes has been examined in a predominantly aqueous solvent leading to homogeneous solutions at equilibrium conditions. Now, it is necessary to extend understanding to more complex conditions, notably multiple-phase and nonequilibrium processes, that more accurately describe biological processes.

Part III
Membranes and Surfaces in Biological Systems

CHAPTER 15

Lipids in Aqueous Solution: The Formation of the Cell Membrane

15.1. The Form and Function of Biological Membranes

So far, this textbook has dealt with the thermodynamics and properties of aqueous solutions. The properties of bulk water have also been considered. However, cells and biological systems are not comprised of simple homogeneous solutions or bulk aqueous phases. One of the most fundamental conceptual as well as structural components of biological organization is the presence of a delimiting membrane. In a very real sense, cells are defined physically and functionally as different from their environment by the cell membrane. The organization of most biological systems, and certainly eukaryotic cells, depends on compartmentalization of cellular functions. This compartmentalization, which allows varied microenvironments to exist in close proximity, is dependent on the membranes that constitute the boundaries of the compartmentalized organelles. Virtually all biological membranes are comprised of lipid molecules arranged in bilayers, with inserted or attached protein and carbohydrate molecules playing a variety of roles at or in the lipid phases. While, in general, the formation of a membrane can occur with a wide variety of polymeric substances including proteins and carbohydrates, biological membranes share the property of depending on lipid molecules to form the membrane structure itself. Although membranes in cells are comprised primarily of hydrophobic lipid elements, they almost always separate two phases whose predominant species is water. Generally, these aqueous-dominated phases are treated as aqueous solutions. Biological membranes are characterized as being permeable to some but not to all of the components of these solutions. The arrangement of these aqueous–lipid–aqueous "phases" leads to a generalized mechanism through which the cell can perform a wide variety of tasks that allow it to sense, judge, and respond to its environment. In other words, the membrane not only defines where the cell as "self" and the environment as "not self" begins and ends, but also it allows the cell to collect information and energy from the environment, process these inputs, and respond in a variety of output patterns. The membrane either is active in or is the site of action for physical separation and protection; mediation of the flow of chemical species into and out of the cell; accumulation and conversion of energy and electric charge for the cell; and information processing and exchange between cellular elements and the environment, both local and remote. Understanding biological systems in any modern sense demands a familiarity with the physics and chemistry that allows a membrane to form and to function in such diverse ways. This section will focus on certain biophysical and biochemical aspects of lipid membrane formation from which a more complete description of the cell can eventually emerge.

15.2. Lipid Structure: Components of the Cell Membrane

Lipids form a very diverse collection of molecules that can be obtained from cells and tissues by treatment with nonpolar solvents such as chloroform and benzene. They all share the characteristic of being water insoluble, a property derived from the extensive hydrocarbon portion of their structure. Lipids can be classified in a variety of ways. The most generally useful scheme is a classification based on their backbone structure. These structures are listed and illustrated in Figure 15.2-1. On the basis of their structure, the families of lipids are considered as fatty acids, nonpolar long-chain alcohols, substituted glycerols, substituted glycerol-3-phosphates, sphingosines, terpenes, prostaglandins, or steroids.

The simplest of the lipids are the single-chain fatty acids, which are similar to the large organic ions discussed in Section 13.3.2. Fatty acids are extremely common as components of the more complex lipids found in most cell membranes, although it is uncommon to find them in the free or unesterified form in biological tissues. All fatty acids are composed of a long hydrocarbon chain that terminates in a carboxyl group. At the appropriate pH, the carboxyl group will be ionized, but more commonly the carboxyl group will esterify with an alcoholic moiety from a suitable backbone such as glycerol to form complex lipids such as the glycerol esters. It is worth noting that a fatty acid is a potential electrolyte, and in the pure state it would be expected to be uncharged. Most fatty acids of biological importance have alkyl chains between 12 and 22 carbons in length. The alkyl chains are either saturated or partially unsaturated. If the hydrocarbon chains contain double bonds, the bonds are almost always found in the *cis* configuration, although certain unusual fatty acids do contain *trans*-oriented double bonds. The length of the hydrocarbon chain and its degee of unsaturation are important determinants of the physical properties of fatty acids. As Figure 15.2-2 indicates, increasing the length of the hydrocarbon chain leads to a rise in melting temperature. The presence of double bonds leads to a significant drop in the melting point when compared to other chains of equivalent length.

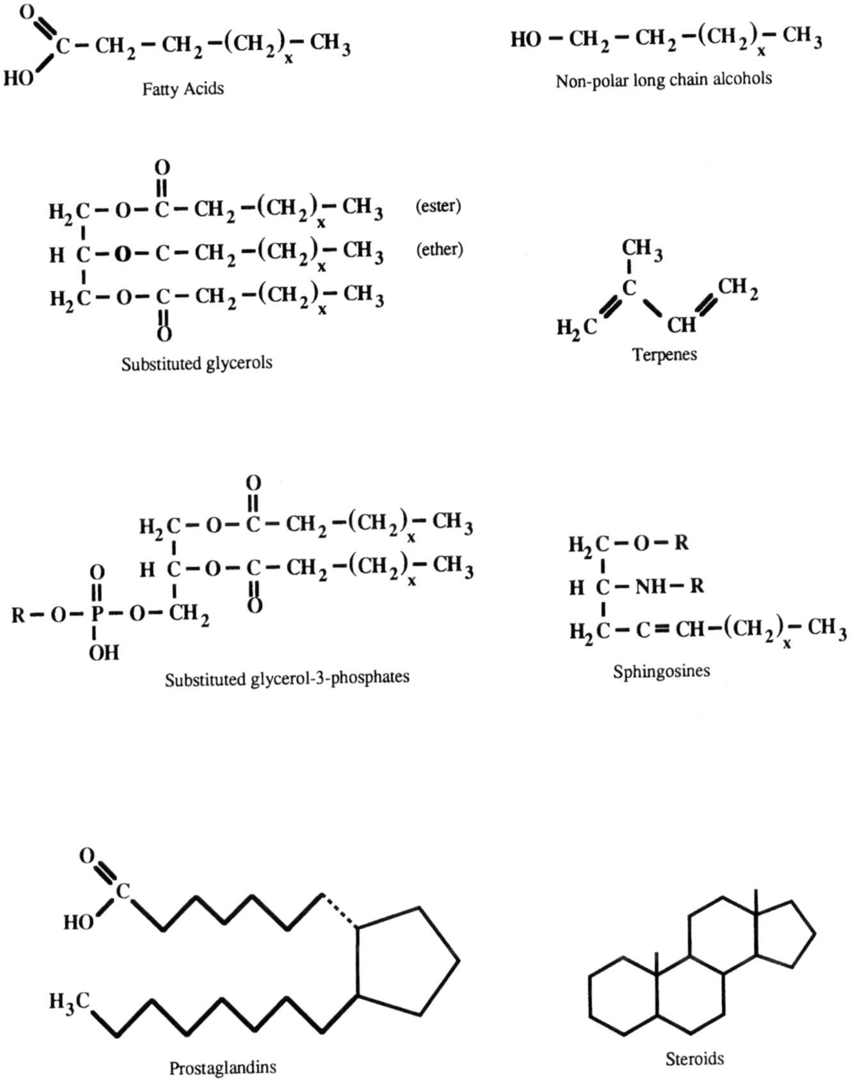

Fatty Acids

Non-polar long chain alcohols

(ester)

(ether)

Substituted glycerols

Terpenes

Substituted glycerol-3-phosphates

Sphingosines

Prostaglandins

Steroids

Figure 15.2-1 Lipids may be classified on the basis of their backbone structure.

The physical reasons for these variations in thermodynamic properties are easily described. As Figure 15.2-3 shows, the saturated hydrocarbon chain has free rotation around all of its carbon–carbon bonds, and hence its minimum energy configuration is a fully extended arrangement of the carbon atoms. Double bonds restrict rotation around the carbons, leading to the formation of rigid kinks in the chains. The maximal interaction between the hydrocarbon chains of the fatty acid is achieved through the

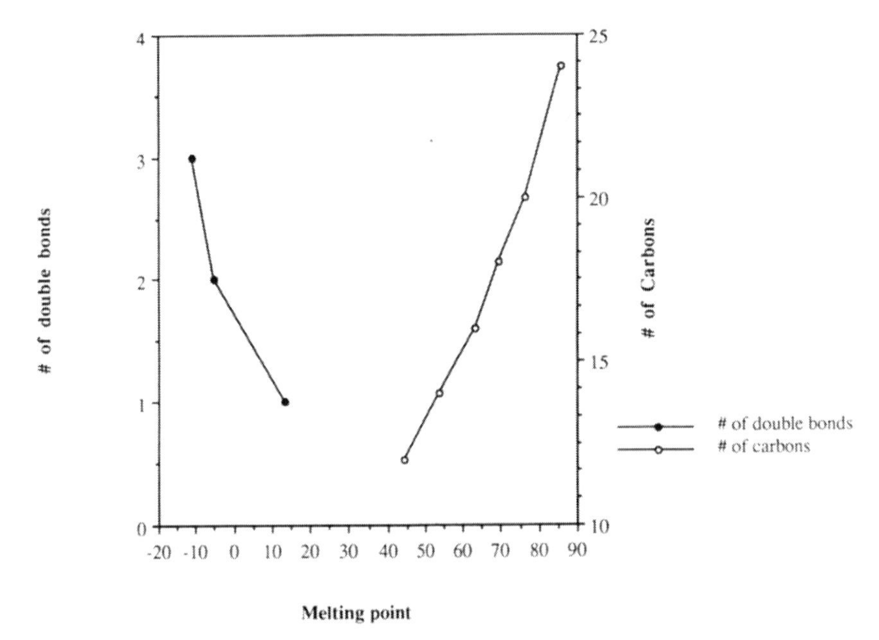

Figure 15.2-2 The melting point for fatty acids increases as the chain length increases, as shown by the curve on the right representing the melting points for saturated fatty acids as a function of chain length. Conversely, the melting point falls as the number of double bonds increases, as shown by the curve on the left, representing the melting points for 18-carbon fatty acids as a function of the number of double bonds.

lateral interactions. The aliphatic chains form layers interacting side- by-side with the resultant structures being stabilized by multiple van der Waals interactions. If the aliphatic chains are saturated, the packing can be tighter with an increased number of van der Waals interactions. Conversely, when the aliphatic chains are unsaturated the kinked chains prevent efficient close packing. Unsaturated fatty acids therefore cannot interact as strongly with each other. As a result, it takes less heat energy to disrupt the interactions in a solid unsaturated fatty acid compared to a solid saturated fatty acid.

In contrast to these simple single-chain lipids, many of the lipids found in cell membranes are **glycerol esters**. If all three carbons of the glycerol backbone are linked to simple fatty acids via an ester linkage, the compounds are called triacylglycerols, neutral fats, or sometimes triglycerides. These lipids are neutral and somewhat polar due to the ester linkages. As a group, they are extractable in nonpolar solvents. The properties of the glycerol esters depend upon the length and degree of unsaturation of the fatty acids attached to each of the three carbons of the glycerol backbone.

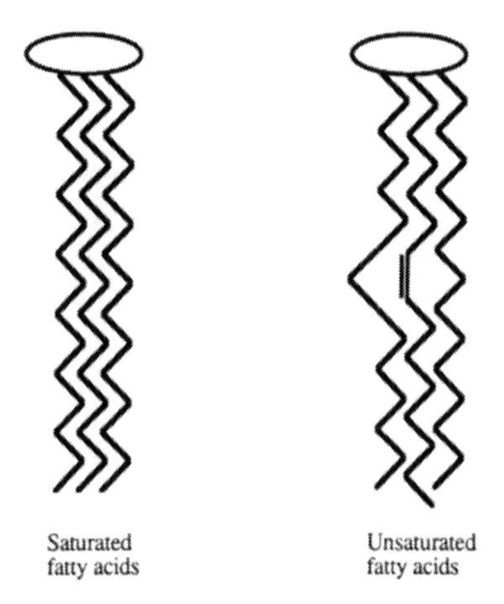

Figure 15.2-3 The freedom of rotation around the carbon–carbon bonds in a saturated fatty acid leads to an extended configuration that allows maximal intermolecular interactions to stabilize the structure. When double bonds are present, the chain becomes fixed in *cis* configurations, which prevents efficient packing of the molecules, leading to a less favorable interactional energy.

If the third carbon of the glycerol backbone is phosphorylated, while the first two each carry a fatty acid ester, the resultant lipid is called **phosphatidic acid**. Because the C-3 position carries a charge, the properties of phosphatidic acid are quite different from those of a neutral fat carrying the same fatty acids on C-1 and C-2. The **phospholipids** are derivatives of phosphatidic acid in which the phosphate group is esterified with an amino acid, an aliphatic amine, a choline residue, a carbohydrate residue, or some other residue (Figure 15.2-4).

In mammalian systems, the glycerol ester-based lipids frequently have a saturated fatty acid on C-1 and an unsaturated one on C-2. Like their parent fatty acids, the nonpolar aliphatic substituents of the glycerol esters prefer not to interact with H_2O but will tend to form hydrophobic interactions with each other. If the third carbon has a polar or charged substituent, these moieties will seek stabilization by hydrogen bond and electrostatic interactions. At physiologic pH, the phospholipids phosphatidylcholine and phosphatidylethanolamine have no net charge; phosphatidylinositol and phosphatidylserine have a net charge of -1; and phosphatidic acid and diphosphatidylglycerol have a net charge of -2.

Other lipid families are present in membranes, notably the **sphingolipids** and the **sterols**. Sphingolipids are complex lipids that are found in many membranes, but especially in membranes of nerve and brain tissues. All of the sphingolipids contain three components: a long-chain amino alcohol called **sphingosine** or one of its derivatives, a fatty acid, and a polar head group. It is perhaps easier to visualize the structure of

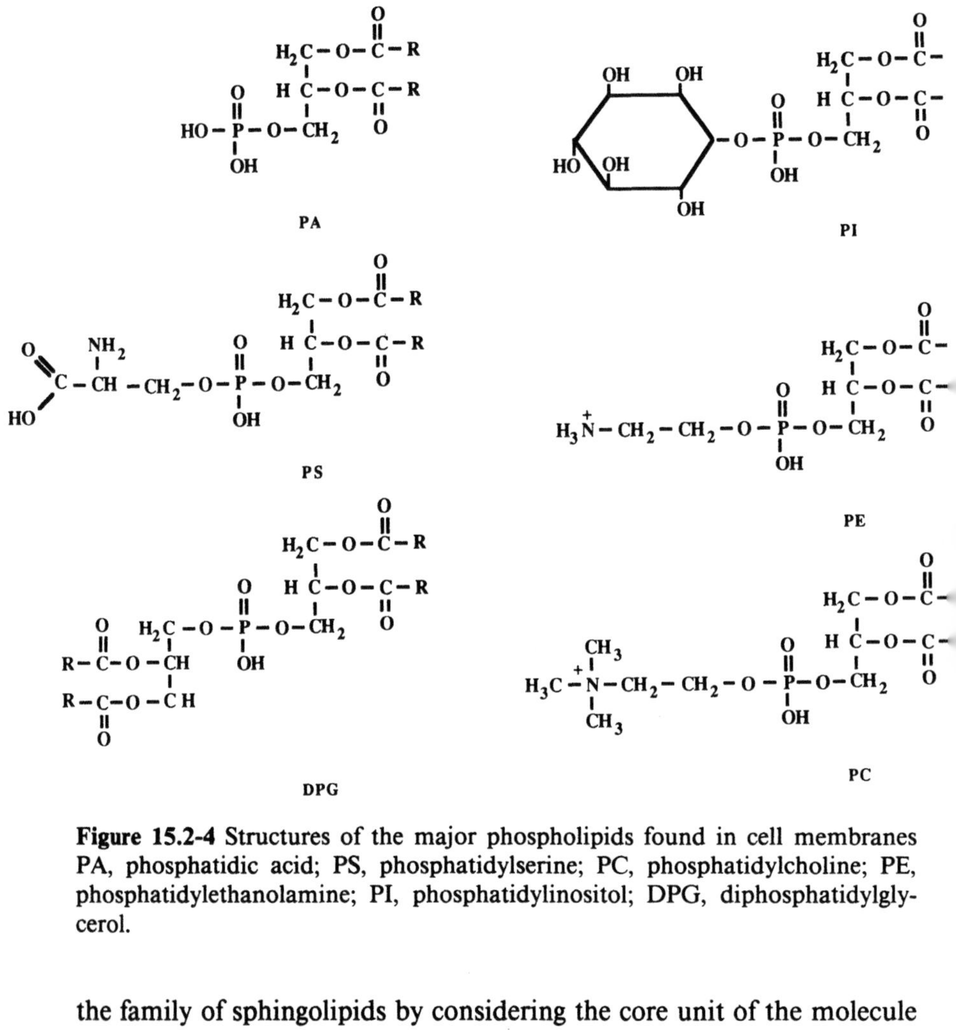

PA

PI

PS

PE

DPG

PC

Figure 15.2-4 Structures of the major phospholipids found in cell membranes PA, phosphatidic acid; PS, phosphatidylserine; PC, phosphatidylcholine; PE, phosphatidylethanolamine; PI, phosphatidylinositol; DPG, diphosphatidylglycerol.

the family of sphingolipids by considering the core unit of the molecule to be 1-amino-2-hydroxyethane, which is then substituted as shown in Figure 15.2-5. The sphingolipids characteristically have two nonpolar tails and a polar head group. The most abundant sphingolipids are the sphingomyelins, which contain either phosphorylethanolamine or phosphorylcholine. Like the phospholipids phosphatidylserine and phosphatidylethanolamine, these molecules are zwitterions at neutral pH. If the polar head group of a sphingolipid is comprised of a neutral saccharide or polysaccharide, a neutral sphingolipid called a **cerebroside** or **ganglioside**, respectively, results. It is interesting to note that three gangliosides differing only in the polysaccharide structure are responsible for the antigenic determinants of the ABO blood groupings.

The **sterols** are quite distinct as membrane lipids, especially in terms of their molecular structure and shape. Cholesterol is the best known of

```
         H
         |
   H2C - C - NH
    |    |    |
    O    R    R
    |
    R
```

A

```
   CH3              O
 + |                ||
H3C - N - CH2 - CH2 - O = P - O - H2C  OH
      |                   |        |   |
     CH3                 OH      H C - CH - CH = CH-(CH2)13 - CH3
                                   |
                                 H N - C - CH2 -(CH2)x - CH3
                                       ||
                                       O
```

B

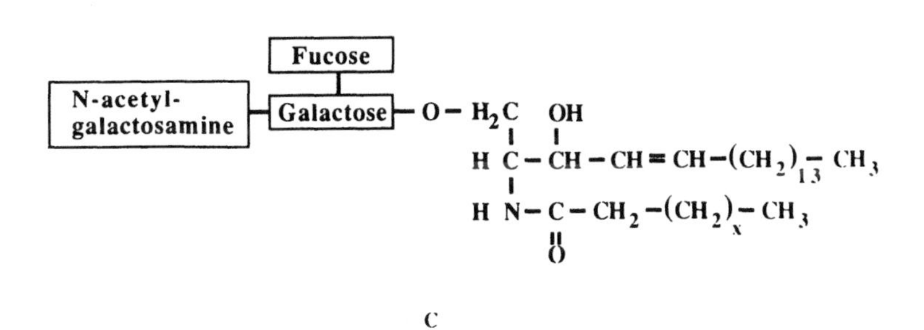

C

Figure 15.2-5 Basic structure of the sphingosine lipids: (A) 1-amino-2-hydroxy-ethane; (B) sphingomyelin; (C) the ganglioside responsible for the group O antigen of the ABO blood group.

the animal steroids and is found extensively in cell membranes. Since it is the precursor of many other steroid derivatives, such as bile acids, and hormones, such as androgens, estrogens, and adrenocorticosteroids, it is found widely dispersed in tissues. While the nonpolar chains of the previous classes of lipids were comprised of predominantly single carbon–carbon-bonded aliphatic chains, the nonpolar structure of the steroids is a planar ring structure of significant rigidity (Figure 15.2-6). When cholesterol inserts into the lipid bilayer of the membrane, the rigid cholesterol nucleus affects the fluidity of the membrane. This effect will be described in Section 15.4.

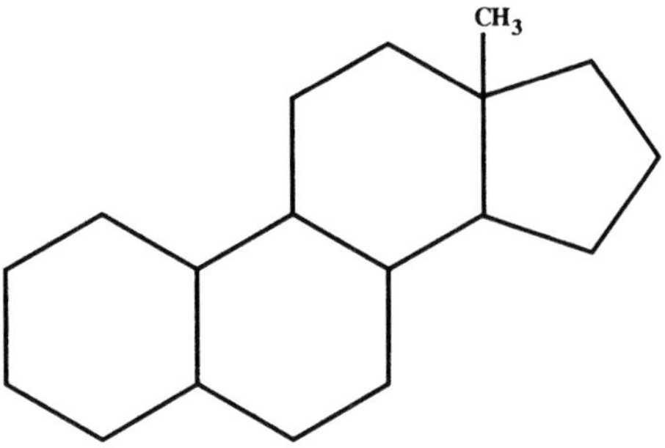

Sterol nucleus

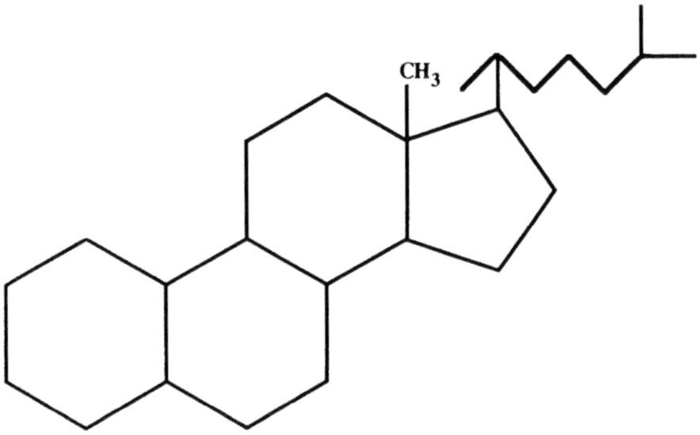

Cholesterol

Figure 15.2-6 Structure of the sterol nucleus of the lipid cholesterol.

15.3. Aqueous and Lipid Phases in Contact

The behavior in an aqueous environment of simple amphiphilic molecules having a long aliphatic hydrophobic chain and a polar head group, such as molecules of a fatty acid, was described in Section 13.3.2. To review, when found in an aqueous environment above the critical micelle concentration, these simple lipids associate hydrophobically side by side so that the aliphatic chains maximize van der Waals interactions and exclude water. The charged ends interact electrostatically, thereby either attracting or repelling one another. Molecules containing aliphatic chains of 10 to 15 carbon atoms will form micelles if the lipid solution is strongly agitated (usually sonicated). The aliphatic chains in a micelle are internal, while the charged ends reside on the surface of the sphere, where maximum hydration can be attained. These same lipids, if gently layered onto an aqueous surface, will form a monolayer in which the polar ends stick

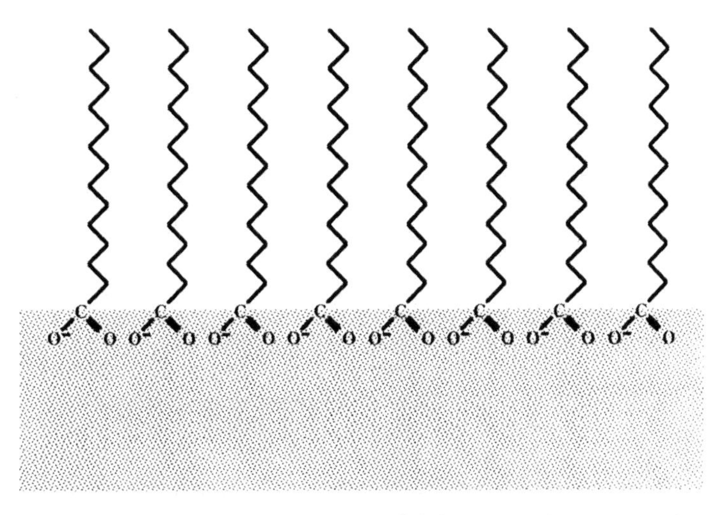

Figure 15.3-1 Formation of a monolayer of lipid molecules occurs when a fatty acid is layered onto an aqueous solution. The polar head group associates with the water molecules, and the hydrocarbon tail orients so that it avoids contact with the aqueous solvent.

into the aqueous medium and the nonpolar aliphatic chains protrude above the solvent (Figure 15.3-1). The surface tension of the water will be markedly reduced.

In similar fashion, if a polar complex lipid such as a phospholipid is added to water in a very low concentration, the lipid molecules may be dissolved by inclusion in clathrates. Above its critical micelle concentration, the lipid will associate into a thermodynamically favored aggregation such as a micelle or a lamellar **bilayer**. The formation of these structures is favored by a ΔG of approximately -60 kJ per mole of lipid transferred from water to a micelle. The types of lipid structures resulting from these interactions are shown in Figure 15.3-2. The interaction of these polar lipid assemblies with water and the structures that result depend strongly upon the nature of the nonpolar and polar substituents. In every case, these structures will orient so as to expose the polar residues to the aqueous milieu and the nonpolar aliphatic regions to each other. If the lipids are deposited gently at the interface between air and water, a monolayer will be produced with the aliphatic chains protruding from the water and the polar ends interacting with the aqueous solvent at the interface just as was described in Figure 15.3-1. If the lipid–water system is strongly agitated, the lipids will be dispersed and will form micelles (Figure 15.3-2) in which a monolayer of lipid forms an array with the aliphatic regions interacting only with each other, while the amount of surface exposure of polar residues to the aqueous solvent is maximized. These structures can be spherical or concave in shape.

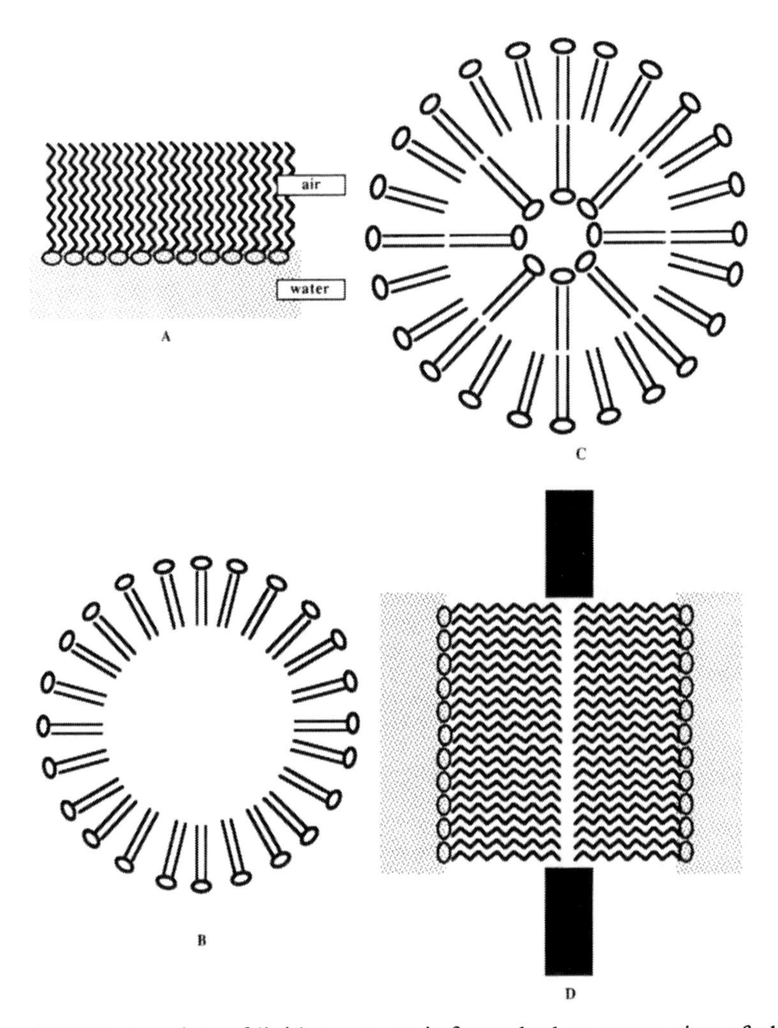

Figure 15.3-2 A variety of lipid structures is formed when aggregation of phospholipids in contact with aqueous phases occurs: (A) a monolayer formed at an air–liquid interface; (B) a micelle; (C) the liposome, a bilayer structure, that is self-enclosing and separates two aqueous phases; (D) a black membrane, a planar bilayer, that separates two aqueous phases.

If a mixture of phospholipids in water aggregates so that water is found both outside and also trapped inside the lipid layer, a bilayer structure called a **liposome** will form. In a bilayer structure such as the liposome, a pair of aqueous phases is separated by a lipid phase. The bilayer forms so that the hydrophilic head groups face outward to form an interface with the aqueous phase, while the hydrophobic hydrocarbon chains are protected from contact with water by remaining on the inside of the two leaflets. The nonpolar aliphatic chains on carbons 1 and 2 of the glycerol backbone interact as closely with each other as their degree of unsaturation permits. The lipid molecules tend to associate side- by- side so as to maximize both the possible van der Waals interactions between the aliphatic chains and the hydrogen bonds and electrostatic interactions between the polar ends. The van der Waals energy accounts for a free energy change of approximately -4 to -6 kJ mol^{-1}, and the electrostatic and hydrogen bonding interactions account for a free energy change of -4 to -400 kJ mol^{-1}, depending upon the environment. Furthermore, there is an entropic component favoring the self-association of lipids, since H_2O is excluded from the contact region between the aliphatic nonpolar ends.

Bilayers of this type are considered to be elementary models for biological membranes and have been used extensively in research on membranes. Multiple physical techniques such as nuclear magnetic resonance, electron - spin resonance, electron microscopy, and low-angle X-ray diffraction studies have all confirmed that lipid bilayers of the type described are a fundamental component of biological membranes. Liposomes have been extensively studied not only because they are a reasonable experimental model of cell membranes but also because of their possible use as drug delivery systems for the treatment of disease. These therapeutic studies are directed at utilizing the tendency of lipids to interact with each other. The principle is based on the idea that the bilayer membrane of a liposome will fuse with the cellular plasma membrane, thus mingling its contents with those of the cell and delivering the drugs, probes, or other entities enclosed in the liposome into the cellular cytoplasm. Liposomes, depending on the phospholipid involved, can be internalized inside other larger ones, forming polylamellar or "onionskin" liposomes. If these structures are sonicated, that is, given enough energy to rupture and re-form, the liposomes will be unilamellar, containing a single bilayer of lipids, and will be of uniform size. This size will be dictated by the surface-to-volume ratio, which is dependent on the nature of the lipid.

An important result from the studies of liposomes is that when artificially mixed phospholipid liposomes are formed, there is an asymmetric distribution of the phospholipid types between the inside and outside leaflets. This observation coincides with a variable but demonstrated asymmetry of phospholipid components in the membranes of cells. The asymmetry of the phospholipid distribution is considered to result from a thermodynamically favored configuration in which strain resulting from the curvature of the surface and the electrostatic interactions of the charged head groups is minimized. It is not yet clear if the asymmetry of the membrane components has a physiologic purpose. Furthermore, while the cytoskeleton and the consumption of ATP seem to be necessary for maintenance of the asymmetry, the rationale and details of these observations are not yet known.

Another type of artificial bilayer can be formed, called a **black membrane**. This is a planar synthetic bilayer that is made by introducing the lipids under water at a small hole in a barrier erected between two aqueous compartments (Figure 15.3-2D). Black membranes have proven to be very useful in modeling mammalian membrane properties, in evaluating the importance of a specific phospholipid and/or of the composition of the C-1 and C-2 fatty acids, and in describing the properties of a membrane. It is clear that the substituents on all three glycerol carbons play a vital role, as do the other membrane-associated molecules such as cholesterol, proteins (including glycoproteins and lipoproteins), and carbohydrates. Among the characteristics of lipids which are important in this role is their structure. In particular, the lipids in mammalian membranes at body temperature are in the **liquid crystalline** form, a partially ordered but nevertheless flexible array which permits the lateral motions demanded by the **fluid-mosaic model** of a mammalian membrane. The membrane lipid structure also affects the conformation of the transmembrane and intrinsic membrane proteins, which in turn affects their function as receptors, channels, carriers, exchangers, recognition sites, etc. Thus, any modulation of membrane lipid structure or any change of state from one state to another (e.g., crystalline to liquid crystalline) will alter the properties of the membrane and may affect the function of the cell. The physical state and properties of lipid bilayers must now be considered.

15.4. The Physical Properties of Lipid Membranes

15.4.1. Phase Transitions in Lipid Membranes

Lipids undergo changes in state just as other compounds do. A distinguishing characteristic of lipids in membranes is the ability to exhibit an intermediate or mesomorphic state called the **liquid crystalline** or the gel state. A liquid crystalline state is intermediate in level of organization between the rigid crystal and the fluid liquid state (Figure 15.4-1). The transition of state between the solid crystalline and the liquid crystalline form in lipid bilayers is quite sensitive to temperature. In pure lipid bilayers, there is a well-defined melting or **transition temperature**, T_m, below which the lipid will be in a solid crystalline phase and just above which the lipid will usually assume a less ordered liquid crystalline arrangement. In the pure lipid bilayer, there is significant cooperativity in the melting process, which leads to the sharply defined transition temperature. While the discussion in this section will be confined to the simpler cases of single-component membranes, the heterogeneous composition of real membranes does not give rise to a sharp transition point, and melting will occur instead over a several-degree range of temperature.

Figure 15.4-1 Comparison of the lipid configurations in the (A) solid crystalline; (B) liquid crystalline; and (C) liquid states.

At a temperature below the melting point, the solid crystalline phase exists in which the nonpolar hydrocarbon tails are rigidly held in an all-*trans* configuration; there is little lateral mobility of each molecule. The all-*trans* arrangement of the nonpolar tails leads to occupation of a minimum volume by the hydrocarbon groups in the membrane. The *cis* configuration that results from the presence of unsaturated bonds in the hydrocarbon tails leads to a greater occupied volume. Because the interactions between the hydrocarbon chains in the *cis* configuration are more limited, the transition temperature is lower than that of chains with an all-*trans* configuration. Below the transition temperature, a saturated hydrocarbon chain (normally in the all-*trans* configuration) will have a very low frequency of **kink** or *cis*-conformation formation, about one kink per 10 acyl chains. As the temperature of the system is increased, there is increasing disturbance of the tightly packed crystalline structure until the transition temperature is reached, where there will be a kink frequency of about one per acyl chain. Above the melting point, the acyl chains will have two or more kinks per chain. The liquid crystalline phase is characterized by a fairly rigid structure near the polar (charged) head groups and a considerably more disorganized and flexible region near the central hydrophobic portion of the bilayer. The loosened organization of the liquid crystal allows for lateral diffusion and a freedom of rotation along the long axis of the hydrocarbon chain. The transition from a stretched all-*trans* conformation to a kinked conformation is accompanied by an increase in membrane volume and a decrease in membrane thickness.

15.4.2. Motion and Mobility in Membranes

The existence of the liquid crystalline state allows a significant increase in the intra- and intermolecular movement of components in the membranes. The phase transition to the liquid crystalline state in membranes affects the ordering and fluidity of the membrane. Measures of these properties are of value when describing membrane behavior. The fluidity is described by a term that refers to the viscosity of the membrane. Viscosity is actually a term appropriately used in macroscopic homogeneous phases without anisotropy. Membranes even of pure lipids are by definition anisotropic and nonhomogeneous, so the concept of local viscosity or **microviscosity** is used. The microviscosity of a membrane or of a lipid bilayer is a measure of the ease with which lateral motion of lipid molecules with respect to one another can occur in that membrane or bilayer. The microviscosity of a membrane in the liquid crystalline phase is highest near its surface and lowest in its center. This is true whether the bilayer contains pure lipids, as is true in liposomes, or whether proteins, glycoproteins, and lipoproteins have been inserted, as in a typical mammalian membrane. This property is important in the fluid-mosaic model of a mammalian membrane because it contributes to the velocity with which transmembrane proteins can travel toward each other in order to form a membrane protein assembly (a "cap") and may control the energy of activation required for the conformational changes occurring in the membrane, such as those associated with receptor-mediated functions. The microviscosity is a function of the liquid crystalline structure of the membrane, and the energy required to achieve a transition from one liquid crystalline state to another as well as the temperature at which such a transition occurs can be calculated from the thermal denaturation curves of a membrane in a fashion similar to that described for macromolecules in Chapter 14. The microviscosity is sensitive to temperature, to pressure applied to the membrane, and to the chemical composition of the membrane.

The microviscosity can be evaluated by various techniques such as:

1) ^{13}C nuclear magnetic resonance (NMR) to measure the freedom of motion around C–C bonds.
2) Electron paramagnetic resonance (EPR) to measure the freedom of rotation of a spin probe, usually introduced into the membrane as a nitroxide attached at the end of a long aliphatic chain.
3) Fluorescence polarization techniques to evaluate via depolarization of the incident polarized light beam the extent to which a lipophilic probe's fluorescence emission is altered.

Each of these techniques has limitations. NMR requires a high concentration of phospholipid and is consequently difficult to apply to cellular systems unless the cells tolerate relatively tight packing. Furthermore, NMR cannot detect carbon-carbon movement whose frequency falls below 10^5 s^{-1}. Since the paramagnetic tag covalently linked to an EPR probe cannot penetrate the membrane, there is a risk that EPR will measure the fluidity of, at best, the outer surface of the cell membrane, and, at worst, the spinning of the probe itself outside the membrane. If fluorescence depolarization is used, the choice of probe dictates the region that is being examined, since some probes are completely embedded in the membrane while others protrude into the extracellular milieu because they are charged. Further details of the actual measurement of microviscosity are beyond the scope of this text and will not be discussed. It suffices for the reader to know that the fluidity of a membrane can be correlated with its phospholipid composition, with the degree to which the fatty acid components of the phospholipids are unsaturated, and with the level of incorporation of intercalating entities, such as cholesterol, in the membrane.

A variety of modes of motion can occur in lipid membranes (Figure 15.4-2). The most common of these is rotation around the carbon–carbon bonds in the long axis of the acyl chain. The speed of these rotational events is on the order of 100 picoseconds. The lifetime of the *cis* kinks in the hydrocarbon chains is about 10 times longer, about 1 nanosecond. The lipid molecules also undergo bending motions. With the exception of the formation of kinks, which is tied to the fluidity of the membrane, these carbon–carbon rotations and bends seem to have no clear functional relevance. Rotation of the entire lipid molecule including the head group also occurs in the membrane with a rotational speed on the order of 2π radians/10 nanoseconds. The parameter describing this rotational motion is called **rotational diffusion**.

Perhaps one of the most important motions that will occur in a membrane is **lateral diffusion**. Lateral diffusion can be thought of in terms similar to those that are the basis of the treatment of Brownian motion by rate theory, which will be covered in some mathematical detail in Chapter 17. For now, lateral diffusion can be thought of as a process

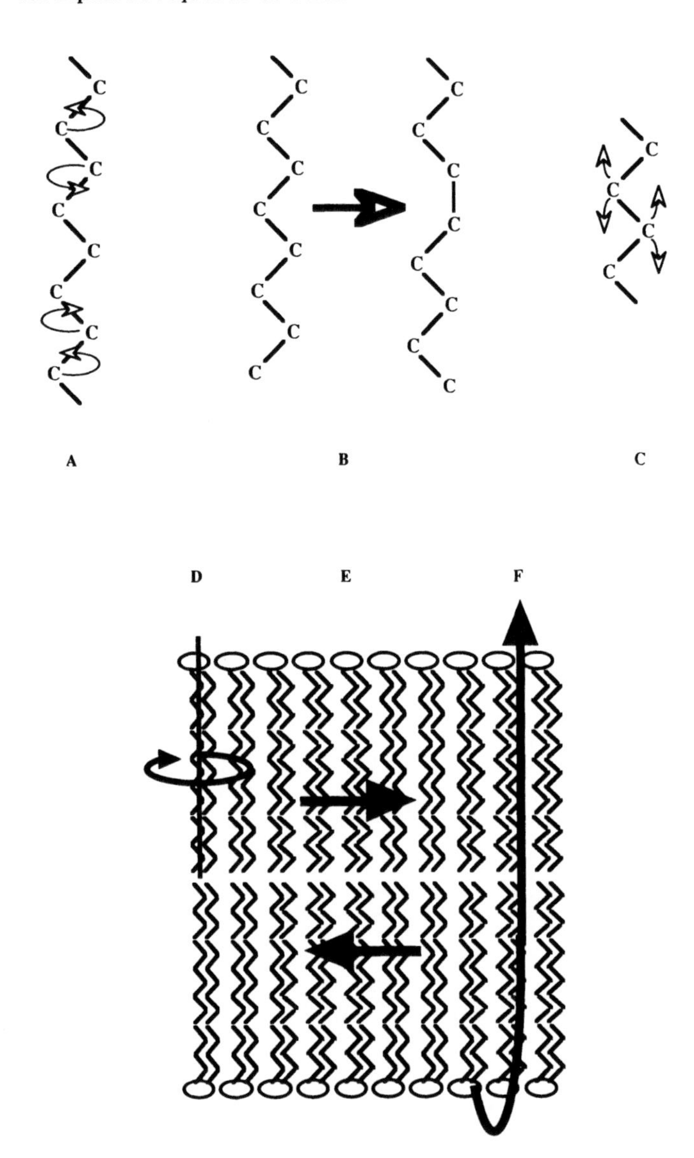

Figure 15.4-2 Various modes of motion found for lipids in membranes: (A) intrachain rotation around carbon–carbon bonds; (B) kink formation; (C) bending of the hydrocarbon chain; (D) rotational diffusion; (E) lateral diffusion; (F) transmembrane flip-flop movement.

where a lipid molecule will experience net translational motion in the plane of the membrane by executing a series of jumps from one vacancy in the liquid crystalline array to another. The rate of lateral jumping in membranes is on the order of 10^7 to 10^8 s^{-1}, while the net movement is much less, on the order of micrometers per second. The rates of lateral diffusion are strongly affected by the presence of membrane-associated proteins and immobilized lipid molecules closely associated with these membrane proteins.

The form of motion with the longest time frame is the **transmembrane flip-flop** of a lipid molecule. This motion requires the complete translocation of a lipid molecule from one half of the bilayer to the the other leaflet. The fastest rate of exchange is reported to be minutes, but frequently the equilibration times for flip-flopping are in the range of hours to weeks. The slowness of the process can be explained by the enormous energy barrier posed by the required movement of the polar head group through the hydophobic interior of the bilayer in the course of the translocation. The kinetics of the flip-flopping can be shortened by the presence of certain membrane proteins, just as other parameters of membrane motion were affected by the nonhomogeneous state of the biological membrane. Before concluding this discussion of the behavior of lipid membranes therefore, a brief look at the complete structure of a biological membrane is in order.

15.5. Biological Membranes: The Complete Picture

A biological membrane acts as a separator between two solutions, permeable to some but not to all components of these solutions. In mammalian cells, several types of membranes exist, including the plasma membrane separating the cytoplasm from the extracellular milieu and different organellar membranes separating the cytoplasm from the inner contents of organelles such as mitochondria or secretory granules. The modern model of the biological membrane is called the **fluid-mosaic model** and is described as a lipid bilayer with an array of various proteins "floating" in the bilayer structure (Figure 15.5-1).

These membranes are phospholipid bilayers which contain membrane-spanning proteins and also, in many cases, membrane-bound proteins which do not traverse the membrane. In the case of the plasma membrane separating the cytoplasm from the exterior of the cell, the external portion of the proteins may be glycosylated. The membrane proteins act as carriers, pumps, channels, and, in general, mediators of cell function and/or sites of cell–cell interactions or cell recognition. In all of these capacities, the conformation of the protein plays a critical role. That conformation is dependent on the interactions between the protein and the membrane with which it comes in contact, and therefore on the structure of the membrane. That structure, as discussed in this chapter, depends on the nature of the lipids, which also controls the microviscosity, the transition temperatures of the membrane lipids, and as well the ease of passage of water, ions, and uncharged entities from one side of the mem-

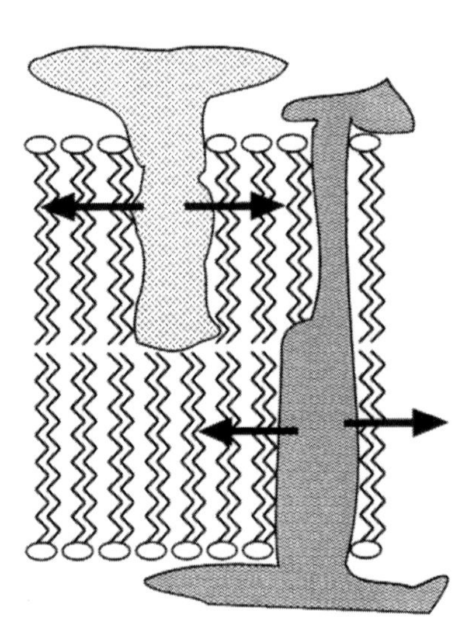

Figure 15.5-1 Fluid-mosaic model of the cell membrane.

brane to the other. Some proteins project into the cytoplasmic space, while other proteins are only inserted into one side or bound to one side of the membrane. Many intrinsic membrane proteins traverse the membrane several times (e.g. seven times for the β-adrenergic receptors and rhodopsin, the main visual protein). Each such traversing region contains a sequence of hydrophobic residues capable of forming an α helix. The role of each exposure on the exterior or cytoplasmic sides of the membrane is not yet clear.

The stage is now set with the players to model the behavior of cells. The cast includes aqueous solutions of polar and nonpolar molecules, both small and large, as well as self-associating lipids arranged into membranes. How do these characters act to create a cellular system? This will be the subject of the remainder of this book.

CHAPTER 16

Irreversible Thermodynamics

16.1. Transport: An Irreversible Process

So far, the focus of this book has been on systems at equilibrium. At equilibrium, systems experience no net flux of heat, work, or matter. Classical thermodynamics treats these systems easily. As pointed out earlier (Chapter 2), the greatest value of thermodynamics is that the behavior of a system can be predicted, even when the mechanistic details are not known. Homogeneous systems, at constant temperature and pressure, such as the solutions of electrolytes and macromolecules described so far, are comprised of molecules that individually experience a variety of forces, both orienting and randomizing. On an instantaneous time scale, this might lead to net movements of mass or energy; however, the time average of the forces leads to the steady-state condition of equilibrium. The activity of a component is the reflection of the time-average molecular forces acting in a system at equilibrium. There are cases in which the time average of a force or forces acting on a system results in the flow of material. When these events occur, transport phenomena results. Transport phenomena and the principles associated with nonequilibrium behavior are extremely important in biological systems because, as has already been suggested, true equilibrium states are achieved only in death. Steady-state systems, which have constant fluxes, are common. These systems are treated by nonequilibrium methods.

There are four phenomena associated with transport. These include diffusion, electrical conduction, heat flow or conduction, and fluid flow or convection. Each of these represents net movement in the direction of a gradient from a higher to a lower potential. The gradients are due to differences in chemical potential, electrical potential, temperature, or pressure, respectively. All of these phenomena are important in biological systems though the primary focus of this chapter will be on diffusion and electrical conduction. The general equation that applies to all forms of transport events is written

$$J_x = -B\frac{\partial A}{\partial x} = -BF_A \qquad (16.1\text{-}1)$$

This equation states that the flow of material in the x direction, J_x, is proportional by some constant B to the gradient of force of type A in the x direction. Similar equations could be written for each coordinate, x, y, or z.

Clearly, transport phenomena are events that do not fall in the realm of the classical thermodynamics that have been described at some length in this text. It is possible to analyze and study these nonequilibrium processes mechanistically, that is, to calculate the forces on each molecule and then relate the combined actions of each and every molecule to the properties of the system. In fact, this approach will be used in this chapter to provide a qualitative picture of the events that lead to transport. There are, however, substantial problems with using only a mechanistic approach. The first problem lies in defining the actual forces that may be acting on a particular molecule. As will be seen, the dimensions of molecules are important parameters in calculating transport properties mechanistically. In a system of macromolecules, this would be a significant difficulty. In many cases, biochemists and biophysicists do not even know all the components that go into making a system, much less their dimensions. Furthermore, the approximations for studying the forces on moving objects such as Stokes' law, which is a mainstay of this approach, assume that the transport occurs in a medium that is a continuum. Such an assumption, especially in the case of aqueous media, ignores the forces that act between a component and its solvent and other components, leading to approximations that can be drastically at variance with reality. Ideally, a parallel (or complementary) set of laws and equations to those applied in the equilibrium studies already described could be found for cases where equilibrium is approached but not yet reached. Such a macroscopic set of empirically (or phenomenologically) derived descriptions of properties (such as transport or kinetic rates) could complement the molecular-mechanistic approach. The study of nonequilibrium or irreversible thermodynamics provides this effective phenomenological approach.

16.2. Principles of Nonequilibrium Thermodynamics

It has been repeatedly stated that the principles of thermodynamics derived in Chapter 2 are universally valid. At no point was it stipulated that only systems at equilibrium could be treated. The reason that only equilibrium systems have been treated to this point has been one of definition and convenience. This has occurred because some of the fundamental variables of state, namely, temperature, pressure, and entropy, were defined at equilibrium. It is more difficult to define them during an irreversible or nonequilibrium process. Other variables of state do not suffer from this limitation and can be successfully used under any circumstances; these include volume, mass, energy, and amount of a component. Recognizing that variables like temperature and pressure are intensive, while volume, mass, and energy are extensive can help explain this difference. An intensive property was defined as one in whose evaluation a small sample was representative of the entire system. This only has meaning if a system is at equilibrium. Consider the example of two heat reservoirs of different temperature connected by a metal bar through which heat travels by thermal conduction. The flow of heat will be irreversible from the reservoir of greater temperature to the one of lower temperature. Choosing small samples at points along the bar will give different measurements for the temperature. Clearly, this does not fit the requirement for an intensive variable. Consequently, the method used to define the parameter of temperature must be different in this system, because it is not at equilibrium. A similar argument could be made for pressure or entropy. Until the variables, such as temperature and pressure, can be defined in an irreversible system, thermodynamic calculations will be unsuccessful.

Through the use of a new postulate, that of **local equilibrium**, this problem can be overcome. The system is divided into small cells, small enough that effectively each point in the system is treated, but large enough so that each cell contains thousands of molecules. At a specific time, t, the cells are isolated from the system and allowed to come to equilibrium in time dt. Therefore, at time $t + dt$, measurements can be made that give an equilibrium temperature or pressure. The variable of state at time t is then considered to be equal to the measurable variable at time $t + dt$. The relationships derived by this postulate are then considered to be equivalent to the relationships derived from equilibrium states. It must be realized that this postulate has its limitations. The presumption is that the variables in system are not changing too rapidly. If time, dt, necessary for the cell to achieve local equilibrium approximates the time during which a change for the whole system may be measured, then the postulate cannot be reasonably applied.

Entropy has been shown to play an important role in aqueous systems and will be further implicated as a crucial driving force in transport phenomena. How does the treatment of irreversible systems work in the case of entropy? Instead of employing the relationship $\Delta S = q_{rev}/T$, it is more convenient to determine ΔS from another relationship, such as that given in Equation (4.4-5):

$$\Delta S = C_p \ln \frac{T_2}{T_1} \tag{4.4-5}$$

Once the entropy of each cell is determined, the free energy for each cell can be determined:

$$dG = V\,dP - S\,dT + \sum \mu_i\,dn_i \tag{16.2-1}$$

The point was made earlier that reversible processes take an infinite amount of time to complete but do not lead to the production of entropy. Irreversible processes, on the other hand, occur in a finite time and create entropy. The rate of a process therefore can be defined in terms of the rate of entropy production with respect to time, dS/dt. This means that as a reaction or process proceeds in an isothermal system, there will be heat flow into and out of the surroundings and system. The differential change in entropy will be given by

$$dS = d_iS + d_sS \qquad (16.2\text{-}2)$$

where d_iS is the entropy change in the system, and d_sS that in the surroundings. dS will always be zero or greater.

Historically, the formulation of irreversible thermodynamics started when Thomson (Lord Kelvin) was investigating the relationship between the flow of electricity and heat flow in thermocouples. If two dissimilar metal wires are twisted together at one end and a voltmeter is used to complete the circuit between the two, a voltage can be demonstrated arising from the contact of the two phases. This is the **contact potential**. If both ends are connected together, there will be two junctions in the circuit. If these two junctions are isothermal and an electric current is passed between them, heat will be absorbed from the surroundings at one junction and an equal amount of heat will be released at the other junction. This heat flow is reversible in that when the direction of the current is changed, the direction of heat flow also changes. This reversible heat is called the **Peltier heat**. A second source of heat is also produced during this process due to the resistance of the metal to the flow of charge, and this heat is called the **Joule heat**. Joule heat is irreversible. If the two junctions are now placed at two different temperatures, then an electromotive force will exist between them. The electromotive force between the two junctions is called the **Seebeck emf**. If a charge is allowed to move around the circuit because of the Seebeck emf, it will be found by experiment that the Peltier heat appearing at the junctions is not sufficient to account for the work accomplished. Thomson therefore proposed a second reversible heat associated with the flow of current, the **Thomson heat**. The Thomson and the Peltier heats are reversible and are accompanied by two irreversible heats in this system, one due to Joule heating and one due to heat conduction. Thomson treated the thermocouple as if it were a reversible heat engine in which only the Thomson and Peltier heats circulated. He described a series of relationships that showed that there was no entropy production, that is, that the two heats were equal and reversible. Thomson's theoretical treatment of this system was experimentally validated even though he ignored the two irreversible terms. Thomson himself recognized that the treatment was incomplete since the process described is an irreversible one and hence the total entropy of the process must be positive. However, his analysis assumed that the entropy increase associated with the Joule heating and the heat conduction would be positive and constant and therefore tested the hypothesis that the Peltier and Thomson heats did not add to the entropy generation of the process, that is, that they were indeed reversible. His result demonstrated that in transport phenomena there are reversible and irreversible processes.

A unifying method for generally treating irreversible systems was given by Onsager in 1931. Onsager based his formulation on the **principle of microscopic reversibility**, which says that at equilibrium any process and its reverse process are taking place on the average at the same rate. He further assumed that for a process that is near equilibrium, equations may be written for the transport process in which the fluxes are linearly proportional to the forces. The theory is valid only for deviations from equilibrium where this linear relationship exists. Processes like the one described above can be generally treated by considering that in a transport process there are a number of flows that occur simultaneously . For example, in the case of thermoelectricity, there is a flux of heat, J_l, and one of current, J_2. The two flux equations take the general form

$$\begin{aligned} J_1 &= L_{11}X_1 + L_{12}X_2 \\ J_2 &= L_{21}X_2 + L_{22}X_2 \end{aligned} \tag{16.2-3}$$

The term X_x represents the force gradient, L_{ij} are the phenomenological coefficients, and L_{ii} are the direct coefficients. In this case, X_1 represents the temperature gradient and X_2 the electrical gradient. The forces represented by X_x are thermodynamic driving forces and have the form

$$\frac{\partial S}{\partial X_x} = F_x \tag{16.2-4}$$

This type of analysis indicates that when more than one gradient is causing flux, there will be a coupling of the flows. The direct coefficients represent the fluxes due to the directly related gradient, that is, the flow of heat due to a thermal gradient. These always increase the entropy of the reservoirs. The cross terms, L_{ij}, are the coupled flows caused by the gradient that is not directly related, that is, the flux of heat caused by the flow of electricity. Onsager showed that the phenomenological coefficients are equal:

$$L_{ij} = L_{ji} \tag{16.2-5}$$

This equality is called the **Onsager reciprocity relation**. The coupling between the flows indicates the interaction of one flow with another. The idea that some fluxes are independent while others occurring simultaneously are interacting and reversibly coupled is an important one in transport phenomena.

In mechanistic terms, transport can be thought of as the balance between the motion of a particle moving directly down a gradient of force and the scattering of the particle away from this direct path because of interactions between other forces or particles. While the relationships of irreversible thermodynamics are probably the most accurate expression of the balance between direct and scattering forces, there are two mechanistic models frequently considered. One is the concept of the mean free path, and the other is based on relaxation time. The focus here will be on the concept of the mean free path and how this approach can be used to examine diffusion in a solution.

CHAPTER 17

Flow in a Chemical Potential Field: Diffusion

17.1. Transport Down a Chemical Potential Gradient

Transport is measured by **flux**, which is defined as the net movement of matter in unit time through a plane of unit area normal to the gradient of potential. In the case of **diffusion**, molecules of specific components will travel down their concentration gradient, $\partial c/\partial x$, until the gradient is removed or neutralized. During **electrical conduction**, mass, as well as charge, is transported in an electric field $\partial\psi/\partial x$. In aqueous solutions, the mass and charges are derived from either electrolyte ions or protons and hydroxyl groups from the autoproteolysis of water. In **convection** or fluid flow, molecules are moved by a streaming process by an external force $\partial\rho/\partial x$, for example, pressure from a pump as in the cardiovascular system or gravitational forces generated in a centrifuge. **Heat conduction**, caused by a difference in temperature $\partial T/\partial x$, is also the result of a net flow, not of the total number of particles, which remains constant in the system, but of the particles with a higher kinetic energy. Each of these transport phenomena has a named relation of the form given in Equation (16.1-1):

$$J_x = -B\frac{\partial A}{\partial x} = -BF_A \tag{16.1-1}$$

For diffusion, it is **Fick's law** and the constant is given as $\boldsymbol{D}$, the **diffusion constant**; for electrical conduction, it is **Ohm's law** with the constant κ, the **electrical conductivity**; fluid movement is given by **Poiseuille's law** and the constant, C, is a **hydraulic conductivity** related to the viscosity; and heat flow is described by **Fourier's law** and κ_T is the **thermal conductivity coefficient.**

The treatment of diffusion and chemical potential gradients can be illustrated by considering a rectangular container of an aqueous solution of i, as in Figure 17.1-1. For simplicity, assume the solution is ideal and so the activity of i can be given in terms of the concentration of i. In this rectangular container, the concentration of i is allowed to vary in

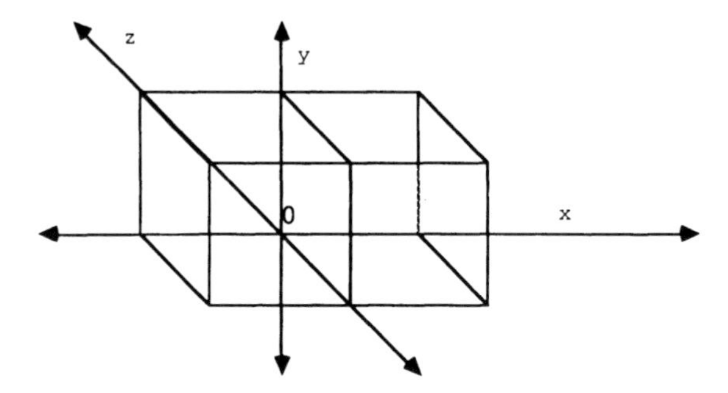

Figure 17.1-1 Container system for the consideration of chemical potential gradients.

the x direction only; that is, there are a series of planes, in the yz plane, perpendicular to the x axis such that the concentration of i is equal within each plane, but may vary from plane to plane. There will be a gradient in concentration measured moving along the x axis from plane to plane. The concentration gradient can be described in terms of the chemical potential as follows. The first plane chosen for description will be at the position $x = 0$. The chemical potential at $x = 0$ can be described by the term

$$\mu_{i\text{-}0} = \mu_i^{\circ} + RT \ln c_{\text{o}} \tag{17.1-1}$$

At the end of the container, $x = f$, there will be a plane where the chemical potential can be given as

$$\mu_{i\text{-}f} = \mu_i^{\circ} + RT \ln c_f \tag{17.1-2}$$

The change in free energy associated with moving a mole of species i from $x = 0$ to $x = f$ will be the difference in the chemical potential of the planes; thus

$$\Delta\mu_i = \mu_{i\text{-f}} - \mu_{i\text{-}0} = RT \ln \frac{c_f}{c_{\text{o}}} \tag{17.1-3}$$

From the earlier discussion of thermodynamics, it is known that the free energy change is equal to the net work done on the system in a reversible process at constant temperature and pressure. Therefore

$$\Delta\mu_i = w \tag{17.1-4}$$

Work is defined as a force acting over a distance, and hence it could be written that the work and the free energy change are related to a force (of diffusion) acting over the distance between 0 and f:

$$w = -F_D(x_f - x_{\text{o}}) = -F_D\Delta x = \Delta\mu_i \tag{17.1-5}$$

The justification for the negative sign will be discussed shortly. The force, F_D, can be written in terms of the free energy change and the distance:

$$F_D = -\frac{\Delta\mu_i}{\Delta x} \tag{17.1-6}$$

The gradient can be written in terms of infinitesimal changes, giving

$$F_D = -\frac{d\mu_i}{dx} \tag{17.1-7}$$

Equation (17.1-7) states that the force leading to diffusion is equal to the negative gradient of the chemical potential. The negative sign can be explained by analogy to the work done in lifting a weight in a gravitational field. If a weight is lifted from $y = I$ to $y = F$, then the difference in potential energy of the weight, U, will be given by

$$w = \Delta U \tag{17.1-8}$$

The work is done in a gravitational field so an expression analogous to Equation (17.1-5) can be written:

$$w = -F_g(y_F - y_I) = -F_g\Delta y = \Delta U \tag{17.1-9}$$

The gravitational force is acting in a downward direction and, because of its vector, will be negative; the displacement of the weight will be upward and so Δy will be a positive quantity. Therefore, the product of F_g and Δy is negative. By convention, the work done on the weight should be positive and therefore the quantity $F_g\Delta y$ needs to be given a negative sign to fulfill these requirements. The analysis can then proceed to give the result that the gravitational force is defined by the gradient of the gravitational potential energy:

$$F_g = -\frac{dU}{dy} \tag{17.1-10}$$

Both of these cases indicate that a force can be described that causes a flux of a material down a gradient. Interestingly, in the case of diffusion, there is actually no directed force like that of gravity or an electric force field that acts on the particles of a species. As will be seen shortly, there is instead an uneven distribution of the particles with respect to position. This phenomenon can be formally treated as if it were a directed force, but in fact it is only a **pseudoforce**. This diffusional "force" will be found shortly to be closely related to the entropy of mixing.

The experimental description of the net transport of materials by diffusion is quite simple. Again, considering the rectangular volume described earlier, the amount of substance i that travels through a yz plane of unit area (1 cm^2 for example) in a given amount of time (usually 1 s) is called the **flux** of species i. Flux in this example would be given in units of mol cm^{-2} s^{-1}. The concentration gradient will determine the flux through the *yz* plane. If there is no concentration gradient, then there will be no flux, and, conversely, when the concentration gradient is steeper, the measured flux will be higher. **Fick's first law** describes these experimental relationships:

$$J_i = -D\frac{dc}{dx} \tag{17.1-11}$$

D is a proportionality constant called the **diffusion coefficient**. Fick's first law indicates that there is a decrease in the number of particles proportional to the concentration gradient; that is, the transport of species is in a direction opposite to the concentration gradient. The concentration gradient will have units of mol m^{-4}, and the diffusion coefficient is written as m^2 s^{-1} . Clearly, the flux of a species i will depend on the apparent force causing the net transport. This force was described in Equation (17.1-7) as

$$F_D = -\frac{d\mu_i}{dx} \tag{17.1-7}$$

This equation can be combined with the general relationship described earlier for transport phenomena under the conditions of a sufficiently small driving force such that the flux is linearly related to the driving force:

$$J_i = -B\frac{\partial A}{\partial x} = -BF_A \tag{16.1-1}$$

The driving force that will cause the transport across the transit plane where flux is being measured will be given by the quantity $c_i\, d\mu_i/dx$ where c_i is the concentration of species i in the plane adjacent to the transit plane. Therefore, Equation (16.1-1) can be written

$$J_i = -Bc_i\frac{d\mu_i}{dx} \tag{17.1-12}$$

Since $\mu_i = \mu_i^o + RT \ln c_i$, Equation (17.1-12) can be rewritten entirely in terms of concentration:

$$J_i = -Bc_i\frac{RT}{c_i}\frac{dc_i}{dx} = -BRT\frac{dc_i}{dx} \tag{17.1-13}$$

and

$$D = BRT \tag{17.1-14}$$

This gives the relationship between Fick's law and the phenomenological theoretical treatment of a transport process under nonequilibrium conditions.

Is the diffusion coefficient, D, constant with respect to concentration? Experimentally, it can be shown that D varies with concentration. The reason for this should be obvious if the assumptions of this analysis are considered. The solution under consideration was assumed to be ideal, and this assumption led to Equation (17.1-13). The discussion in Part II of this book makes it clear that solutions of biological importance are not ideal and that their activities vary considerably with concentration. Taking this into account gives the following result:

$$J_i = -Bc_i \frac{d\mu_i}{dx} \tag{17.1-12}$$

Now explicitly taking into account the activity of i, $\mu_i = \mu_i^\circ + RT \ln \lambda_i c_i$, leads to

$$J_i = -Bc_i \frac{d}{dx} (\mu_i^\circ + RT \ln \lambda_i c_i) \tag{17.1-15}$$

Solving the equation gives the following result:

$$J_i = -BRT \left(1 + \frac{d \ln \lambda_i}{d \ln c_i}\right) \tag{17.1-16}$$

Now rewriting Equation (17.1-14) in terms of Equation (17.1-16) gives

$$D = BRT \left(1 + \frac{d \ln \lambda_i}{d \ln c_i}\right) \tag{17.1-17}$$

The diffusion coefficient does depend on concentration though the dependency is only significant when the activity coefficient varies in the range of concentrations causing diffusion in the first place.

Fick's first law applies to diffusion events where the rate of transport is steady state; that is, even though there is net flux in the system, the rate of flux is constant and the driving force, i.e., $-dc/dx$, remains constant. In physical terms, this means that the movement of species i through the transit plane of the rectangular volume is constant and that at any time the same amount of material is leaving each yz plane as is entering that plane. If the flux through a small volume, with the dimensions of a yz plane by an $x + dx$ coordinate, is such that less material is leaving one side of the volume than is coming into the volume from the other side, then the concentration in the volume is increasing with respect to time. Hence, the following partial derivative can be written:

$$\left(\frac{\partial c}{\partial t}\right)_x = -\left(\frac{\partial J}{\partial x}\right)_t \tag{17.1-18}$$

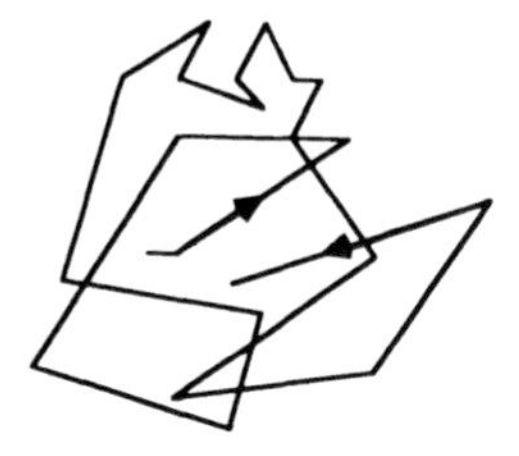

Figure 17.2-1 Random wandering of a particle in a solution at equilibrium.

Using Fick's first law, the change in concentration in a small volume that varies with respect to time can be found:

$$\left(\frac{\partial c}{\partial t}\right)_x = \mathrm{D}\left(\frac{\partial^2 c}{\partial x^2}\right)_t \tag{17.1-19}$$

This is **Fick's second law** and is accurate if the diffusion coefficient is independent of the concentration. Fick's second law is a second-order partial differential equation, and the solution of such equations can be difficult. Its solution will not be covered here. Fick's second law is the basis for the treatment of many non-steady-state or time-dependent transport problems. Experimental use of Fick's second law is often made to determine diffusion coefficients. It is important to be sure that the experimental conditions and analysis of the data take into account the dependency of the diffusion coefficient on the concentration of the measurement system.

17.2. The Random Walk: A Molecular Picture of Movement

What is a reasonable molecular picture of the movement of the particles making up a solution at equilibrium? A clue can be gained by observations on the motions of small colloidal particles of dye. These particles, if observed under a microscope, show a haphazard irregular motion of each particle that ultimately leads nowhere. The behavior of each colloidal particle, and in fact of each molecule in the solution, can be qualitatively described and depicted as shown in Figure 17.2-1. The irregular zigzagging path of the particle is due to the multiple collisions of that particle with other particles in the solution. The average distance a particle will travel between collisions is called the **mean free path**. The mean free path, l, is given by dividing the average distance a particle will travel in unit time if unimpeded, $<u>$, by the number of intermolecular collisions the particle will experience in that time, z, i.e.,

$$l = \frac{<u>}{z} \tag{17.2-1}$$

One measure of the average velocity of molecules is the **root-mean-square velocity** given as follows for an ideal gas or solution:

$$<u> = \left(\frac{8RT}{\pi M}\right)^{1/2} \tag{17.2-2}$$

where M is the molecular weight. The number of collsions, z, depends on the concentration of the particles N/V, their diameter, σ, and their mean velocity. Expressions of the following form may be written:

$$z = 7.09 \frac{N}{V} \sigma^2 \left(\frac{RT}{M}\right)^{1/2} \tag{17.2-3}$$

Consideration of Figure 17.2-1 suggests that although the net distance a particle may travel from its origin will be zero, the total distance traveled by the particle in a particular time, t, may be significant. Knowledge of the distance traveled is necessary for a molecular understanding of the events of diffusion. In a single dimension, the average distance from the origin will be determined by summing the distance covered by all of the steps the particle takes in a positive or negative direction from the origin in a large number of tries, i, and then dividing by the number of tries:

$$<x> = \frac{\sum_i x(i)}{\sum i} \tag{17.2-4}$$

Obviously, for a large enough number of tries, the number of moves a particle makes away from the origin to the left will be equal to the number of moves to the right and $<x>$ will equal zero. This is the mean progress. This result is the consequence of adding the positive and negative forays from the origin together. If each of the jumps is squared and then summed, a value that is positive will result. This is the **mean square distance,** $<x^2>$. Since the distance that is traveled with each step or start is the same (i.e., the mean free path of the particle) and the number of steps is increasing at a constant rate with time, the mean square distance traveled increases in a linear fashion with time; that is, $<x^2>$ is proportional to time.

Now, it is possible to point out the relationship between the random-walking particles and diffusion. Recalling the rectangular box used earlier, consider the random walk of some solute molecules, for example, ions in solution. A yz plane is dropped at the origin of the x axis. Two more yz planes are now established on either side of the origin at $-(<x^2>)^{1/2}$

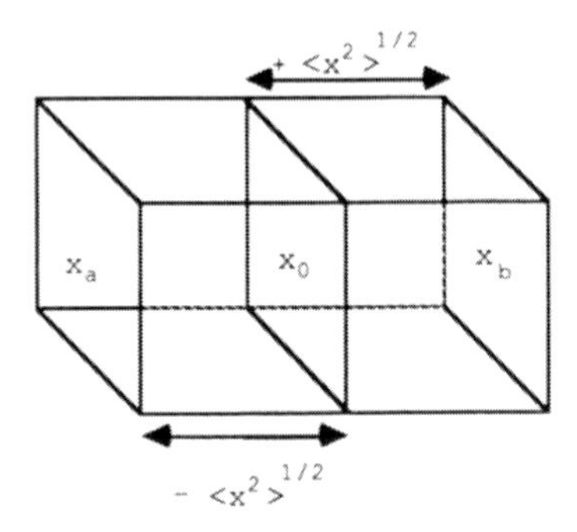

Figure 17.2-2 System to study the random walk approach to diffusional forces.

and $+\langle x^2\rangle^{1/2}$. Therefore, two equal volumes, *a* and *b*, now exist on each side of the plane at the origin, each with a concentration of the ion species *i*, c_a and c_b (see Figure 17.2-2). Both volumes, *a* and *b*, start with the same number of molecules of *i*. The ions move in a random walk along the *x* axis. In a particular time, *t*, all of the ions in compartment *a* that are moving in the left-to-right direction will cross the *yz* plane at the origin (this is ensured by choosing the dimension $\langle x^2\rangle^{1/2}$ for the volume). Since there is an equal probability of an ion going left to right or right to left across the plane at the origin, the flux across the transit plane from *a* to *b* will be given by $\frac{1}{2}\langle x^2\rangle^{1/2}c_i$. An equivalent analysis of the flux across the transit plane from *b* to *a* via a random walk can be made. If the number of ions in each volume is equal at the end of *t*, there will be no net flux, which is the result that is expected. The flux from side *a* to *b* can be written as follows:

$$J = \frac{1}{2}\frac{\langle x^2\rangle^{1/2}}{t}(c_a - c_b) \tag{17.2-5}$$

If the number of ions of species *i* found in the two volume elements, *a* and *b*, is not the same, there will be a measurable net flux simply because of the random walk mechanism. On a molecular basis then, it is clear that diffusion occurs without a specific diffusive force that acts to move each particle. As was mentioned previously, this is why diffusion can be formally treated as a force but in fact is different in nature from an electrostatic or gravitational force. The relationship of the random walk-based molecular model to Fick's laws can be shown as follows. The concentration gradient from side *a* to *b* is evaluated:

$$\frac{\partial c}{\partial x} = \frac{c_b - c_a}{\langle x^2\rangle^{1/2}} = -\frac{c_a - c_b}{\langle x^2\rangle^{1/2}} \tag{17.2-6}$$

Writing this result in terms of the difference $(c_a - c_b)$ gives

$$(c_a - c_b) = -(\langle x^2\rangle)^{1/2}\frac{\partial c}{\partial x} \tag{17.2-7}$$

This result can be directly substituted into Equation (17.2-5), which after simplifying gives

$$J = -\frac{1}{2}\left(\frac{\langle x^2\rangle}{t}\right)\left(\frac{\partial c}{\partial x}\right) \tag{17.2-8}$$

Equating Equation (17.2-8) with Fick's first law gives

$$-\frac{1}{2}\left(\frac{\langle x^2\rangle}{t}\right)\left(\frac{\partial c}{\partial x}\right) = -D\left(\frac{\partial c}{\partial x}\right) \tag{17.2-9}$$

Simplifying yields

$$\frac{\langle x^2\rangle}{2t} = D \tag{17.2-10}$$

Rearrangement gives the **Einstein–Smoluchowski equation**:

$$\langle x^2\rangle = 2Dt \tag{17.2-11}$$

The factor 2 in this equation is a result of the one-dimensional derivation. A three-dimensional derivation will have a coefficient of 6.

Diffusion processes in biological systems often occur between phases of significantly different structure and composition, such as from the generally aqueous environment of the blood through the bilipid layer of the cell membrane. The relationship of the diffusing species to the solvent clearly must be considered. The diffusion coefficient has been treated as a phenomenological proportionality coefficient, but clearly, if considered on a molecular basis, it contains information about solute–solvent interactions. If the random-walking ion is intuitively considered, the mean square distance would be expected to depend on the distance moved with each jump and the number of jumps that occur in a unit of time, t. These values have been discussed already—the mean free path, l, and the number of collisions, z. It can be shown that

$$\langle x^2\rangle = z\, l^2 \tag{17.2-12}$$

Combining Equations (17.2-11) and (17.2-12) yields the result for a one-dimensional random-walking ion:

$$z\, l^2 = 2Dt \tag{17.2-13}$$

In the case where $z = 1$, that is, only a single jump is considered, this expression reduces to

$$D = \frac{l^2}{2\tau} \tag{17.2-14}$$

where τ is equal to the mean jump time for the particle under consideration to cover the mean jump distance, l. τ is the period of the cycle of the jump including time between the jumps. The frequency of the jump is simply $1/\tau$. The diffusion coefficient then is dependent on the mean jump distance and the frequency of the jumps.

What is the physical consequence of Equation (17.2-14)? First of all, the distance that a particle will be able to jump is obviously limited by the physical structure of the solvent in which it is jumping. Generally, this means that the particle is going to jump from an occupied site in the solvent to an unoccupied site in the solvent. After the jump, there will have been an exchange of occupied and unoccupied sites. The mean free path or mean jump distance will be dependent on the structure of the solvent at the instant of the jump.

While the mean free path is essentially dependent on the solvent structure, the frequency of the jumps is a question of rate. Application of rate process theory, in which the free energy change from the prejump to postjump coordinate goes through an activated state of higher free energy, provides an approach to this problem. Since the particle must cross an energy barrier to go from site to site, the frequency of the jumps will be given by a rate constant, k_{forward}:

$$k_{\text{forward}} = \left(\frac{kT}{h}\right) e^{-\Delta G\ddagger/RT} \tag{17.2-15}$$

k is Boltzmann's constant, h is Planck's constant and $\Delta G\ddagger$ is the free energy of the transition state.

The diffusion coefficient can therefore be considered in terms of the mean jump distance and the frequency of the jumps. Because all jumps are of interest in a problem of diffusion, the coefficient in the denominator from the Einstein–Smoluchowski equation can be taken as unity, giving

$$D = \left(\frac{l^2kT}{h}\right) e^{-\Delta G\ddagger/RT} \tag{17.2-16}$$

Both molecular and empirical sketches of the transport phenomenon of diffusion have been presented so far. The driving force that causes movement down a chemical potential gradient was shown to be a pseudoforce. Next, the movement of particles in an electric field will be discussed. While the nature of the forces in these two cases is significantly different, it will be found that the formal treatment of these forces is quite similar.

CHAPTER 18

Flow in an Electric Field: Conduction

18.1. Transport in an Electric Field

To this point, the discussion has focused on the transport of molecules simply by the random process of diffusion. What events change when the system is placed under the external force of an electric field? This is an extremely important aspect of cellular systems, because so many biologically active chemical species, including proteins, carbohydrates, lipids, and ions, carry a net charge and will respond with a net flow of current to the electric fields that are ever present in cellular systems. The focus in the following derivation will be on the net movement of charge in solutions of inorganic ions, because the majority of charge transfer in solution is accounted for by these ions.

An electrolyte solution has been shown to contain a certain amount of charge, given by its ionic strength, but balanced precisely so that the condition of electroneutrality is met. Under the conditions of only thermal randomizing forces (i.e., no chemical, gravitational, electric, or other force fields), the ions and solvent molecules are in constant motion without any net movement or transport. What will the effect be when an electric field is imposed across this system? The most obvious and easily measured consequence will be the conduction of electric current by the electrolyte solution. It can be demonstrated that the conductivity of the electrolyte solution is due primarily to the addition of the ionic species to the solvent and that conduction of electricity in solution is due to the net transport of ions. The discussion of electrical conduction in electrolyte solutions will be approached by first developing a qualitative sketch; subsequently, empirical and quantitative descriptions will be used to provide more detail.

18.2. A Picture of Ionic Conduction

When an external electric field is imposed on the electrolyte solution, ions will experience an electrical force superimposed on the thermal randomizing forces that lead to the random walk behavior of the molecules.

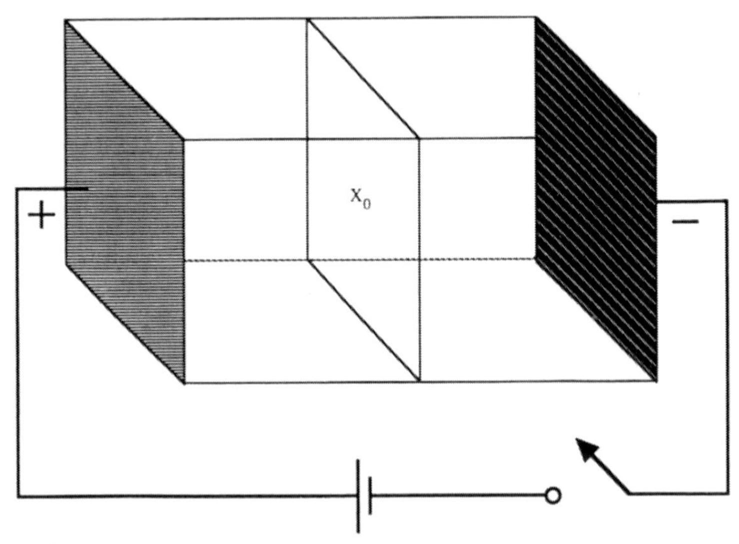

Figure 18.2-1 Experimental system for the description of electrical conduction.

To understand the effects of the external electrical force on the ions in solution, it is necessary to have a description of the electric field that is created in the electrolyte solution upon the application of an external field. The system to be placed under study will be similar to the rectangular box used earlier for the examination of diffusion except that now a pair of parallel plate electrodes are placed at the ends of the box. These electrodes are connected through a circuit to an external power source, for example, a battery. A switch is in the circuit to connect and disconnect the power source and the box. Figure 18.2-1 illustrates this system.

An electrolyte solution fills the box between the electrodes, and the switch connects the battery to the circuit. Charge in the form of electrons flows from the negative terminal of the battery through the wires of the external circuit into the negative electrode or **cathode**. At the cathode–electrolyte interface, a charge transfer reaction occurs in which the electron, which is responsible for carrying charge in the external circuit, is transferred via an **electronation** or **reduction reaction** to an ion in the electrolyte phase. In the electrolyte phase, the ions carry the charge down the potential gradient to the positive electrode, the **anode**. At the anode, the charge carried by the ion will be transferred to the anode via an **oxidation** or **deelectronation reaction** and will complete the trip around the external circuit as an electron. This description of charge transfer in a circuit comprised of both ionic and electronic charge carriers leads to the classifications used by electrochemists to discuss the system. The study of the conduction and behavior of the electrons in the external circuit is called **electronics** while the study of the charge transfer by ions in solution is called **ionics**. The extremely interesting and important study of the charge transfer reactions at the electrodes is called **electrodics** and is the focus of modern electrochemistry. The focus of this chapter will be on ionics, but the study of electrodics as related to biological systems is emerging as a fascinating and important field in bioelectrochemical research. The basic principles of bio-electrodics are discussed in the following chapters.

If conduction is taking place in the electrolyte solution, a driving force or potential gradient field must be present down which the ions move. A potential map of the system in Figure 18.2-1 can be drawn by finding the work necessary to bring a test charge from infinity to any point x in the system. This work will define the potential, ψ, at each point. Such a potential map is illustrated in Figure 18.2-2. It is notable that the largest potential drops in the system occur at the two electrode–electrolyte interfaces. This is not suprising given the events of electronation and deelectronation occurring at the electrodes, since the work done to bring a test charge through the interface involves transfer between two phases and also the generation or consumption of chemical species. The focus here however is on the potential map between the two electrodes, that is, the potential gradient in the electrolyte solution, for it is this gradient that the ions in the bulk of the electrolyte solution are seeing. One of the advantages in choosing an experimental design with parallel plate electrodes is that the electric field, X, is a linear force field as the graph in Figure 18.2-2 indicates. The strength of the electric field can be easily found by taking two equipotential planes such as x_0 and any other plane x_j, finding the potential difference between them, and dividing by the distance separating them, d:

$$E = -\frac{\Delta\psi}{d} \tag{18.2-1}$$

The negative sign occurs because E is the negative of the gradient of the potential. A series of equipotential planes parallel to the electrodes will exist for this experimental system, providing a linear gradient in a fashion similar to the example developed for diffusion. Although in this case the gradient of the field responsible for transport is linear, this does not have to be the case, and the potential field can generally be written

$$E = -\frac{d\psi}{dx} \tag{18.2-2}$$

The molecular picture of movement in a solution has already been described. Considering the experimental circuit of Figure 18.2-1 before the switch is thrown, and focusing at this point on the movement of a single ion in an electrolyte solution, the ion is found in position in the quasi-crystalline water and is vibrating about an equilibrium point with-

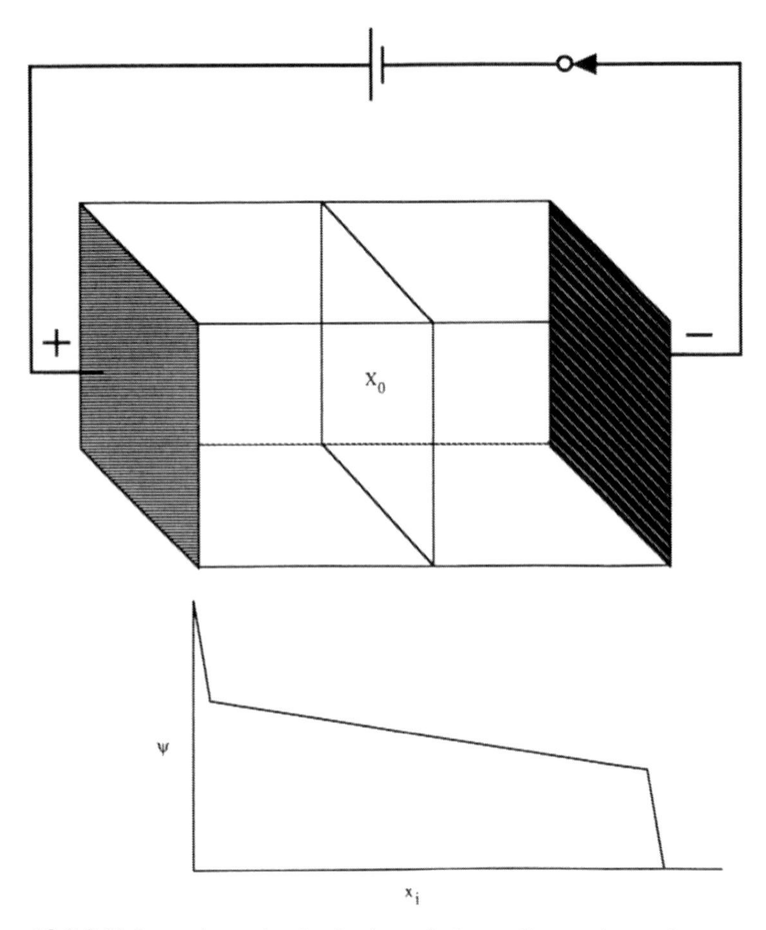

Figure 18.2-2 Voltage drops in the ionic and electrodic portions of the experimental circuit.

out any net jumping. At a certain time, the ion will have attained sufficient energy to jump to a free site in the water. The rate of the jumping is on the order of 10^{10} to 10^{11} s^{-1} for systems of biological interest. The jumps occur with a frequency that depends on both the energy needed to separate the ion from its first site and the energy needed to form a hole into which the jumping ion can go. The electric field in the solution is now switched on and the ions experience an electrical force, so that positive ions are attracted to the negative electrode (cations) and negative ions are attracted to the positive electrode (anions). This added force acts to nudge the jumping ions in the preferred direction of the appropriate electrode, leading to an overall net flux of current. This flux is ionic conduction. Figure

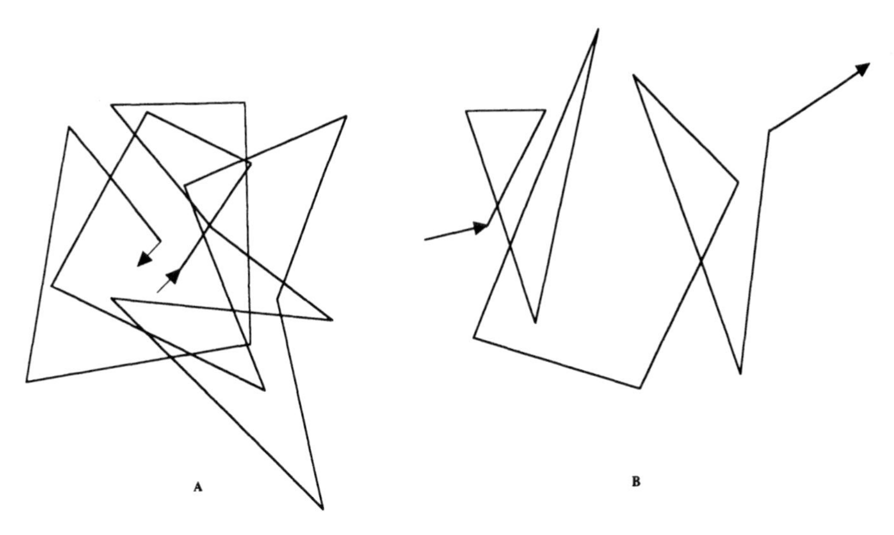

Figure 18.2-3 Comparison of the path of an ion on a random walk (A) versus that of an ion in an electric field (B).

18.2-3 compares the motion from the random walk to the motion of an ion being acted upon by an electric field. The velocity of motion imparted to the ions by the electric field is small compared to the instantaneous velocity resulting from the thermal collisions. For example, the ionic velocity for a 1 V m^{-1} field will be approximately 50 nm s^{-1}, while the thermally induced instantaneous velocity can be 100 m s^{-1}. The molecular picture of ions in an electric field is like that of a crowded lobby during intermission at the theater. Before the bell rings to summon the crowd for the second act, people are milling about with constant motion, but no actual net movement into the theater itself. At the call for the second act, people begin to move toward their seats, certainly not in a direct manner. Ultimately, there is a net flux from the lobby to the house, and each individual may be tracked zigzagging and starting and stopping along the way.

Conduction of charge will continue as long as the net electroneutrality of the solution is maintained. The charge added to the solution at one electrode must be removed at the same rate at the opposite electrode. If this were not the case, the electrolyte would almost immediately attain a net charge and, like the capacitor discussed previously (cf. Chapter 11), a counter field would rapidly develop that would prevent the addition of any further charge to the solution, and flow of current would cease. The reactions at the electrodes therefore are the vital link to the continued flow of charge in these systems.

18.3. The Empirical Observations Concerning Conduction

After the electric field has been applied to the electrolyte solution for some time, a steady-state flow of ions will be observed. As already discussed, the steady-state flow through the ionic conductor depends on the preservation of electroneutrality in the solution because of charge transfer reactions at the two electrodes. **Faraday's law** expresses the quantitative relationship for the transfer of electronic charge to ionic charge that occurs at the electrodes:

$$\frac{m}{\text{FW}} = \frac{it}{|z_i|F} \tag{18.3-1}$$

where m is the mass of an element of formula weight FW liberated or consumed at an electrode, i is the current passed in amperes in t seconds. z_i is the charge on a given ion, and F is the Faraday, equal to 96,484.6 C mol^{-1}. When an electrical potential field is imposed across a conducting material, the amount of current flow depends on both the potential and the resistance to flow of current, the **resistance**. This is the relationship quantified in **Ohm's law**:

$$I = \frac{E}{R} \tag{18.3-2}$$

where I is the current expressed in amperes, E is the electrical potential field expressed in volts, and R is the resistance expressed in ohms (Ω). The resistance of a sample of conducting material is dependent on both the geometry and the intrinsic **resistivity** of the material to conduction of current. For a conductor of resistivity ρ, the resistance will increase as the length, l, of the current path increases and the resistance will fall as the cross-sectional area, A, of the conductor increases:

$$R = \rho \frac{l}{A} \tag{18.3-3}$$

The units of ρ are Ω m^{-1} or Ω cm^{-1}. Alternatively, the reciprocal of resistivity is a measure of the ease with which current flows in a conducting medium; it is quantitated as **conductivity** and is given the symbol κ. Conductivity or specific conductance can be expressed therefore as follows:

$$\kappa = \frac{1}{RA} \tag{18.3-4}$$

The units of κ are Ω^{-1} m^{-1} or Ω^{-1} cm^{-1}. Resistivity or conductivity is determined in a cubic volume element of 1 m^3 or 1 cm^3. The conductivity of a variety of materials is listed in Table 18.3-1.

Table 18.3-1. Conductivity values (κ) for selected materials at 298 K.

Material	Conductivity (Ω^{-1} m^{-1})
Silver	6.33×10^{7}
Copper	5.80×10^{7}
KCl (0.1 *M*)	1.33×10^{0}
Acetic acid (0.1 *M*)	5.20×10^{-2}
Water	4.00×10^{-6}
Xylene	1.00×10^{-17}

Just as the conductivity is the reciprocal of the resistivity, it is often convenient to use the reciprocal of the resistance, the **conductance, *G***, to express the relationship between the current and voltage of a sample. Ohm's law may be rewritten

$$I = GE \tag{18.3-5}$$

The unit of G is the mho, the reciprocal ohm (Ω^{-1}).

While electrolyte solutions obey Ohm's law (at least for small potential fields and steady-state conditions), the current flux depends on the concentration of the ions making up the solution, as well as their identity and charge. Consequently, the conductivity of an electrolyte solution varies with concentration (see Figure 18.3-1). Because of the variation in

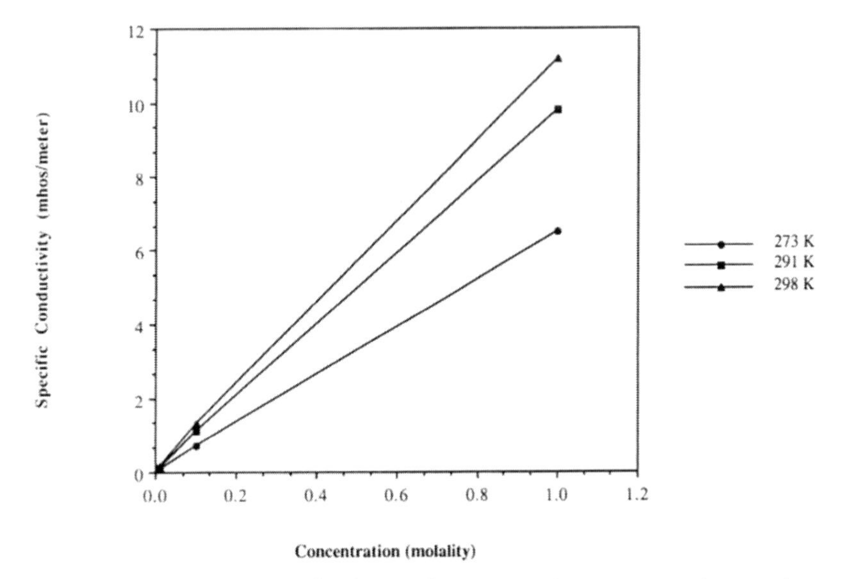

Figure 18.3-1 Conductivity of KCl solutions at varying concentrations and temperatures.

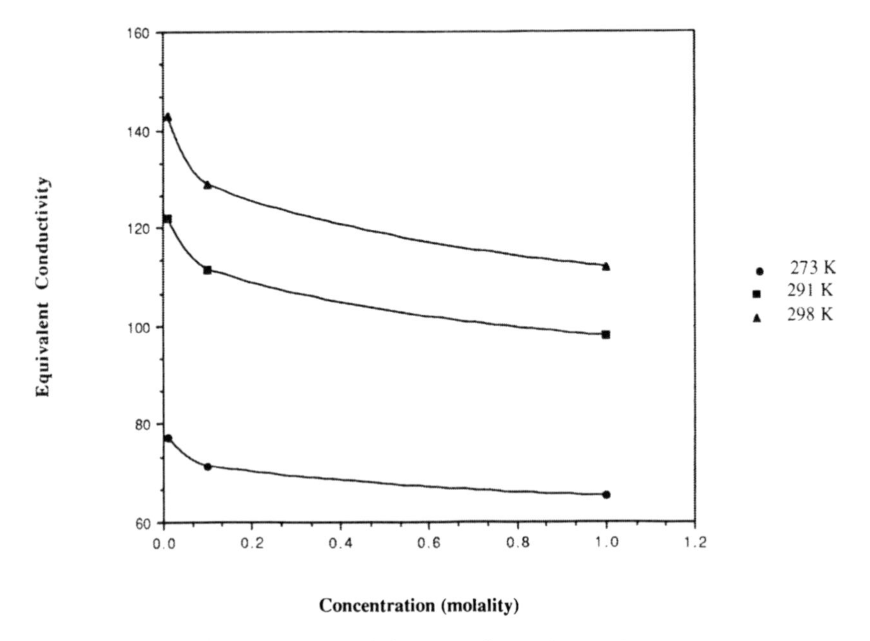

Figure 18.3-2 Equivalent conductivity, Λ, of solutions of KCl at various temperatures.

conductivity with concentration, it is important to determine the conductance on a per particle basis and to introduce the concept of **molar conductivity, Λ_m:**

$$\Lambda_m = \frac{\kappa}{c} \tag{18.3-6}$$

The units of Λ_m are Ω^{-1} cm^2 mol^{-1} or Ω^{-1} m^2 mol^{-1}. The molar conductivity is valuable when comparing the conductivities of solutions that may not have the same concentration. The comparison of molar conductivities however is complicated by the fact that different types of electrolytes, i.e., 1:1, 1:2, etc., will contain anions and cations of different charge. To normalize for this variable, the **equivalent conductivity**, Λ, can be used:

$$\Lambda = \frac{\kappa}{cz_i} \tag{18.3-7}$$

What is apparent when the molar or equivalent conductivities are graphed against the concentration of the solution is that the Λ actually falls as the concentration is increased. In Figure 18.3-2, the data from the graphs in Figure 18.3-1 are expressed as Λ rather than as conductivity, κ, of the solution. This result may be surprising, since it would seem at first glance that the increase in potential charge carriers should lead to an increase

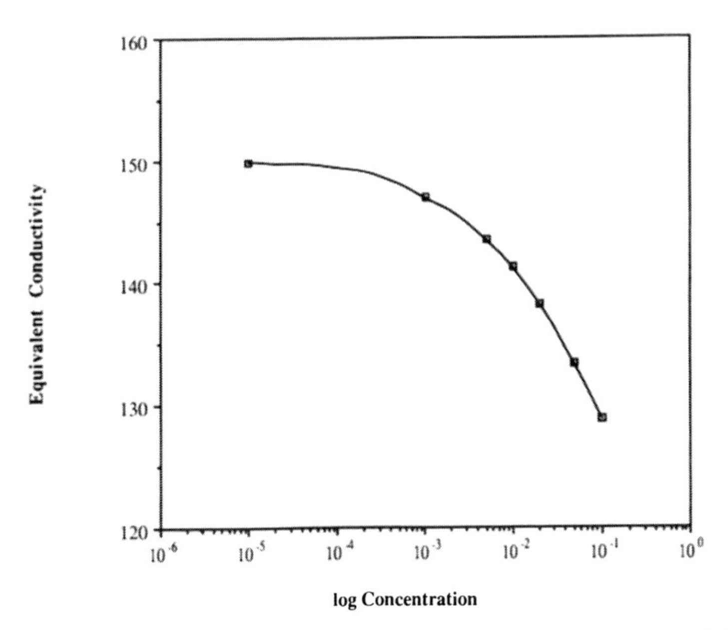

Figure 18.3-3 Method for determining the equivalent conductance at infinite dilution, λ°, for KCl.

in the flow of charge in the conductor, as is shown in Figure 18.3-1. However, the equivalent conductivity data indicate that the increase in the concentration of charge carriers leads to a diminished ability of each ion to transport charge. The shape of the Λ versus concentration curve suggests that the maximum conductivity for each ion occurs in the limit of an infinitely dilute solution. This observation invites two conclusions. If the maximum conductivity per ion occurs under the conditions of an ideal solution, then the molecular reason for the observed decrease in equivalent conductivity will probably lie in interactions between the ion and other ions, as well as the ion and its solvent, as in the case of the activity coefficient. It also suggests that if a series of experimental points are obtained, and then an extrapolation is made to the infinitely dilute solution, a reference state of maximum equivalent conductivity can be defined that will be useful for comparisons between any electrolytes. In Figure 18.3-3, the method of determining this reference state, Λ°, the **equivalent conductivity at infinite dilution**, is illustrated. Table 18.3-2 lists the equivalent conductivity at infinite dilution for a variety of electrolytes.

The dependence of the equivalent conductivity of true electrolytes such as KCl on the concentration was determined by Kohlrausch, who showed that Λ varied with the square root of the concentration at low concentrations:

$$\Lambda = \Lambda^\circ - kc^{1/2} \qquad (18.3\text{-}8)$$

Table 18.3-2. Equivalent conductivity at infinite dilution, $\Lambda°$, for selected electrolyte solutions at 298 K.

Electrolyte	$\Lambda°(\Omega^{-1}\ m^2\ equiv^{-1} \times 10^{-4})$
HCl	426.16
LiCl	115.03
NaCl	126.45
KCl	149.86
NaOH	247.80
$MgCl_2$	129.40
$CaCl_2$	135.84
$LaCl_3$	145.80

where k is an empirical constant which is found as the positive slope of the straight line obtained by graphing Λ against $\sqrt{c}$ as shown in Figure 18.3-4.

Kohlrausch also showed by experiment that the value of $\Lambda°$ for any true electrolyte depends on the sum of the equivalent conductivities at infinite dilution of the anions and cations making up the true electrolyte. These ionic conductivities at infinite dilution are given the symbols λ_+°

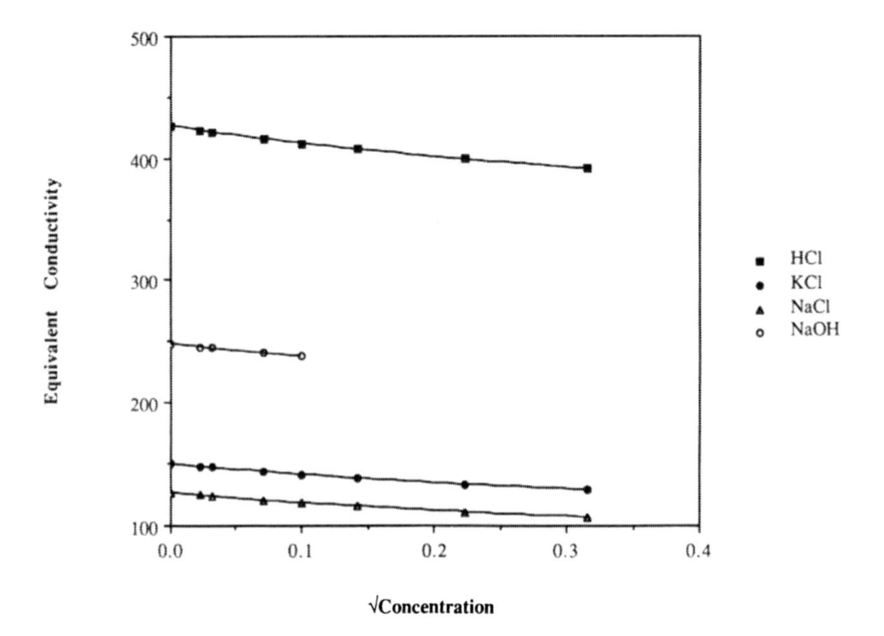

Figure 18.3-4 Straight lines showing the validity of the Kohlrausch law, indicating the dependence of conductivity on the concentration for true electrolytes.

Table 18.3-3. Ionic conductivities at infinite dilution for selected ions at 298 K.

Cation	λ_+° (Ω^{-1} m^2 × 10^{-4})	Anion	λ_-° (Ω^{-1} m^2 × 10^{-4})
H^+	349.80	OH^-	197.60
Li^+	38.69	Cl^-	76.34
Na^+	50.11	Br^-	78.40
K^+	73.50	$CH_3CO_2^-$	40.90
1/2 Mg^{2+}	53.06		
1/2 Ca^{2+}	59.50		

and λ_-°. This observation is known as **Kohlrausch's law of the independent migration of ions:**

$$\Lambda^\circ = \nu_+\lambda_+^\circ + \nu_-\lambda_-^\circ \qquad (18.3\text{-}9)$$

where ν_+ and ν_- account for the stoichiometry of the cations and anions, respectively, that are carrying the charge; for example, for HCl, $\nu_+ = 1$ and $\nu_- = 1$ while for $CaCl_2$, $\nu_+ = 1$ and $\nu_- = 2$. Equation (18.3-9) can be used to predict the equivalent conductivity of a true electrolyte at infinite dilution from the equivalent conductivities at infinite dilution of the ions. The values of Λ° in Table 18.3-2 could be found by using Table 18.3-3 and Equation (18.3-9). This result is consistent with the idea put forth earlier that the fall in equivalent conductance of solutions of electrolytes with increasing concentration is related to the interactions between the moving ions during transport in an electric field, since at infinite dilution it would be expected that no interaction can occur and the conductivities would be truly independent of other ions. How the experimental evidence presented here fits a molecular model will be the subject of the next section.

Before proceeding to develop a model, it is important to realize that up to this point the discussion has dealt with true electrolytes only. Empirical studies show significantly different behavior for potential electrolytes. It is necessary to consider the behavior of potential electrolytes in addition to that of true electrolytes, because of the prominent role played by potential electrolytes such as the bicarbonate ion, in some biological systems. Figure 18.3-5 shows the dependence of Λ on the concentration for both a true electrolyte, HCl, and a potential electrolyte, CH_3COOH. The potential electrolyte clearly does not behave in a fashion similar to that of the true electrolyte, in that it has a very low equivalent conductivity at even low concentrations, but as the concentration of the electrolyte approaches infinite dilution, the equivalent conductivity increases dramatically toward the levels of true electrolytes. The reason for this behavior is that potential electrolytes are not fully ionized until very low concentrations, and therefore the equivalent conductivity is dependent

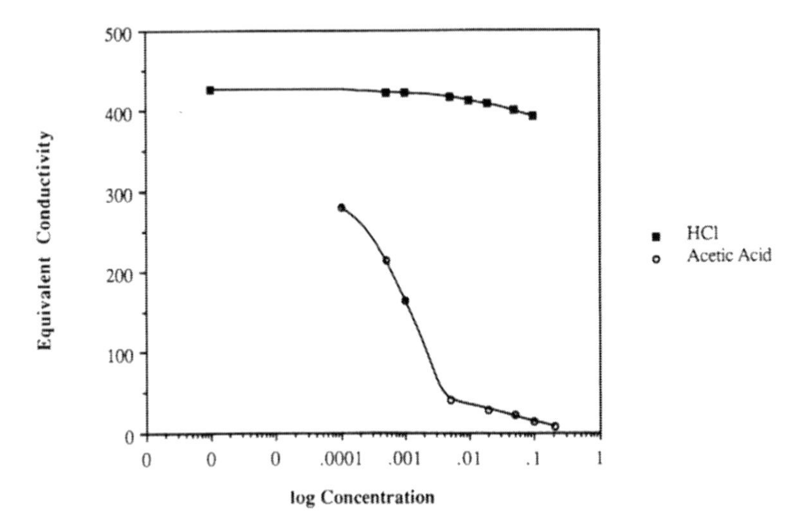

Figure 18.3-5 Comparison of the dependence of Λ on concentration for a true and a potential electolyte solution.

on the equilibrium between the unionized and ionized forms of the potential electrolyte at higher concentrations.

It would be expected that Kohlrausch's relationship of the dependency of Λ on the $\sqrt{c}$ would not hold in the case of the potential electrolyte, and Figure 18.3-6 confirms this. The conductivity of a potential electrolyte solution depends on the degree of ionization, α, of the electrolyte. The equilibrium constant of apparent dissociation is related as follows (c is the concentration of the species in the following equations) :

$$K = \frac{(c_{M^+})\,(c_{A^-})}{(c_{MA})} \tag{18.3-10}$$

$$= \frac{(\alpha c)\,(\alpha c)}{(1 - \alpha c)} \tag{18.3-11}$$

$$= \frac{(\alpha^2)}{(1 - \alpha)}\,c \tag{18.3-12}$$

Because the conductivity depends on the degree of ionization, the measured conductivity, Λ', can be related to the equivalent conductivity of a fully ionized solution:

$$\Lambda' = \alpha\Lambda^{\circ} \tag{18.3-13}$$

It is noted that Kohlrausch's law of independent migration is true for both potential and true electrolytes, since at infinite dilution α will equal

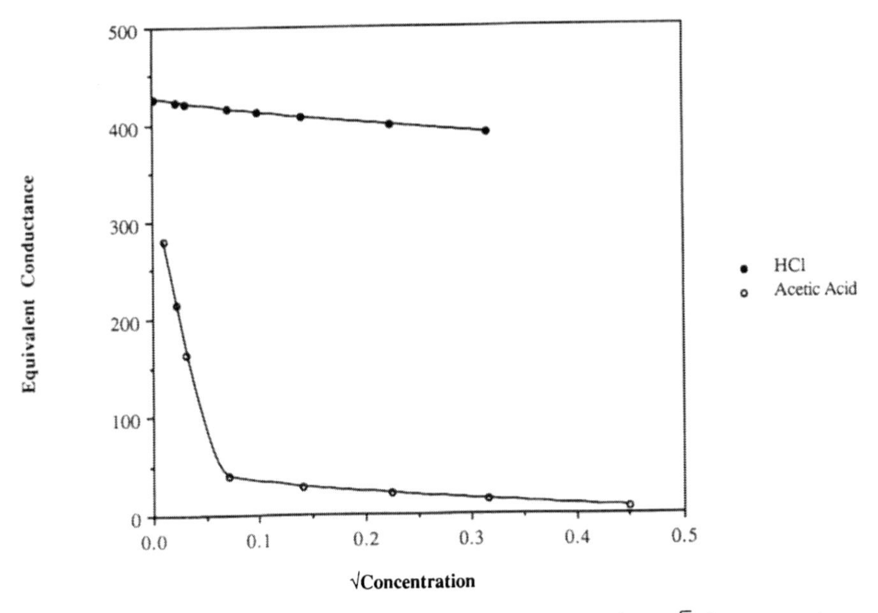

Figure 18.3-6 Plotting the equivalent conductance against $\sqrt{c}$ does not give a linear relationship for a potential electrolyte.

1. Solving Equation (18.3-13) for α and substituting into Equation (18.3-12) gives **Ostwald's dilution law**:

$$K = \frac{\Lambda^2 c}{\Lambda^\circ(\Lambda^\circ - \Lambda)} \tag{18.3-14}$$

It is now necessary to develop a a molecular model that can explain the empirical observations given here. As was discussed in Part II, it will be found that the unexpected alterations in ionic conductivity are related to a variety of interactions. These interactions lead to a loss of independence as a consequence of the increased concentration.

18.4. A Second Look at Ionic Conduction

The most obvious suggestion made by the law of independent migration is that the total current-carrying capacity of an electrolytic solution depends on the vectorial motion of oppositely charged ions in a solution. This picture (Figure 18.4-1) is of anions traveling toward the anodic (positive) electrode, while cations are going in the opposite direction toward the cathode. Since the current is the sum of all of the charge carriers, this picture is perfectly consistent with the empirical evidence presented so far. A moment's consideration of Figure 18.4-1 leads to

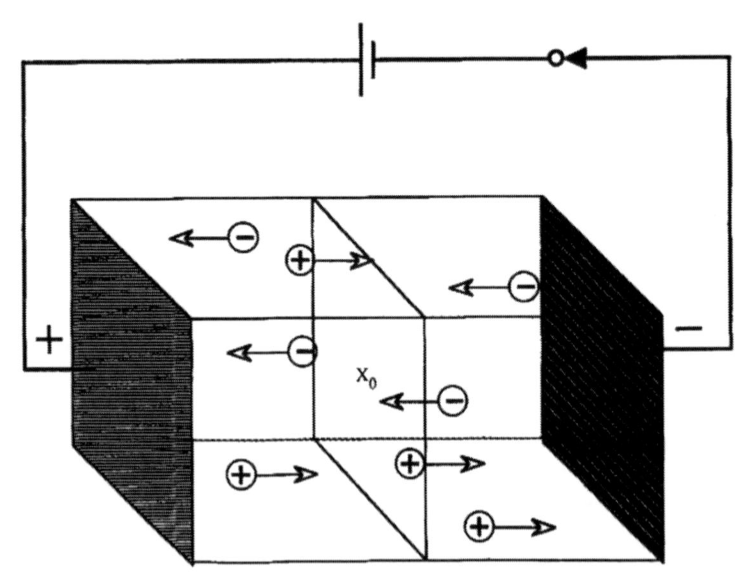

Figure 18.4-1 The total current is found by the vectorial sum of all of the charge carriers.

several valuable questions. Is it reasonable to assume that under any condition except the most ideal, that is, infinite dilution, the ions flowing past one another have no contact or interaction? Even ignoring the question of interionic interaction, why do some ions have a higher equivalent conductivity (cf. Table18.3-3) at infinite dilution when compared to others under the same conditions? These questions must be satisfactorily answered before a suitable molecular model of ionic conduction can be proposed as valid.

Consider, first, the case where the law of independent migration is operating. Any interaction between ions that will prevent their independence will be ignored. As already described (cf. Section 18.2), an ion under an electrical potential field experiences a force given by

$$F = z_i e_o E \tag{18.4-1}$$

where the electric field, E, was described by the negative gradient of the potential:

$$E = -\frac{d\psi}{dx} \tag{18.4-2}$$

The force on the ion that is generated by the potential field will accelerate the ion up to a maximum velocity, the drift velocity, v. The fact that the ion reaches a maximum velocity depends on the drag that acts to retard the movement of the ion. This drag or frictional counter force in a so-

Table 18.4-1. Comparison of the ionic mobilities with $\lambda^{o}_{\pm}$ in H_2O at 298 K for selected ions.

Ion	μ ($m^2\ s^{-1}\ V^{-1} \times 10^{-8}$)	$\lambda^{o}_{\pm}$ ($\Omega^{-1}\ m^2 \times 10^{-4}$)
H^+	36.25	349.82
K^+	7.62	73.52
Na^+	5.19	50.11
Li^+	4.01	38.69
OH^-	20.48	198.50
Br^-	8.13	78.40
Cl^-	7.91	76.34
$CH_3CO_2^-$	4.24	40.90

lution is the viscous force. The viscous force for a spherical object is often calculated using **Stokes' law:**

$$F_v = 6\pi r \eta v \tag{18.4-3}$$

where r is the radius of the object, η is the viscosity of the solvent, and v is the **drift velocity**. At the drift velocity, the forces acting to accelerate the ion, F_{electric}, and to retard it, F_{viscous}, should be equal:

$$z_i e_o E = 6\pi r \eta v \tag{18.4-4}$$

The drift velocity can be written in terms of this balance:

$$v = \frac{z_i e_o E}{6\pi r \eta} \tag{18.4-5}$$

The drift velocity is related to an index of the ease with which an ion can move, the **mobility,** μ. The absolute mobility is a proportionality constant that defines the drift velocity with respect to a unit force:

$$\mu = \frac{v}{F} \tag{18.4-6}$$

The mobility of an ion is related to its ability to contribute to carrying a charge in an electric field. Table 18.4-1 demonstrates the relationship.

The relationship between conductivity and mobility is of fundamental importance and can be related to current flux through Ohm's and Faraday's laws. These equations are as follows, for the individual conductivity:

$$\lambda_{\pm} = \mu_{\pm}\, |z_{i\pm}|\, F \tag{18.4-7}$$

and for a 1:1 electrolyte, where $z_+ = |z_-| = z_i$:

$$\Lambda = z_i(\mu_+ + \mu_-)F \tag{18.4-8}$$

Table 18.4-2. Comparison between the crystallographic radii and $\lambda°$ for selected ions.

Ion	Crystallographic radius (pm)	Hydration number	$\lambda^{\circ}_{\pm}$ (Ω^{-1} m^2 $\times$ 10^{-4})
H_3O^+	133.0	3	349.82
Li^+	60.0	5	38.69
Na^+	95.0	4	50.11
K^+	133.0	3	73.52
Cl^-	181.0	2	76.34
Br^-	195.0	2	78.40
I^-	216.0	1	76.80

In the case of the conduction of ions, the product of the mobility and the electric field will give the drift velocity:

$$v = \mu E \tag{18.4-9}$$

Since the equivalent conductivity of an ion must be related to the ease of moving it in an electric field, that is, to the mobility, it would be predicted on the basis of Equations (18.4-5) through (18.4-9) that the equivalent conductivity of an electrolyte should increase as the viscosity decreases. Since the viscosity of aqueous solutions decreases as the temperature rises, it would be reasonable to expect the equivalent conductivity to increase with increasing temperature. This is indeed the case as Figure 18.3-2 shows. Further examination of Equation (18.4-5) would suggest that the equivalent conductivity will fall as the ionic radius increases. This raises the same question that was discussed previously in the development of an ion–solvent model (cf. Chapter 11). Is the radius the crystallographic radius or is it the solvated radius of an ion? Table 18.4-2 lists the crystallographic radii and equivalent ionic conductivities at infinite dilution for a variety of ions. Clearly, the conductivity falls with decreasing crystallographic radius. When the association of the hydration sheath with the ion is included however, it appears that as the hydration sheath becomes larger the conductivity falls, as would be expected. The association of a larger hydrodynamic radius with smaller ions is expected, because, as was seen in Chapter 11, the formation of the hydration sheath is essentially electrostatic in nature. Electrostatic theory predicts a stronger electric field to be associated with the ions of smaller radii, and consequently, a larger association of solvent molecules and a larger hydration sheath. The behavior of ions in an electric field so far is consistent with the picture described in Chapter 11, that is, the mobility of the ion will be dependent on its relationship with the solvent in terms of the solvent's properties, such as viscosity, and on its association with the solvent, as in the hydration sheath. If these considerations of ion–solvent interactions are important here just as they were in Chapter

11, what about the effects of ion–ion interactions? These interactions will be the next focus of attention. Before moving forward however, it should be noted explicitly that the equivalent conductivity of the proton is exceptionally high, especially when the fact that the proton exists as a hydrated hydronium ion in solution is considered (its hydration number of three suggests that it is part of a tetrahedral array with other water molecules). It will be necessary to account for this anomalously high conductivity before this analysis is completed (cf. Section 18.5).

18.5. How Do Interionic Forces Affect Conductivity?

The consideration of ion–solvent interactions has provided a reasonable explanation for the differences found in the equivalent conductivities of ions at infinite dilution. As in the earlier discussion of the chemical potential and activity coefficient, it seems likely that the interactions between the ions will lead to effects on the measured conductivity. Ions in solution have been described as being surrounded by a cloud of ionic countercharge (cf. Chapter 12). It is not hard to imagine that as the concentration of electrolyte in solution increases, the ionic clouds may interact with one another; that is, the migration of the ions is no longer independent of other ions. As Figure 18.4-1 shows, because the net flux, and hence the conductivity, of an electrolyte solution depends on the sum of the movement of positive charge and negative ions, each moving in opposite directions, the resistance to current flux will increase as the ion clouds drag past one another. This effect of an increased viscous force due to the interaction of the ionic clouds is termed the **electrophoretic effect**. Figure 18.5-1 illustrates this effect.

The picture of the spherically symmetrical ionic cloud that was detailed earlier was developed under static conditions. Consider the dynamic events that occur when the central ion begins to move due to the electrical potential field. As the ion begins to move down the gradient of force, the ionic cloud will also need to move along with the ion. However, the ionic cloud is composed of a time-average arrangement of other ions that are drawn to the central ion. As the central ion moves, two asymmetric fronts develop. The asymmetry of the dynamic ionic cloud is a result of the finite time it takes for the old cloud to decay and the new cloud to form. Because the ions can neither dissipate behind the moving ion nor form a new cloud in front of the ion instantly, the ion experiences a separation of charge within its own cloud. The center of charge is in fact behind the ion, thus generating a local electric field that acts to retard the movement of the ion with respect to the external field. This retarding force due to the finite time it takes for ion cloud rearrangement to occur is called the **relaxation effect**. The relaxation time for dilute solutions is on the order of 10^{-6} s. The relaxation effect is illustrated in Figure 18.5-2.

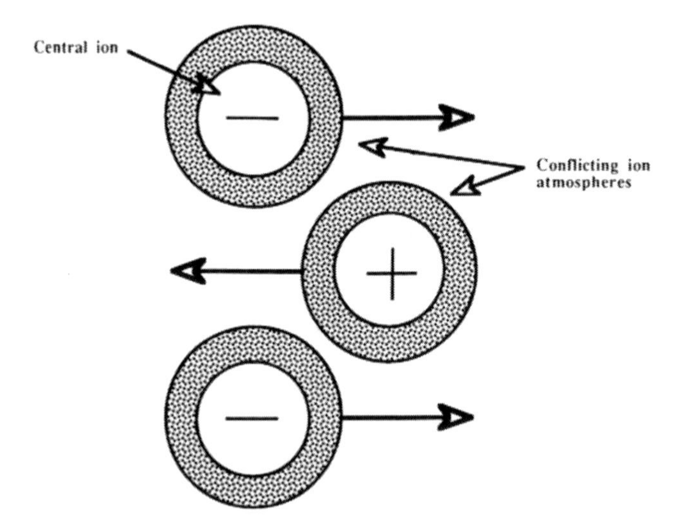

Figure 18.5-1 Ionic cloud interference leads to an increased viscous force in the electrophoretic effect.

Both the electrophoretic and the relaxation effects depend on the density of the ionic atmosphere and can be shown to increase with $\sqrt{c}$. This is consistent with the observations made by Kohlrausch as described earlier (cf. Section 18.3). A model theory based on these effects has been developed and is called the **Debye–Hückel–Onsager** theory. A rigorous development of the theory is beyond the scope of this volume, but the conclusion can be given. This model has the form

$$\Lambda = \Lambda^\circ - (A + B\Lambda^\circ)c^{1/2} \tag{18.5-1}$$

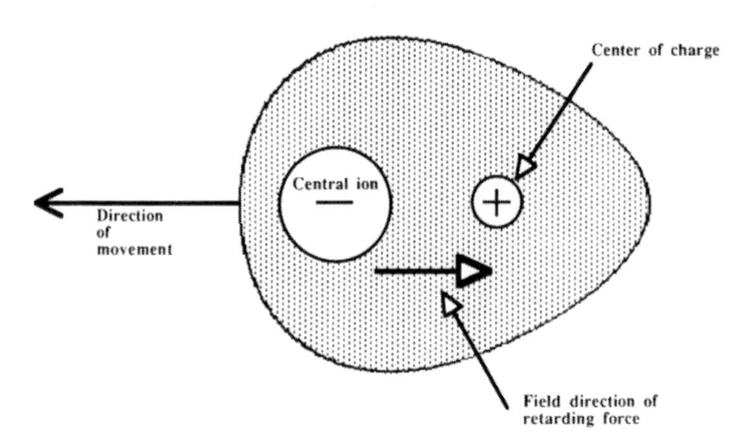

Figure 18.5-2 The dynamics of the relaxation effect.

Here, the constant A describes the electrophoretic effect and has the form:

$$A = \frac{z_i e_o F}{3\pi\eta}\left(\frac{2z_i^2 e_o^2 N_A}{\epsilon_o \epsilon kT}\right)^{1/2} \tag{18.5-2}$$

B describes the relaxation effect and can be written

$$B = \frac{e_o^2 z_i^2 q}{24\pi\epsilon_o \epsilon kT}\left(\frac{2z_i^2 e_o^2 N_A}{\pi\epsilon_o \epsilon kT}\right)^{1/2} \tag{18.5-3}$$

where q is approximately 0.5 for a symmetrical electrolyte. This equation provides reasonable agreement with the empirical conductivity data, thus providing good evidence for the ion–ion interactions described so far. It is accurate for concentrations up to only approximately 0.002 M in the case of aqueous solutions of 1:1 electrolytes. Further modifications have been suggested for more concentrated solutions, but the assumptions and development of these theories are both complex and controversial and will not be discussed.

Two further lines of evidence that support the model of ion–ion interaction as described above are worth mentioning. The first is the **Debye–Falkenhagen effect**, in which conductivities are studied with a high-frequency electric field. In a high-frequency field, the ionic cloud will not have time to form, and hence the displacement of the center of charge will not occur. Without the separation of the charge, the retarding force responsible for the relaxation effect would not be expected to occur, and conductivities should be higher under these conditions. Experiment confirms this effect. The Debye–Falkenhagen effect is probably of no significant consequence in biological systems. A second line of evidence may well have biological relevance. For electric fields of low strength, i.e., 10^2 to 10^3 V m^{-1}, the molar conductivity of a true electrolyte is independent of field strength. However, for fields of high strength, i.e., 10^6 to 10^7 V m^{-1}, the conductivity is found to be increased. This is the first **Wien effect** and is due to the fact that at high field strengths, the movement of the ion is so rapid that the ion moves an order of magnitude faster than the relaxation time of the ionic cloud formation. Consequently, the ion travels without its ionic cloud and the retarding force due to the charge separation is eliminated. It is interesting to note that the transmembrane field strengths of many cells are on the order of 10^6 V m^{-1} or greater, and the ion fluxes across these membranes may well be subject to this Wien effect.

18.6. The Special Case of Proton Conduction

The equivalent conductivity of protons in aqueous solutions is significantly higher than would be predicted on the basis of the previous discussion. Protons have been shown to exist as hydronium ions and not

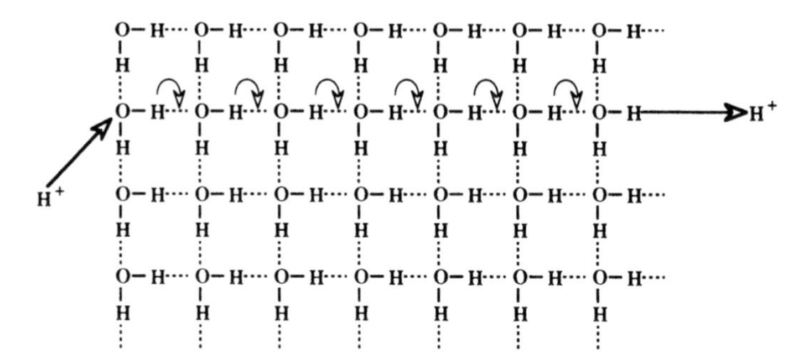

Figure 18.6-1 Schematic of the movement of a proton through liquid water by a bucket-brigade mechanism.

as free protons in aqueous solution. As Table 18.4-2 shows, the ionic radius of the H_3O^+ ion is close to the radius of a potassium ion. Furthermore, the hydronium ion has been shown to exist in a tetrahedral complex as $(H_9O_4)^+$. Taken on these terms, there does not seem to be any explanation for the surprising mobility of the proton. Could the structure of the aqueous solvent be responsible for the high mobility? Experimental evidence that this is the case is found in studies where the equivalent conductivity of H^+ falls to the level of that Λ of K^+ when the water is substituted with another solvent. What is the unique interaction between protons and water that make this phenomenon so dramatic?

Protons in water are part of a dynamic network of hydrogen-bonded water clusters, as has been well described at this point. The existence of this extensive hydrogen-bonded structure is part of the reason that proton mobilities are so high in water. Rather than actually being physically transported down the electrical potential gradient, the proton is passed through the lattice from molecule to molecule with the net effect being the movement of charge. This mechanism is illustrated in Figure 18.6-1. The schematic nature of the Klotz bucket-brigade mechanism shown in Figure 18.6-1 should be approached carefully so that it is not misleading. While liquid water and ice are both highly structured due to the hydrogen bonding of the molecules, this structure is not static in nature. The idea that the net transport of protons occurs by a "simple" brigade mechanism should be further scrutinized.

In Section 17.2, a picture of mobility based on the ideas of mean free path and rate processes was described and was summarized in Equations (17.2-12) through (17.2-16). It was noted that the mean free path for a particle making a jump in solution depends greatly on the structure of the solvent and that the rate or frequency of the jumps depends on the size of the energy barrier between the pre- and postjump coordinates. Can the exceptional mobility of the proton moving in a brigade fashion be understood in these terms? First, consider the question of the potential

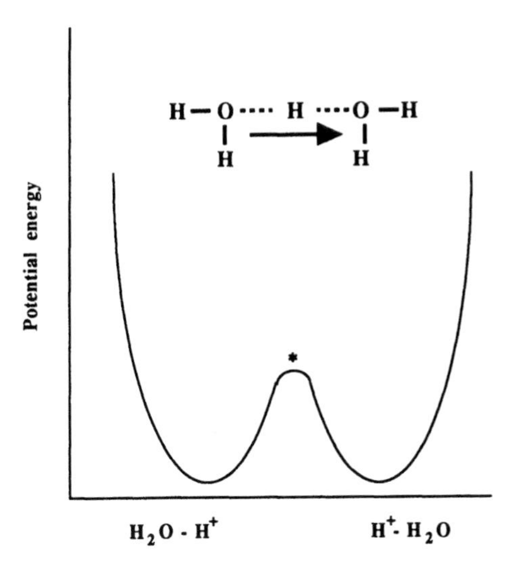

Figure 18.6-2 Potential energy graph showing the potential energy barrier that occurs as a proton is abstracted from an oxygen center (left well) and is moved to a new acceptor center (right well).

energy barrier that will exist between the pre- and postjump water molecules on the brigade line. The process of proton exchange in the brigade can be written as follows:

$$H_2O\text{–proton} \rightarrow H_2O\text{–proton}^* \rightarrow \text{proton–}H_2O \qquad (18.6\text{-}1)$$

This process is symmetric and it would be expected that a potential energy diagram could be drawn for it. This diagram (Figure 18.6-2) would be comprised of two mirror image plots of the potential energy associated with removing a proton from an H_3O^+ molecule. The potential energy barrier associated with the proton transfer is on the order of 40 kJ mol^{-1}. When this value is used in an analysis like that outlined in Equations (17.2-12) through (17.2-16), there is reasonable agreement between the predicted and measured mobilities. However, the analysis of proton mobility is not really complete at this point. Although the concurrence between theory and experiment is good for the proton mobility, other experimental tests suggest that the model as proposed needs further adjustment. One of these conflicts is in the prediction that the mobility of H^+ in H_2O will be significantly increased over that of D^+ in D_2O; this, it turns out, is not the case (the ratio of mobilities is 1.4). A second conflict lies in the prediction that with increasing temperature the distance between the donor and acceptor H_2O molecules will increase, leading to a higher energy barrier and hence a higher heat of activation. In fact, the energy of activation falls with increasing temperature. What further considerations should be included in this analysis?

Protons, although not as small from the standpoint of mass as electrons, are sufficiently light that they may be sometimes better treated by quantum mechanics than by classical approaches. One of the results of a quantum mechanical treatment of a particle in a potential energy well with finite barriers is that a distinct probability exists that the particle may be found outside the potential barrier without having climbed up and out of the well. In effect, the particle tunnels through the barrier. This effect is called **quantum mechanical tunneling** and leads to the physical consequence that particles may cross a potential energy barrier, even if they do not have sufficient energy to go over the barrier in classical fashion. The consideration that protons may tunnel in aqueous solution is an important one. It has been shown that tunneling is the most likely mode of transport in water, and it is preferred over the classical approach given above. However, comparing the mobilities derived from the tunneling calculations with the experimental measurements leads again to trouble—the mobility predicted by a tunneling approach is simply too fast. Considerations of speed however are only meaningful when there is somewhere to go. If a proton can rapidly tunnel through a barrier, where does it end up? Attention must be turned to the ability of the acceptor H_2O molecule to receive a proton.

A water molecule is able to assume only a limited number of positions where its unbonded electrons can form a bond with the transferring proton. There is a finite time that is required for the water molecule to assume a position where it can receive a proton. Furthermore, since the concern is with conduction of the proton, the same water molecule must now reorient and assume a position in which it can pass on the proton to another acceptor molecule. Even if the actual transfer is accomplished by a rapid tunneling process, the rate of transfer will depend on the time necessary to position and reposition the acceptor/donor water molecules along the way. If calculations are made based on the random movement of the H_2O molecules induced by thermal forces, the reorientation occurs too slowly to account for the experimentally found mobility. However, it is not reasonable to treat this system as being dependent only on thermal forces for its interactions. An H_3O^+ molecule is positively charged and will exert an attractive coulombic force on the partial negative charge of the acceptor dipole molecule. This coulombic interaction will act to increase the speed of reorientation; hence the generation of a local electric field will assist the reorientation of the acceptor molecule. This orienting force significantly increases the rate of reorientation over that found for a strictly thermally induced motion and allows for a mobility consistent with that found experimentally. The complete picture of proton conduction in H_2O is a combination of fast tunneling through, rather than over, potential energy barriers but with a rate-limiting step, the orientation of the acceptor molecule. While the correct orientation of the acceptor molecule is the rate-limiting step, this step is accelerated by the ion–dipole interaction of the approaching hydronium ion whose electrostatic field assists and speeds the reorientation of the acceptor molecule. Not only does this model give proton mobilities that are consistent with experiment, but as well it predicts the surprising fall of activation energy with increased temperature and gives the correct ratio of H^+/D^+ mobility (incorrectly forecast by the earlier model). The activation energy drop is predicted, because in the reorientation model, the rate of orientation will be related to the number of hydrogen bonds that must be broken and reformed. As the temperature increases, the number of hydrogen bonds is decreased, reorientation is easier, and the activation energy falls.

CHAPTER 19

The Electrified Interface

19.1. When Phases Meet: The Interphase

Earlier, it was discussed that while homogeneous systems are more easily described, many cellular processes are heterogeneous. Because so many processes occur that require the exchange of components across at least one phase, it is extremely valuable for the biological scientist to have an understanding of the forces and structures that act in the zone of transition between phases. When different phases come in contact with each other, an interface between them occurs. This interface is a surface, and the properties of a surface are different from those of either of the phases responsible for creating it. Additionally, the changeover between phases is never instantaneously abrupt, but instead there is a zone of transition extending from the surface for a finite distance into the bulk of each of the phases where the properties are representative of neither bulk phase. The surface and the regions immediately adjacent are termed the **interphase**, a very useful distinction. The properties of interphases will be the the subsequent focus of this section.

The development of interphase regions is a thoroughly general phenomenon and applies throughout nature. The approach here will be to develop a qualitative picture of these interface regions in general and then to relate this picture to cellular systems. Much of the seminal work in understanding the nature of the interphase has been done in aqueous electrolyte–mercury drop studies, and, though not directly convertible, there is much valuable information that can be gained from considering this knowledge. Although the cell membrane is probably better considered an insulator or a semiconductor than a conductor like mercury, the picture of the interphase region developed through the study of the mercury electrode is certainly a good place to start for an understanding of the cellular interphase. In Chapter 22, a somewhat different model of the double layer will be described for membranes whose only charge carriers are electrolytes. This model is called the **interface of two immiscible electrolyte solutions (ITIES).**

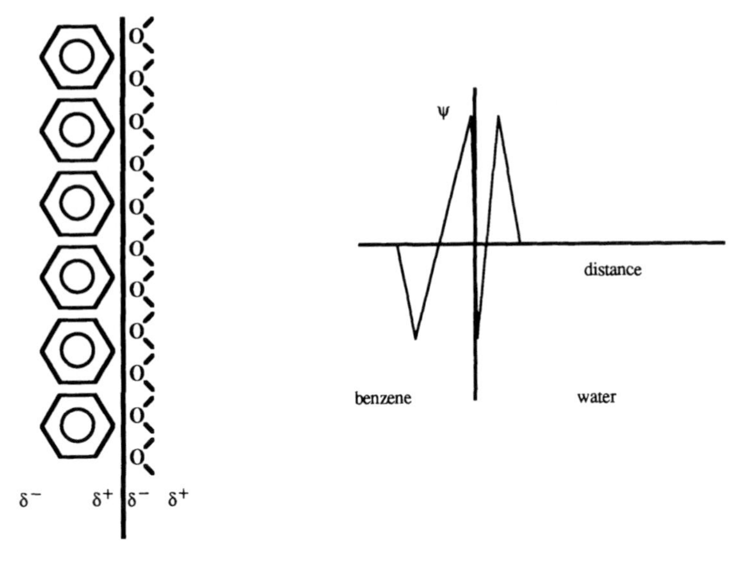

Figure 19.1-1 The electrification of the interphase region due to the orientation of charges between the two phases, water and benzene. The graph is a representation of the change in potential, ψ, with respect to the distance into each phase.

What happens when two phases are brought into contact with one another? Perhaps an easy way to consider this question is to take a short thought trip like the one in Chapter 8. Consider, first, the environment experienced by a single water molecule moving from a phase of pure bulk water to the interface with a phase of benzene (Figure 19.1-1). Initially, deep inside the aqueous phase, the water molecule is free to look around and find the monotonous uniform tug (at least on a time-average scale) of other water molecules. No matter which way it looks it will see (feel forces exerted on it) the same thing. Specifically, there will be no net dipole orientation, and electroneutrality will prevail in any reasonably chosen sample of the bulk phase. For the convenience of the argument, a lamina is chosen parallel to the benzene–water interface. Now, if the water molecule was to move randomly toward the interface between the benzene and water and suddenly was to find itself looking out not on another watery neighbor, but instead now on a benzene molecule, it would be profoundly changed in its view of the world. It has gone from an **isotropic environment** where direction is inconsequential to an **anisotropic environment** where the forces experienced depend on magnitude as well as direction. In fact, the water molecule will begin to experience a gradient of altered forces produced by the benzene–water interface before coming face to face with a benzene molecule as neighbor. The behavior of the molecules that experience the anisotropic environment will no longer appear random on a time-average scale. Water molecules located near the interface will experience orienting forces that depend on the nature of the interactions with the benzene molecules. In this case, it would be found that the water molecules will arrange themselves with their oxygen atoms facing the boundary, and their hydrogens facing into the aqueous phase. The layer of water apposing the benzene phase therefore is no longer randomly oriented, and the arrangement of the water dipoles results in a lamina of charge. Because the water dipoles are no longer randomly oriented, which is the condition in the bulk phase, a charge separation has occurred in association with the interface. The charge derived from the aqueous dipoles is sensed across the phase boundary, and the benzene phase responds by producing a countercharge of equal but opposite magnitude. (In fact, the thought experiment could have been considered from the point of view of the benzene molecules with focus on a different mechanism for generating the benzene charge but yielding the same overall result.) Across the phase boundary, a potential difference exists, usually with a magnitude of no more than 1 V. However, the distance of the separation is generally quite small (1 nm), and therefore the resulting electric field has a magnitude of approximately 10^8 to 10^9 V m^{-1}. Such pairs of separated charges across the boundary between two phases is what is called the **electrical double layer** or the **electrified interface**. The region affected by this electric field extends not only across the phase boundary, but also for a distance into each phase. In this region, the forces acting on the molecules are significantly different from those in the bulk of either phase, and the behavior of the molecules therefore deviates significantly from that of their bulk phase. The region where such deviation from the bulk properties occurs is defined as the **interphase**.

The interphase includes both sides of the phase boundary, and if the entire interphase is considered, the electroneutrality principle will prevail. As previously mentioned, the existence of this electrified interface is a completely general phenomenon when phases meet. Potentials are developed by dissimilar metals in contact at liquid–liquid junctions, at liquid–solid junctions, etc. Even when a metal is in contact with a vacuum, there is a tendency for the electrons to protrude out of the metal surface, leading to a negative charge above the boundary countered by a positive layer inside the surface.

A range of mechanisms lead to charge separation at phase boundaries, and while not all have equivalent relevance in biological systems, some of the mechanisms will be presented here by way of a preview before a more detailed consideration of the interphase is undertaken. These mechanisms can be separated into categories that are somewhat artificial but can nonetheless be valuable in understanding the forces that lead to electrification of interphases (Table 19.1-1).

Table 19.1-1. Classification of the variety of mechanisms leading to interphase electrification.

Type I	Charge established or imposed on one phase
	Ionizable surface groups
	Charged metal electrode surfaces
	Physical entrapment of charge
Type II	Charge generated by differential affinity between phases
	Difference in electron affinity between phases
	Preferential attraction of ionic species to surfaces
	Differential distribution of ions between phases
	Arrangement of permanent dipoles to generate net charge separation
	Arrangement of inducible dipoles to generate net charge separation

The interactions categorized as type I are similar in that one of the phases contains an identifiable charge when compared to the other phase. This should not be taken to suggest that the so-called "charged" phase is not electroneutral but rather that definable charged entities (ions and electrons) can be identified as associated with a specific phase whether or not the interface exists. For example, a surface containing carboxyl groups will become charged when the pH of the environment surrounding these groups rises above their pK_a. Although the development of this surface ionizable charge clearly depends on the pH of the associated phase, the charge must remain resident on the surface of the phase that contains the carboxyl groups. This particular case is especially applicable to biological systems and includes the ionizable charges associated with cell membrane components, the extracellular matrix, and the cellular glycocalyx. The entrapment of charge in a phase can occur in cases where the charge is associated with a component that cannot diffuse through a semipermeable membrane. This can also be an important effect biologically since many membranes are impermeable or variably permeable to different cations and anions. When a metal electrode is connected to an external power source, electrons can be either forced into or removed from the electrode, resulting, respectively, in a negatively or a positively charged electrode. The charge associated with the altered electron density of the electrode secondary to the external source is different from a charge separation that also results from the interaction of the electrode phase with its contiguous phase. The treatment of these two separate charges applies for all the cases type I. These forces will be examined in some detail shortly.

Type II mechanisms are invoked in cases where the charge separation results from an interaction that depends to a great extent on the interface being present. For example, the difference in affinity for electrons between two dissimilar metals leads to the contact potential difference between them. The charge separation is the result of the greater attraction of conductivity band electrons from one metal into the other metal. This case is probably more important in metal–metal and metal–semiconductor junctions. In cases where an aqueous electrolyte constitutes one of the phases, as in all biological systems, specific anionic and cationic species will be adsorbed on surfaces in a differential fashion leading to charge separation. This chemically specific absorption is often referred to as **specific adsorption.** Anions and cations will also partition across a phase boundary in a differential manner because of preferences for one phase or the other. A special case of this is the difference in potential, called the **diffusion potential**, that results when two similar solutions (but of different concentration) are brought together and, due to differences in the rates of diffusion of the ions, a potential difference results. The diffusion potential will be discussed further in Section 22.1. The case of phases that contain permanent dipoles, or may be induced to contain dipoles, has already been presented.

19.2. A More Detailed Examination of the Interphase Region

If an electric field is to be expected at every surface and interface between two phases, what will be the arrangement of the molecules and forces interacting in this region? To answer this question, an examination of the situation at the metal–electrolyte interface is valuable since it is reasonably well described. The principles can be generalized to situations relevant to the biological scientist, marking a solid starting place for the discussion.

A highly polished metallic electrode with the geometry of a flat plate is brought into contact with an aqueous solution containing only a true electrolyte (consideration of organic materials in solution will come later). The bulk electrolyte solution has the behavior already described in earlier chapters. The electrode can be connected to an external power source that will add electrons to or withdraw electrons from the electrode so that the electrode may have a negative, a positive, or no charge associated with it. The metal phase will be considered as a crystalline lattice of positive ions that are immobile with a cloud of electrons free to move in response to an electric field. An excess charge on the electrode, whether negative or positive, can be considered to be localized to the surface of the metal or to the electrode surface of the solution–electrode interface. The setup is depicted in Figure 19.2-1.

A positive charge is placed on the electrode. Except for the geometry, the electrode now appears to the ions and water molecules making up the bulk electrolyte as a large cation. As in the case of the Debye–Hückel central ion, the electrolyte will respond by orienting its molecules in such a way as to neutralize the electrode charge. In other words, a cloud of countercharge will form parallel to the electrode surface, similar to the one formed around the central ion. By moving molecules so as to coun-

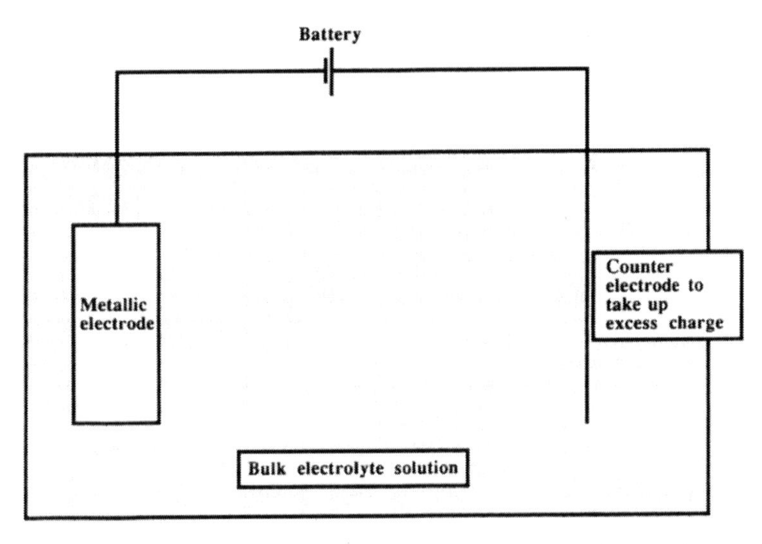

Figure 19.2-1 System for the study of electrode–electrolyte interactions.

teract the charge on the electrode, the electrolyte phase itself now becomes charged in the region of the electrode. Sufficiently far into the bulk phase of the electrolyte, the charge on the electrode is no longer felt and the charge distribution in the solution phase becomes independent of the electrode charge. The properties of the interphase end and those of the bulk phase take over. What is the arrangement of the molecules making up the solution phase in the interphase region?

The structure of the ionic side of the interface has been, and continues to be, an area of intense study. There is good agreement at this time as to the stuctural elements of the charged ionic phase, due in large part to the work of Grahame. Rather than a highly detailed and mathematical treatment of the interphase region, a historical and qualitative approach will be used to describe the arrangement. Appreciation of the historical aspects of the development of the model of the aqueous portion of the electrified interface or double layer is useful because a great deal of the terminology relating to the science of the interphase is historically derived. It is also worth noting that similarities between the structure of electrolyte solutions around electrodes and ions were not lost on the researchers working on these problems in the early part of this century. In fact, as mentioned in Section 12.1, it was the work of Gouy in 1910 and Chapman in 1913, over a decade before Debye and Hückel's central ion model, that suggested the idea of the ionic cloud as a model for their work on the electrified interface.

19.3. The Simplest Picture: The Helmholtz–Perrin Model

Just as the molecular structure around an ion was found by determining the charge density of the ionic cloud with respect to the distance r from the central ion, the structure of the electrified interface can be considered by evaluating the location and density of charge excesses with respect to the distance from the electrode. In a fashion similar to that for the ion, the parameter describing the charge excess is the potential resulting from the charge separation. The charge associated with a metal electrode has been described above as localized in a single layer on the surface of the electrode. The mission of the aqueous electrolyte is to neutralize this charge, and in the earliest model of the electrified interface, developed independently by Helmholtz and Perrin at the turn of the century, the entire countercharge to the electrode was considered to reside in a single rigid layer of counterions. This compact and immobile layer of ions was thought to completely neutralize the charge on the electrode. Were the model of the interphase to stop at this point, the electrified interface would be precisely analogous to a parallel plate capacitor composed of two plates of opposite charge, a double layer system. This is the historical derivation of the term **the double layer**. While the Helmholtz model is generally not adequate for describing the electrified interphase region, the term has remained. If the Helmholtz model of the double layer were correct, then, since the charge on the electrode is completely neutralized by a single plane of ionic charge, a graphical plot of potential versus distance from the electrode would be linear as shown in Figure 19.3-1.

It should be noted that no mention has been made of the arrangement of water and its structure in the interphase so far. Water plays an important structural role in the interphase and will be considered shortly, but the Helmholtz–Perrin model considers only electrostatic type interactions, and therefore only the ion–electrode relationships are examined.

19.4. A Diffuse Layer Versus a Double Layer

One of the fundamental problems with the Helmholtz–Perrin model lies in its lack of accounting for the randomizing thermal effects everpresent in liquids. Gouy and Chapman suggested that the capacitor plate-like arrangement of counterions be replaced by a diffuse cloud of charge that was more concentrated near the electrode surface and extended out into the bulk solution. Their formulation was based on Maxwell–Boltzmann statistics and is similar to the picture already developed for the Debye–Hückel cloud relationship of charge density in the cloud to the potential, ψ_r, from the ion. In the **Gouy–Chapman model**, the decay of the potential derived from the electrode versus the distance, x, into the bulk electrolyte depends in part on the charge, z_i, on the ion and on the ionic strength

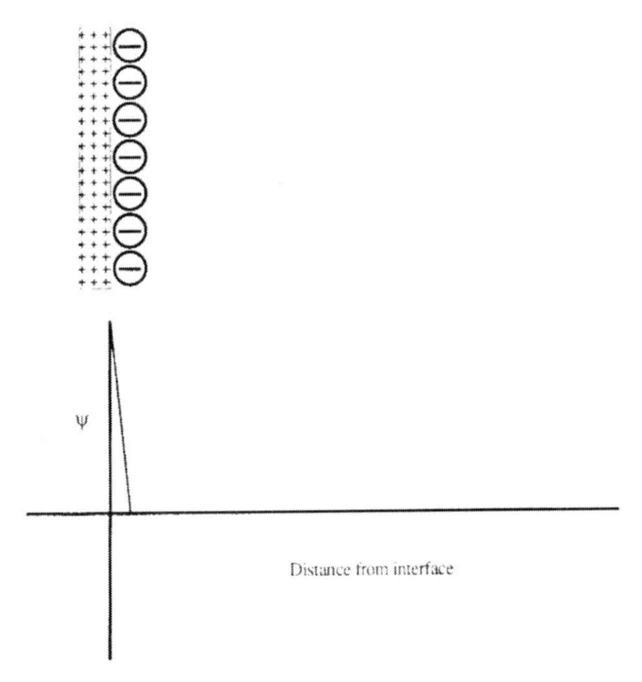

Figure 19.3-1 Structure and associated potential versus distance curve of the Helmholtz model of the electrical double layer. Note that the arrangement of counterions forms a parallel layer that acts like a parallel plate capacitor, and hence the potential curve is linear.

of the solution. In fact, a thickness for the Gouy–Chapman diffuse layer exactly analogous to the reciprocal length parameter, κ^{-1}, of the Debye–Hückel model can be described:

$$\kappa^{-1} = \left(\frac{\epsilon_o \epsilon kT}{2N_A z_i^2 \epsilon_o^2 I}\right)^{1/2} \tag{19.4-1}$$

This means that the charge contributed by the diffuse layer can be simulated, and treated as if it were a capacitor plate placed in parallel a distance from the electrode equal to the reciprocal length. As in the Debye–Hückel model, only electrostatic interactions are considered; the ions are treated as point charges, and the solvent is considered as a structureless continuum. There is no consideration of the concept of closest approach of an ion to the electrode, ion size, or water interaction. The structure and potential versus distance relationships of a "pure" Gouy–Chapman diffuse layer are illustrated in Figure 19.4-1. Figure 19.4-1 is based on the relationship that y varies with respect to distance, x, from the electrode according to the following equation:

$$\psi_x = \psi_o e^{-\kappa x} \tag{19.4-2}$$

where $\psi_x \to \psi_o$ as $x \to 0$.

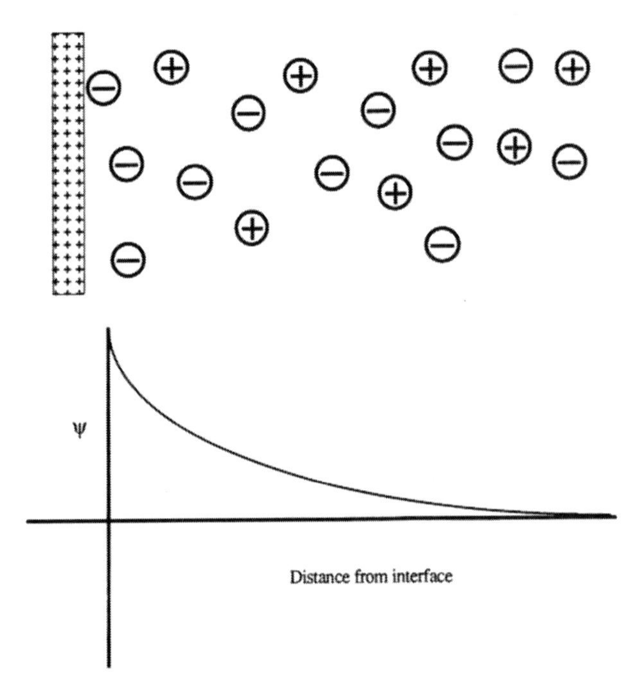

Figure 19.4-1 Structure and associated potential versus distance curve of the Gouy–Chapman model of the interphase region.

19.5. Combining the Capacitor and the Diffuse Layers: The Stern Model

The Gouy–Chapman treatment of the double layer ignores the effect on the dielectric constant of the high potential fields (approximately 10^8 V m^{-1}) present at the interface, the fact that ions have finite size, and the tendency of molecules to adsorb on surfaces through forces other than electrostatic interactions (specific adsorption). In 1924, Stern presented a model that incorporated the questions of finite size and adsorption onto the surface into the structure of the interphase. The mathematical approach of Stern will not be discussed, but a qualitative picture can be sketched. When the finite size of an ion (whether hydrated or not) is considered, it becomes obvious that the distance between a countercharge comprised of ions and an electrode will be greater than if the countercharge were comprised of point charges. The inclusion in the model of interactions at the surface similar to those described by the Langmuir isotherm added an important dimension for the molecules that are on the inside of the interphase region. Because of the adsorptive forces (often described as the "chemical" aspect of the interaction) acting on certain ionic species, the number of ions, and hence the charge excess, in the

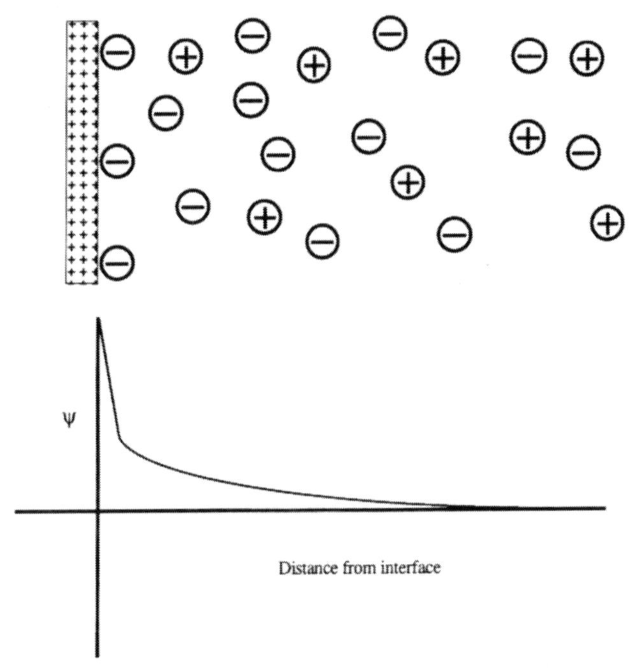

Figure 19.5-1 Structure and associated potential versus distance curve of the Stern model. Note that the model contains elements of both the Helmholtz and Gouy–Chapman formulations.

layer closest to the electrode will be altered when compared to that predicted by a model in which only electrostatic interactions are taken into consideration. The number of ions populating the innermost layer, where adsorption is important, and the distance of closest approach of this layer to the electrode are both finite. Because there is a limit to the nearest approach of the ion cloud, a region of linear potential decay with respect to distance results. Such a region has the behavior of the molecular capacitor suggested by Helmholtz. This inner layer is called the **Stern layer**. Beyond the inner layer, the Stern model predicts that the remainder of the charge needed to neutralize the electrode charge is arranged like a Gouy–Chapman diffuse layer. The **Stern–Gouy–Chapman model** is illustrated in Figure 19.5-1. A very important point that derives from the Stern model is that the amount of charge excess found in the Stern layer or the diffuse layer differs under various conditions of electrolyte valence and concentration. In dilute solutions, the countercharge will be found almost completely in the Gouy–Chapman diffuse layer, while in solutions sufficiently concentrated the charge may instead be localized almost completely on the Stern layer, giving rise to a Helmholtz-type structure.

19.6. The Complete Picture of the Double Layer

Further developments have been made with respect to understanding the structure of the double layer region, and the role of water and its dielectric properties have been taken into account. Building on the historical basis laid to this point, a modern picture of the double layer region will now be drawn (Figure 19.6-1). To develop this picture, imagine moving from the surface of the metal electrode (actually the starting point will be just inside the charge layer of the electrode). As in the model for the ion developed earlier, water molecules will be the first components of the solution encountered. A monomolecular layer of water molecules will be lined up immediately adjacent to the electrode surface. In the case of the positive electrode, the negative end of the water dipole will be preferentially facing the electrode, but this orientation will reverse itself if the charge on the electrode is reversed. The electrode will be hydrated in a fashion analogous to that for an ion. Unlike an ion however, the metal

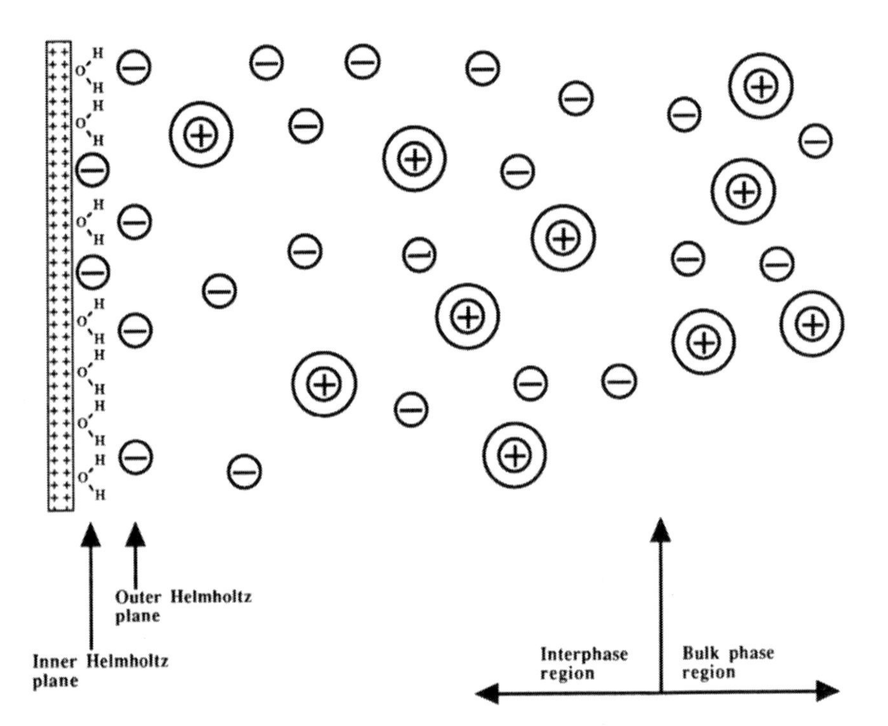

Figure 19.6-1 Schematic of the structure of the interphase, including orientation of aqueous dipoles, specific adsorption of unhydrated anions on the electrode surface, and hydrated cations. Note that the outer Helmholtz plane is the division between the Helmholtz-like molecular capacitor and the diffuse Gouy–Chapman layer.

Table 19.6-1. ΔG values for the movement of selected ions to the inner Helmholtz plane.[a]

	Interaction				
	Ion–electrode	Water–electrode	Ion–water		Total
Ion	ΔG	ΔG	ΔH	ΔS	ΔG
Na^+	−207.5	113.0	165.3	46.9	56.5
K^+	−204.2	79.9	141.8	21.7	10.9
Cl^-	−208.8	76.6	74.1	−69.5	−37.2
I^-	−210.5	69.0	45.6	−136.4	−54.8

[a] ΔG and ΔH values are in kJ mol^{-1}; ΔS values are in J mol^{-1} deg^{-1}.

electrode can be forced to change its charge, and the direction of the water dipole can also be induced to change to a varying degree with the electrode charge. The arrangement of the dipoles and the energy of their attraction to the electrode have been shown to play an important role in allowing or preventing adsorption of organic molecules on electrodes.

On moving through the hydration sheath, a second layer of molecules is encountered, comprised primarily of hydrated ions of appropriate countercharge. In the case of the positive electrode, this layer is populated by anions, and for the negatively charged electrode, cations. This layer of ions, an important constituent of the molecular capacitor, is called the **outer Helmholtz plane** (OHP). The outer Helmholtz plane is the dividing line in the modern view between the outer diffuse Gouy–Chapman layer and the inner compact region of the interphase. Attention will be turned now to the structure of the inner compact region. The electrode is considered to be covered with a monomolecular layer of water with a net dipole orientation determined by the sign of the charge on the electrode. In some cases, it is possible that ions can displace these adsorbed water molecules. When the electrode is positively charged, the counter-field ions will be comprised primarily of anions. Anions differ from cations in their surface activity because they are usually not hydrated. Just as is the case with a water–air interface where the anions are found preferentially at the surface, anions will move in toward the hydration sheath of the positively charged electrode and displace a water molecule. The plane of closest approach to these unhydrated anions is called the **inner Helmholtz plane (IHP)**. Whether an ion will leave the outer Helmholtz plane and move to a position on the inner Helmholtz plane depends on the free energy change associated with the jump from one plane to the other. If the ΔH and ΔS values for the ion–electrode, water–electrode and ion–water interactions are all considered, ΔG for the jump can be found. As Table 19.6-1 shows, ΔG for this jump is positive for Na^+ and K^+, while it is negative for Cl^- and I^-. Consequently, for these selected ions of biological importance, the anions will adsorb at the IHP while the cations

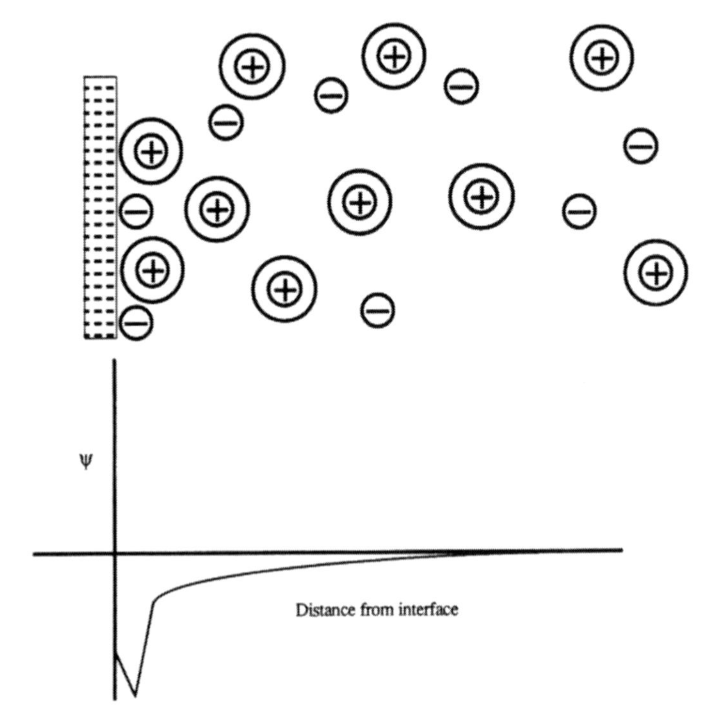

Figure 19.6-2 Structure of the interphase and the potential versus distance curve for a case of specific adsorption of anions on a negatively charged electrode.

will remain in the OHP. Interestingly enough, the specific adsorption of anions on the electrode surface occurs even when the electrode is negatively charged. The tendency of anions to be surface active, and therefore present at the inner Helmholtz plane when simple electrostatic forces would not predict their presence, leads to the interesting case shown in Figure 19.6-2.

These concepts of interphase structure are important in a large number of phenomena associated with biological processes. In the next chapters, the effects of this structure on physiochemical behavior of macromolecular structure and membranes will be explored.

CHAPTER 20

Electrokinetic Phenomena

20.1. The Cell and Interphase Phenomena

Although the cell can be thought of as a pair of phases, one intracellular and one extracellular, separated by a membrane, the membrane itself can be considered a phase. When the membrane is conceived as a phase, the cell is described by an extracellular aqueous phase in contact with a lipid phase that contacts another aqueous phase. However, the intracellular environment should probably not be considered as a unitary aqueous phase except in the most unusual of circumstances. The structure of virtually all cells, except the mature mammalian red cell, is comprised of multiple membranes and intracellular organelles packed densely together. Each membrane has associated with it two interphase regions, one on each side, representing its association with the phases. The nature of the interphase region that is associated with biological membranes is an extremely important aspect of the surface chemistry in biological systems and is also exceedingly complex.

The exact role of the interphase in biological systems is still under investigation, but it surely plays an important role in the transfer of charge in the electron transport chain, the behavior of cell–substrate interaction, the adsorption of hormones and other chemical messengers on the cell surface, and the process of generating, maintaining, and altering the membrane potentials in both excitable and nonexcitable cells, to name just a few examples. The structure of the double layer region gives rise to a number of important effects associated with motion relative to the double layer, called **electrokinetic effects**. These will be discussed next.

20.2. Electrokinetic Phenomena

Charge located in the compact layer is bound tightly and, like the irrotational water, is essentially part of the same kinetic entity as the interface. In other words, it will move in lockstep with the movement of the interface. On the other hand, charge found in the diffuse layer is not held as tightly to the surface. In a sense, the ions making up the countercharge in the Gouy–Chapman diffuse layer have split loyalties to the interphase region and the bulk solution. When the solution phase and interphase are at rest (relative to one another), the ions making up the diffuse layer are held by the forces from the electrified interface (moderated of course by thermal randomizing forces). However, if either the solution or the surface move relative to the other, the ions located in the diffuse layer can be torn away from the interphase region and move with the bulk phase. What are the consequences of removing this charge from the interphase region? First of all, a portion of the charge that was needed to neutralize the field associated with the electrified interface is now gone. If a measurement were made across the complete interphase region, the interphase would no longer be electrically neutral. Furthermore, the charge that was removed from the diffuse region is now associated with the bulk phase; it is also no longer electroneutral. When charge is separated, a potential difference is the result. The motion of the bulk phase with respect to the surface therefore has resulted in a new potential field, the **electrokinetic potential.**

The electrokinetic potential is named the **zeta potential** and given the symbol ζ. The magnitude of ζ is found by measuring the potential difference between the portion of the diffuse layer that can be removed by a shearing force and the bulk of the solution (which represents the reference zero or reference ground). The entire diffuse layer is generally considered to be able to be removed under a kinetic shearing force, so the shear plane is considered to be located just outside the outer Helmholtz plane. The zeta potential therefore has the same value as ψ at the OHP. Knowledge of ζ can be used to indirectly infer some information about the surface charge excess and the compact layer. However, caution must be used, since the potential at the OHP is by no means specific for a particular interface structure, or even for the charge on the surface, as Figure 19.6-2 indicates.

There are four electrokinetic phenomena. Two of these occur when an electric field is applied to the system and cause movement of either the particle or the solvent associated with the double layer; these are **electrophoresis** and **electro-osmosis**, respectively. The other two phenomena, the **streaming potential** and the **sedimentation potential**, are instances of the generation of an electric field when the solvent or the particle, respectively, generates a potential secondary to the shearing forces during its movement. Each of these phenomena will be summarized briefly below.

When a pressure differential is applied to an electrolyte (or any fluid) in a pipe or capillary, the fluid will move through the opening with laminar flow. As the electrolyte begins to flow, its motion is opposed by the viscous force of the solution. The velocity of flow will increase until the force exerted by the pressure gradient is opposed equally by the viscous force. The velocity of flow for any given lamina increases with the distance away from the wall of the vessel. The lamina of fluid nearest the surface of the capillary will not move and will remain fixed to the capillary wall. This is consistent with the interpretation that this layer is represented by the solution phase inside the immobile outer Helmholtz plane. The first lamina of solution that is free to slip past the immobile lamina includes a portion of the diffuse Gouy–Chapman layer. The movement of the volume elements that contain Gouy–Chapman charge leads to the condition where the flowing lamina of electrolyte carries a charge. A separation and accumulation of charge occurs, and as a result, an electrical potential is generated. This voltage is the **streaming potential.** If the amount of charge drawn away from the double layer is measured as it moves through a plane normal to the surface, then a streaming current can be found. The presence of an electric field in the electrolyte causes a current to flow in the opposite direction of the streaming current. At steady state, the streaming current and the conduction current that opposes it are equal but opposite in sign. The streaming potential, E, is found to depend on the viscosity (η) and conductivity (κ) of the electrolyte, the pressure gradient (P) in the capillary, and the zeta potential (ζ), which is usually considered to be synonymous with y at the OHP. The relationship can be written

$$E = \frac{\zeta\epsilon_o\epsilon P}{\eta\kappa} \tag{20.2-1}$$

The streaming potential will be expressed in volts if the zeta potential is also expressed in volts.

Electro-osmosis is essentially the paired opposite of the streaming potential phenomenon. A similar system to that described for the streaming potential is considered. If an electric field is applied parallel to the surface, there will be an induction of charge transport through the electrolyte solution. The charge will be carried principally by the ions in the solution, including some of the ions that are involved in providing the diffuse Gouy–Chapman layer at the surface. As the ions are stripped from the diffuse layer and experience net flux in the parallel electric field, they carry water with them. The movement of water and ions results in a net flux of solvent whose flow is countered by a viscous force identical to that experienced when the driving force is a pressure differential. Consequently, the electrical potential gradient behaves like a gradient of pressure in a capillary or pipe causing net flow. The mathematical relationship for a pressure differential causing current to flow (the streaming potential) and that for a current causing a net flux of fluid (electro-osmosis) can be shown to be equivalent, thus demonstrating the reciprocity of these two processes. The reciprocal equivalence of these phenomena is important because it fulfills the Onsager reciprocity relationship for irreversible processes. The relationship of the electric field to the volume of fluid flow (V) is described by

$$V = \frac{\zeta\epsilon_o\epsilon I}{\eta\kappa} \tag{20.2-2}$$

where V is the volume transported in unit time, and I is the current flowing in the test system. The electro-osmotic mobility, μ_E, is defined as the velocity of flow of the solvent with respect to the external field and is given as

$$\mu_E = \frac{v}{E} = -(4\pi\epsilon_o)\frac{\epsilon\zeta}{4\pi\eta} = -\frac{\epsilon_o\epsilon\zeta}{\eta} \tag{20.2-3}$$

The negative sign indicates that the flow of solvent is toward the electrode of the same polarity as the zeta potential. This occurs because the charge in the diffuse layer is opposite in sign to the potential at the OHP. Electro-osmotic processes are important in the behavior of ion-exchange membranes and may well play important roles in behavior across biological membranes as well.

Perhaps for the biochemist the most familiar phenomenon associated with the electrified interface is the process of **electrophoresis**. In electrophoresis, a particle must be large enough to have a double layer; most macromolecules, organelles, and cells fit this requirement. Electrophoresis is like electro-osmosis in that an external electric field causes relative motion of the surface and the diffuse layer; it differs in that the solvent is now stationary and the particles move. Consider a system of small particles (just how small will be considered shortly) in solution that are subjected to an electric field. Streaming of the liquid secondary to electro-osmotic effects is prevented. The particles are found to move in the field. Shortly after the application of the electric field, a steady-state velocity of the particles is established with the particles moving toward one of the electrodes. The particles move because of the electrostatic interaction between the field and their own charge. They reach steady-state velocity when the electrostatic forces are opposed equally by the viscous forces associated with their movement. The velocity of the particles with respect to field strength is the electrophoretic mobility, μ, and has units of $m^2\,s^{-1}\,V^{-1}$:

$$\mu = \frac{v}{E} \tag{20.2-4}$$

Electrophoresis refers to particles that are large enough to have an electrified interface associated with them. Most quantitative analyses are derived based on the condition that the particle is large compared to the dimension of the diffuse double layer as given by κ^{-1}, which makes the conditions similar to those for the derivation of the electro-osmotic effect.

Table 20.2-1. Electrophoretic mobilities for selected colloids at pH 7.0.

Colloidal species	Mobility ($m^2 s^{-1} V^{-1} \times 10^{-8}$)
Human erythrocytes	−1.08
Human platelets	−0.85
Colloidal gold	−3.2
Oil droplets	−3.1

The particle and the compact layer of the electrified interface move in concert as a kinetic entity, and consequently it is once again the potential at the OHP or ζ that determines the charge associated with the particle. This leads to several interesting effects. First, since the diffuse layer has a sign opposite to that of the particle it will be attracted to the opposite electrode. Since the diffuse layer of charge wants to stay associated with the particle (which is being accelerated in one direction), yet is itself being accelerated in the opposite direction, the ion cloud opposes the motion of the particle. Ultimately, it is the particle that experiences net motion, but at the cost of decreased mobility. Like ions in an electric field, the particle does not actually drag its diffuse layer with it but leaves a portion behind and continually rebuilds the diffuse layer in front as it moves. Second, although most particles of biological interest (e.g., cells, proteins, etc.) have a net charge, a particle with no net charge can also experience electrophoretic effects, because the electrophoretic mobility is determined by the charge at the OHP, and the specific adsorption of ions at the IHP (usually anions) can lead to a net charge at the OHP. The negative mobilities of colloidal gold and oil droplets as shown in Table 20.2-1 are examples of this effect.

The derivation of a simplified expression for electrophoretic behavior can provide a glimpse of some of the complexities in a biological system. The starting point is highly simplified. A system is imagined where a spherical charged particle can have its electrified interface switched on and off and the solvent can be switched from a nonconducting solution to a conducting electrolytic one at will. Starting initially in the nonconducting solvent without an electrified interface, the particle is subjected to an electric field, E. The force experienced by the particle is given by

$$F = z_i e_o E \tag{20.2-5}$$

The velocity of the particle, v, depends directly on the electrostatic force and inversely on the frictional coefficient of the particle. Therefore, the following can be written:

$$v = \frac{z_i e_o E}{f} \tag{20.2-6}$$

Earlier, the electrophoretic mobility, μ, was defined as v/E; this can be combined with Equation (20.2-6) to give

$$\mu = \frac{z_i e_o}{f} \tag{20.2-7}$$

Since the particle is spherical, the frictional coefficient may be calculated from Stokes' law, and Equation (20.2-7) may be rewritten

$$\mu = \frac{z_i e_o}{6\pi\eta r} \tag{20.2-8}$$

where r is the radius of the particle. This is a general equation for the simplified system described. In the more realistic case, it will be found that two approaches to the same problem give two answers. Understanding this apparent discrepancy leads to some important insights.

The switch is thrown and the solvent becomes a conducting electrolyte and the electrified interface is allowed to form. What will have changed? First of all, an electrified double layer exists that contributes its own field to that of the external field. Use of ζ rather than the charge $z_i e_o$ on the particle will yield more accurate results. Second, the size of the diffuse layer and the contribution it makes to the apparent viscous drag on the particle will vary. As the ratio of the reciprocal length to the radius of the particle changes, the importance of the drag contribution by the diffuse layer will vary. It has been indicated that electro-osmosis and electrophoresis share fundamental similarities. One of the most important is the derivation of the electro-osmotic and electrophoretic mobilities. It can be shown that if the radius of the particle is large relative to the reciprocal length of the diffuse layer, then the electrophoretic and electro-osmotic mobilities differ only in sign. The Smoluchowski equation for electrophoretic mobility gives the result

$$\mu = (4\pi\epsilon_o)\frac{\epsilon\zeta}{4\pi\eta} = \frac{\epsilon\epsilon_o\zeta}{\eta} \tag{20.2-9}$$

In 1924, Hückel derived an expression for the electrophoretic mobility which gave the result

$$\mu = \frac{2\epsilon\epsilon_o\zeta}{3\eta} \tag{20.2-10}$$

These are two significantly different answers for the same problem. Interestingly, it turns out that both are correct for a specific set of conditions. How can this apparent discrepancy help in the understanding of the processes in electrophoresis?

It is entirely feasible to derive an equation that sets the electric force on the particle in this system equal to the viscous force when the particle is at steady-state conditions. In like fashion to Equation (20.2-8), the viscous force, f_v, can be found through Stokes' law:

$$f_v = 6\pi v a \eta \tag{20.2-11}$$

where v is the velocity of the particle and a is the particle radius. The electrical force, f_e, is given simply as

$$f_e = QE \tag{20.2-12}$$

Q is the charge and E is the electric field. Combining these equations in a manner analogous to that described for Equations (20.2-6) through (20.2-8) gives the following result for μ:

$$\mu = \frac{Q}{6\pi a \eta} \tag{20.2-13}$$

Q is evaluated to take into account the charge on the double layer and the radius of the particle, including the compact layer, and is written

$$Q = (4\pi\epsilon_o)\epsilon a\,(1 + \kappa a)\zeta \tag{20.2-14}$$

Combining Equations (20.2-13) and (20.2-14) gives the following result:

$$\mu = \frac{(4\pi\epsilon_o)\epsilon\zeta}{6\pi\eta} \tag{20.2-15}$$

The term κa relates the radius of the particle and the thickness of the diffuse layer. In the limit where κa becomes very small (small particles with a relatively thick double layer), then Equation (20.2-15) reduces to

$$\mu = \frac{2\epsilon_o\epsilon\zeta}{3\eta} \tag{20.2-16}$$

whereas in the other limit, as ka approaches infinity (large particles with very thin double layers), the equation becomes the same as Equation (20.2-9). The Smoluchowski equation is derived with similar assumptions as the electro-osmosis mobility equation, and the treatment therefore assumes that the major force retarding the movement of the particles is derived from streaming of the diffuse layer opposite the movement of the particle, also called **electrophoretic retardation**. Hückel's treatment assumed conditions such that the value of κa was small and the major retarding force was the frictional resistance of the medium rather than electrophoretic retardation. The various conditions can be taken into account by applying a correction factor deduced by Henry which allows Equation (20.2-15) to be written

$$\mu = \frac{2\epsilon_o\epsilon\zeta}{3\eta}\,[f(\kappa a)] \tag{20.2-17}$$

where $f(\kappa a)$ is **Henry's function**. The function $f(\kappa a)$ depends on particle

Table 20.2-2. Selected values for Henry's function.

κa	$f(\kappa a)$
0	1.0
1	1.025
10	1.26
100	1.46

shape and, for a sherical particle, varies between 1.0, when $f(\kappa a)$ is 0, and 1.5, when $f(\kappa a)$ is ∞. Table 20.2-2 gives some values for Henry's correction function.

This treatment shows that small particles of diverse size and double layer structure can behave quite differently when conditions are changed just slightly. It is a valuable lesson to see how the size of a particle and its related double layer must be taken into account in understanding the behavior of many molecules of biological importance. Finally, it is worth noting that while the dielectric constant and viscosity of the medium must all be taken as equal to the bulk solution values in these derivations, it is likely that this is a gross simplification. For all of these reasons, in studies of electrophoretic mobility and in analytical techniques involving electrophoresis, most researchers use a constant set of conditions and reference their results to those conditions. However, before the role of electrophoretic forces in cellular systems can be quantitatively understood, most of these assumptions would need to be extended.

The last of the electrokinetic phenomena is the **sedimentation potential** (**Dorn effect**). Mention of it is included here for the sake of symmetry and completeness. When a particle moves in a stationary solvent, a potential will arise, again due to the separation of the charges, in this case derived from the diffuse layer surrounding the particle. Sedimentation potentials are rarely studied, and their measurement and interpretation can be fraught with significant difficulty. Consideration of the effects of sedimentation potentials may be of some importance in ultracentrifugation studies of sedimentation rates.

CHAPTER 21

Colloidal Properties

21.1. Colloidal Systems and the Electrified Interface

Electrokinetic effects are generally considered in the framework of interactions between an electrolyte solution and an insulating surface or in colloidal systems, but since they are derived from the mechanical interactions with the double layer, they may be found acting at any electrified interface including metal–electrolyte systems. Macromolecules and organelles in the cell have many of the characteristics of colloidal systems, and the structure and behavior, for example, of the cytosol of the cell is very much affected by the interphase properties of the electrified interface. The electrokinetic effects are not only restricted to very small systems. For example, the flow of blood past the surfaces of the blood vessel walls is capable of producing a streaming potential. Whether this potential has a role in the normal development or in pathologic processes of the vessel wall remains to be determined.

It is worth remembering that colloidal dispersions are comprised of particles whose size is on the order of 10^{-6} to 10^{-9} meters, and consequently an enormous surface area per unit volume is exposed. The large surface area means that surface behavior is a powerful and frequently dominant property of these systems. The behavior of colloidal systems must be interpreted to a large extent on the properties and phenomena deriving from the electrified interface, such as surface and interfacial tensions, and the electrokinetic potential.

Colloids are frequently categorized as either lyophilic or lyophobic. Lyophilic refers to colloids that are considered "solvent-loving" and lyophobic to those that are "solvent-hating." **Lyophobic colloids** form systems that are called sols, and **lyophilic colloids** form systems called gels. Many aspects of cellular and biochemical systems involve the behavior of sols and gels. Gels have in fact already been extensively covered in this text, since solutions of macromolecules are three-dimensional gels. Many lyophilic colloidal materials can be formed into two-dimensional networks or matrices. These gels are comprised of intertwined lyophilic molecules in a continuous mass with fine pores dispersed throughout the gel, constituting a membrane. The behavior associated with the pores in these membranes is greatly dependent on double layer effects (as is the case with electro-osmosis in ion-exchange membranes). The cytosolic dispersion of organelles and suspensions of many cells including bacteria have properties similar to those of sols. A knowledge of colloidal systems will often help in understanding cellular and biochemical behavior.

Lyophobic colloids, such as colloidal gold or droplets of lipid in an aqueous solvent, exist as a dispersion of particles suspended in the solvent. The formation of micelles as described in Section 13.3 form dispersion of colloidal ions. Interactions of cells may be reasonably considered using these colloidal particles as models. It may be correct to treat organelles in the cytosol with the same model, although the properties of the cytosolic "solvent" are much less clear in this case. The particles prefer not to associate with the solvent and, if conditions are right, will associate with one another, and flocculation or coagulation occurs. Whether a lyophobic colloid exists in a dispersed or flocculated state depends greatly on the presence of the double layer. What are the forces acting on a pair of identical lyophobic particles as they approach one another in the dispersed state? Since each of the particles is surrounded by an identical electrified interphase region, it is this charge layer that each particle sees as it approaches the other. The force that exists between the two particles is an electrostatic one, and, since the charges are the same, the force will cause the particles to repel one another. The potential energy of the interaction will increase progressively as the double layers of the particles approach one another. Because it is the two double layers that are interacting, the electrostatic force is approximated by

$$F_{\text{double layer}} = U_{\text{electrostatic}} = \psi_{o} e^{-\kappa x} \tag{21.1-1}$$

A graph of this relation is shown in Figure 21.1-1.

As the particles approach one another, other interactive forces come into play, given by a curve like the Lennard-Jones potential (cf. Chapter 10). Earlier, this curve was described as including interactions due to the attractive van der Waals forces and repulsion of the electron cloud:

$$U_{i} = -\frac{A}{r^{6}} + \frac{B}{r^{12}} \tag{21.1-2}$$

The Lennard-Jones potential curve is redrawn in Figure 21.1-1. The complete potential energy curve for the interaction of the lyophobic particles is given by combining the electrostatic interactions with the Lennard-Jones potential:

$$U_{\text{colloid}} = \psi_{o}\, e^{-\kappa x} - \frac{A}{r^{6}} + \frac{B}{r^{12}} \tag{21.1-3}$$

As can be inferred from Figure 21.1-1, when the thickness of the double

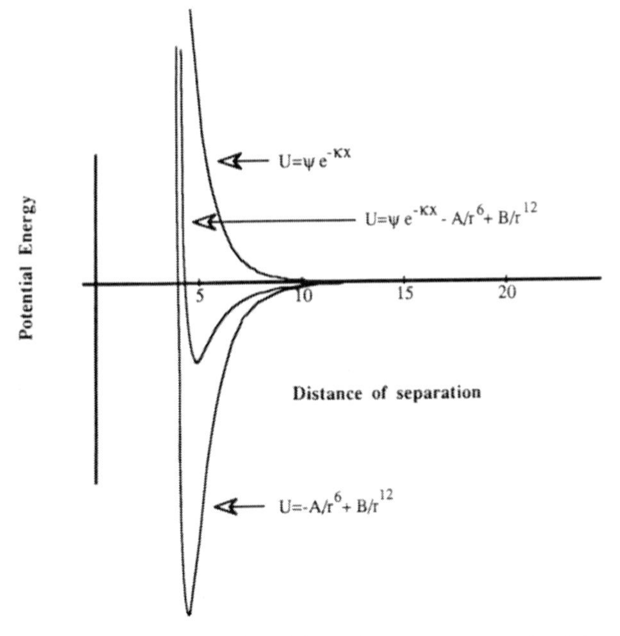

Figure 21.1-1 Plot of the interactional force between two colloidal systems, each having a double layer of identical charge. The top curve represents the energy of interaction of the double layers. The bottom curve is a Lennard-Jones potential curve for the system, and the middle curve is the interactional energy that results from the superposition of the upper and lower curves.

layer is sufficient, the potential energy curve of interaction will not favor the approach of the particles. They will stay dispersed under these conditions.

However, the thickness of the double layer can be made to vary by altering the ionic strength of the electrolyte, and the electrostatic potential curve can be moved. As can be seen from the family of curves in Figure 21.1-2, increasing the ionic strength, reflected by changes in κ^{-1}, moves the double layer in toward the surface of the particle and leads to a significant change in the interactional potential energy with respect to separation distance. As the double layer becomes more compressed, the van der Waals attractive forces become more important, and the potential energy for the flocculation of the dispersed particles becomes favorable. Hence, by changing the thickness of the double layer, the behavior of the colloidal dispersion is affected. Changing the ionic strength can be accomplished both by changing the concentration of a particular type of electrolyte, for example, increasing or decreasing the concentration of a 1:1 salt, or by changing the types of electrolytes that go into making a solution, for example, changing from a 0.10 *M* solution of a 1:1 electrolyte to a 0.10 *M* solution of a 1:2 or 2:2 electrolyte.

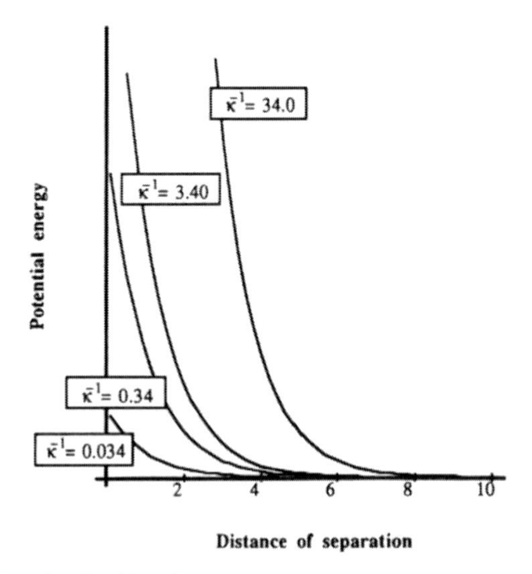

Figure 21.1-2 Plot of a family of curves relating the double layer thickness to the interactional energy with respect to distance.

Manipulations of the double layer thickness by varying concentration can be cumbersome, and that is probably not the method of choice in a biological system. Another type of manipulation of the electrified interface can lead to significant changes in the interactive potential energy curve. Specific adsorption of ions at the inner Helmholtz plane can lead to a reduction in the magnitude of ψ_o, and hence the potential energy of interaction can be made to favor flocculation through the addition of these ions to the system. These alterations are illustrated in Figure 21.1-3.

21.2. Salting Out Revisited

It is now instructive to reflect on the precipitation of a colloid from solution. The precipitation of a lyophilic colloid was discussed in Chapter 14. In the discussion of solutions of ions, it was shown that when the activity of the aqueous solvent was reduced by the addition of electrolyte to a solution of macromolecules (a gel), an oversaturated solution was produced with the subsequent precipitation of the macromolecules. This process is mechanistically explained by the structural consequences of removing the hydration layers from the colloidal material. Relatively large amounts of electrolyte must be added to the gel to salt out the protein. In this chapter, it has been shown that the addition of electrolyte can also lead to the precipitation or salting out of a lyophobic colloid, a

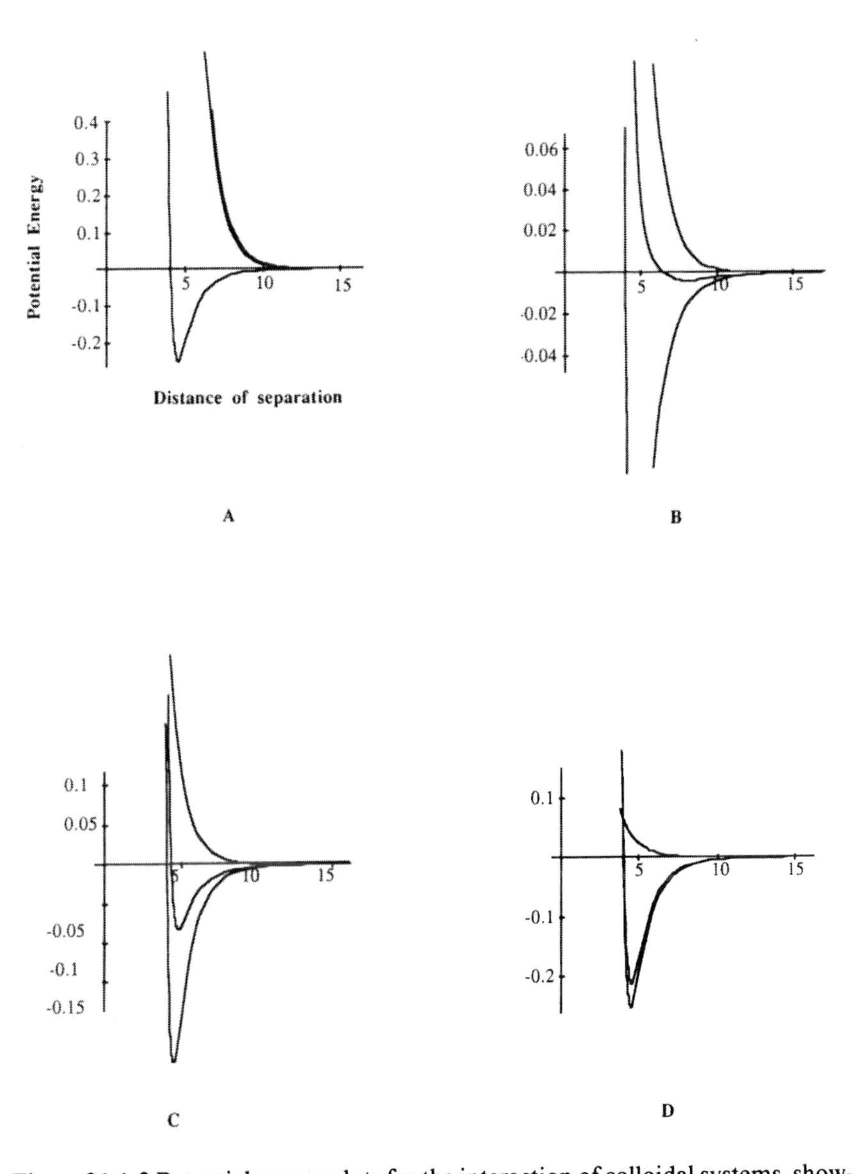

Figure 21.1-3 Potential energy plots for the interaction of colloidal systems, showing alterations that are caused by the adsorption of high-valency ions at the inner Helmnoltz plane. In each panel, the bottom curve is a Lennard-Jones curve, the top curve is the interactional energy between the double layers, and the middle curve is the superposition of the upper and lower energies. In panel A, ψ_o is given an arbitrary magnitude of 100. The interactional energy corresponds to a case where there is no IHP neutralization of charge, and the electrostatic field is quite strong for a significant distance into the interphase region. The colloidal systems are therefore unable to have an attractive interaction. In panel B, the adsorption of counterions at the IHP has decreased ψ_o to a magnitude of 20, and a quite small potential energy minimum is found. Panels C and D represent the reduction in ψ_o to 5 and 1, respectively, and show dramatically the effect that increasing the valency of the counterion can have on the interactional energy of the system. As the electrostatic field is reduced in intensity, the attractive interactions at close distance become much more favorable and association can easily occur.

sol. However, now the salting out occurs by changing the interactional energies that result from the double layer structure. Relatively small amounts of electrolyte are required for these effects. The point to be remembered is that biological systems are comprised of both sols and gels, often existing side by side. The differences in behavior stem to a great degree from the differences in interaction with the aqueous solvent. In the case of the sol, water avoids interaction with the colloid for the reasons described in earlier chapters, and so the only forces acting to keep the sol dispersed are the electrical forces derived from the interactions of the double layers. For the lyophilic colloids, the water associates to a much higher degree with the solute, and stabilization of the dispersion occurs because of forces above and beyond those associated with the electrified interface. It is clear that knowledge of the relationships described here, though complex, is vital to a complete understanding of the cell. Research in this area will continue to be a frontier of great biological importance.

CHAPTER 22

Forces Across Membranes

22.1. Energetics, Kinetics, and Force Equations in Membranes

This book will conclude with a discussion of the forces operating at the biological membrane. Much of the chemistry and functional physiology of the cell, including energy production, protein synthesis, hormone and antigen binding, stimulus–response coupling, and nutrient adsorption, occurs at the cell membrane. The membrane is an anisotropic nonhomogeneous matrix of lipids, proteins, and, in some cases, carbohydrates that is in intimate contact with aqueous-dominated interphases. The nature and treatment of aqueous solutions, membrane properties, and interphase structure have all been presented in previous chapters, as well as the thermodynamics and mechanics of both equilibrium and transport phenomena. The task ahead is to highlight this knowledge by examining some limited aspects of the behavior of cells with respect to the role of their membranes. This chapter will focus first on describing the forces operating at and across the membrane, and then will examine the role of the membrane in modulating the flow of material charge.

22.1.1. The Donnan Equilibrium

The determination of the equilibrium in a homogeneous solution of several components and the approach to equilibrium via diffusion and conduction have been discussed. The addition of a semipermeable membrane will have the effect of creating a heterogeneous system in which the equilibrium between two phases can be considered (Figure 22.1-1). This membrane is freely permeable to simple electrolytes such as Na^+ or Cl^- but is impermeable to a colloidal electrolyte. In this example, a protein of molecular weight 82,000 daltons will be the polyelectrolyte to which the membrane is impermeable. Initially, aqueous solutions of only NaCl are added to each side. Since the membrane is freely permeable to each component of the solutions, the chemical potential of each component

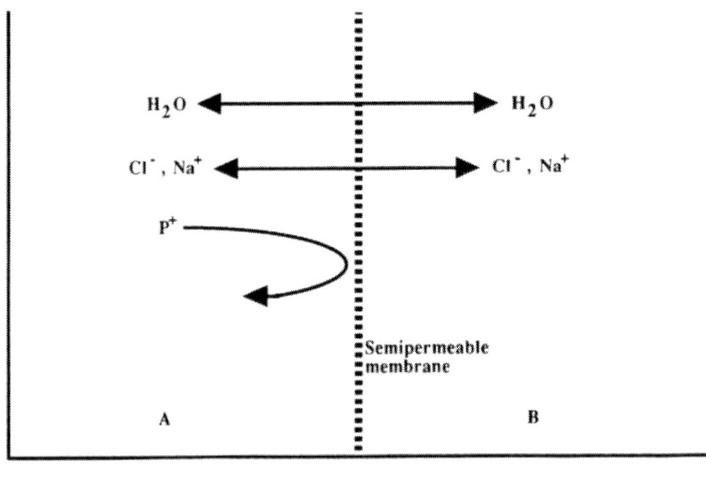

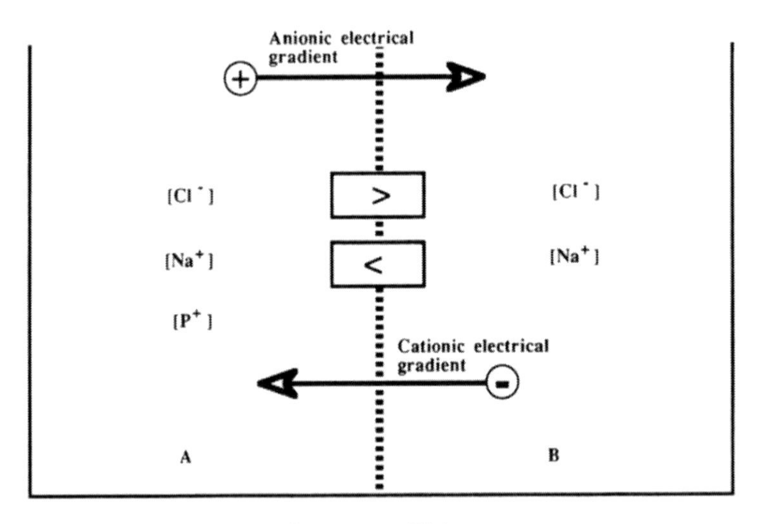

Figure 22.1-1 The Gibbs–Donnan equilibrium describes the case of two phases separated by a semipermeable membrane that allows free passage of H_2O and ions, but not macromolecular polyelectrolytes.

in each phase will be equal at equilibrium, even if the initial concentrations of these solutions are different. The following equations can be written to represent the equilibrium condition:

$$\mu^{A}_{Na^+} = \mu^{B}_{Na^+} \tag{22.1-1}$$

$$\mu^{A}_{Cl^-} = \mu^{B}_{Cl^-} \tag{22.1-2}$$

$$\mu^{A}_{H_2O} = \mu^{B}_{H_2O} \tag{22.1-3}$$

Assuming ideality in this example, it can also be written that the concentrations of Na^+ and Cl^- are equal in each compartment; this fulfills the requirement for electroneutrality:

$$c^{A}_{Na^+} = c^{A}_{Cl^-} = c^{B}_{Na^+} = c^{B}_{Cl^-} \tag{22.1-4}$$

The equilibrium is now perturbed by the addition of the polyelectrolyte to side A. Since virtually all biological macromolecules have some polyelectrolyte nature, this case is a thoroughly general one. The protein will have a net cationic charge. Through the use of a thought experiment, the charge on this protein may be switched on or off at will. Initally, the charge is turned off. What is the effect of the addition of the uncharged protein? The addition of the protein to side A leads to a fall in the mole fractions of Na^+, Cl^-, and H_2O with a concurrent drop in their activities. The osmotic pressure on side A necessary to raise the activity of the permeable components is now higher because of the addition of the protein. If the system is allowed to proceed toward equilibrium, there will be a shift in the concentrations of the permeable components just as was described in the case of the equilibrium dialysis (Appendix 7.1). Ultimately, there will be a movement of water, Na^+, and Cl^- toward side A so that the chemical potentials in each phase are once again the same.

Now, the charges on the protein are switched on. What is the effect? The phases are no longer at equilibrium because electroneutrality has been violated. Side A now contains a charge that is not able to freely partition between the two solutions. What is the effect on the final equilibrium for this system? In this case, the presence of an unbalanced cationic charge on side A makes it electrically positive with respect to side B. There is an electrical driving force caused by the presence of the polyelectrolyte, and also a chemical potential driving force to push the macromolecule down its concentration gradient. The electric field causes the repulsion of the permeable cations (Na^+) from side A and attracts the permeable anions (Cl^-) from side B. The equilibrium direction attempts to reestablish electroneutrality at the expense of establishing a chemical potential gradient. As the Na^+ and Cl^- ions accumulate in different concentrations across the membrane in response to the electric field, a chemical potential difference is generated where [in terms of Equation (22.1-4)]:

$$c^{A}_{Na^+} < c^{B}_{Na^+} \tag{22.1-5}$$

$$c^{A}_{Cl^-} > c^{B}_{Cl^-} \tag{22.1-6}$$

Finally, at equilibrium, there will be offsetting electrical and chemical potentials. Although ΔG will be 0, there will be a persistent electrical gradient driving the anions to move by conduction to side A and for the cations to move by conduction to side B. On the other hand, there will be a persistent chemical potential gradient driving the anions to move by diffusion to side B and the cations to move by diffusion to side A. This electrochemical gradient is called the **Gibbs–Donnan** or **Donnan potential.**

The mathematical derivation of the Gibbs–Donnan effect is as follows. Ideality is assumed in this derivation. The salt is denoted as M^+N^-, the protein as P^+, and the charge on the protein as z_i. The requirement for chemical potential equality is given by

$$\mu^{A}_{M^+}\mu^{A}_{N^-} = \mu^{B}_{M^+}\mu^{B}_{N^-} \tag{22.1-7}$$

The equation for the electroneutrality condition on side A is given by

$$z_i c^{A}_{P^+} + c^{A}_{M^+} = c^{A}_{N^-} \tag{22.1-8}$$

The equation for electroneutrality on side B is given by

$$c^{B}_{M^+} = c^{B}_{N^-} \tag{22.1-9}$$

Equations (22.1-7) to (22.1-9) are combined, with the following result:

$$(c^{B}_{M^+}) = (c^{A}_{M^+})\left(1 + \frac{z_i\, c^{A}_{P^+}}{c^{A}_{M^+}}\right)^{1/2} \tag{22.1-10}$$

and

$$(c^{B}_{N^-} = (c^{A}_{N^-})\left(1 - \frac{z_i\, c^{A}_{P^+}}{c^{A}_{N^-}}\right)^{1/2} \tag{22.1-11}$$

This is the mathematical equivalent of the thought experiment above. There will be a higher concentration of the cation on side B and a higher concentration of the anion on side A at equilibrium. The free energy secondary to the chemical potential gradient will be offset exactly by an opposite electrical potential gradient. The effect of the Donnan equilibrium on the osmotic pressure in the system is often important and can be significant in cellular systems.

22.1.2. Electric Fields Across Membranes

Biological membranes have interesting electrical properties that can be derived from measurements of the dimensions and electrical capacitance of the membranes. Biological membranes have a capacitance of approximately 1 μF cm^{-2} and are about 7.5 nm thick. The measured transmembrane potential is on the order of 0.050 to 0.100 V. The bilayer membrane can be considered roughly as a parallel plate capacitor. Therefore, the electric field strength across the membrane will be between 7×10^6 and

1×10^7 V m^{-1}. The cell membrane is capable therefore of sustaining a large potential without breaking down, a property quantitated by a measure called the **dielectric strength**. The dielectric constant of the membrane can also be estimated from the formula for capacitance of a parallel plate capacitor:

$$C = \frac{\epsilon\epsilon_o A}{d} \qquad (22.1\text{-}12)$$

where C is the capacitance, A is the cross-sectional area, and d is the distance separating the two plates. ϵ is the dielectric constant, and ϵ_o is the permittivity of free space. Using this formula and the values given above, the membrane has a calculated dielectric constant of approximately 8.

22.1.3. Diffusion Potentials and the Transmembrane Potential

Two NaCl solutions of different concentrations are brought into electrical contact either through a salt bridge or a membrane, and a voltmeter of very high impedance is used to measure the potential difference between two Ag/AgCl electrodes immersed in the solutions (Figure 22.1-2). A

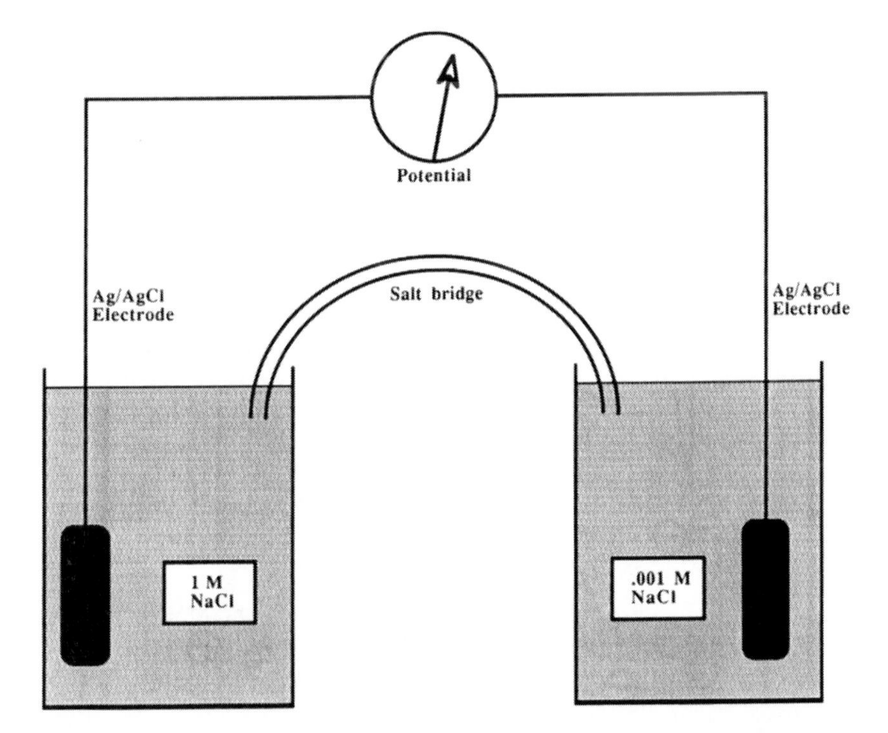

Figure 22.1-2 An example of a system generating a concentration potential.

potential will be found between the two solutions. The magnitude of the potential will be given by the Nernst equation:

$$E = -2.303 \frac{RT}{nF} \log \frac{c_2}{c_1} \tag{22.1-13}$$

The emf is generated by the concentration difference between the two solutions and represents the difference in chemical potential between them, as sensed by the electrodes, which in fact are responding to the difference in chloride ion concentration. A potential generated in this fashion is called a **concentration potential.**

A different kind of potential can be shown in a system where a semipermeable membrane that will allow the free exchange of water and ions is placed between two NaCl solutions of different concentrations and the ions are allowed to progress down their chemical potential gradients toward equilibrium. Each ion will move down its gradient toward equilibrium but, because the mobilities of Na^+ and Cl^- are different, the Na^+ ions will diffuse faster than the Cl^-ions. As a consequence, there will be a front of positive charge that moves ahead of the diffusing negative charge. There will be a charge separation across the membrane with the more concentrated solution negative with respect to the more dilute solution. A potential of this type, illustrated in Figure 22.1-3, is called a **diffusion potential.**

The diffusion potential depends on both the concentration difference between the two phases and the difference in mobilities of the ions carrying the charge:

$$E = -\left(\frac{\mu_+ - \mu_-}{\mu_+ + \mu_-}\right) 2.303 \frac{RT}{nF} \log \frac{c_2}{c_1} \tag{22.1-14}$$

When the mobilities of the anions and cations are identical, there will be no diffusion potential. The emf generated by a diffusion potential reaches a steady state because as the charge separation increases, the potential field that is generated between the positive and negative fronts causes the leading positive ions to be retarded and the lagging negative ions to be accelerated. The concept of the diffusion potential is important because it helps to explain the physical basis for the generation of the transmembrane potentials in biological systems.

22.1.4. Goldman Constant Field Equation

In biological systems, electrical potentials are generated across membranes because of a combination of chemical potential gradients that are established by active ion transport systems that are coupled with a differential permeability of the membrane to the movement or mobility of the ions down their chemical potential gradients. Although the molecular

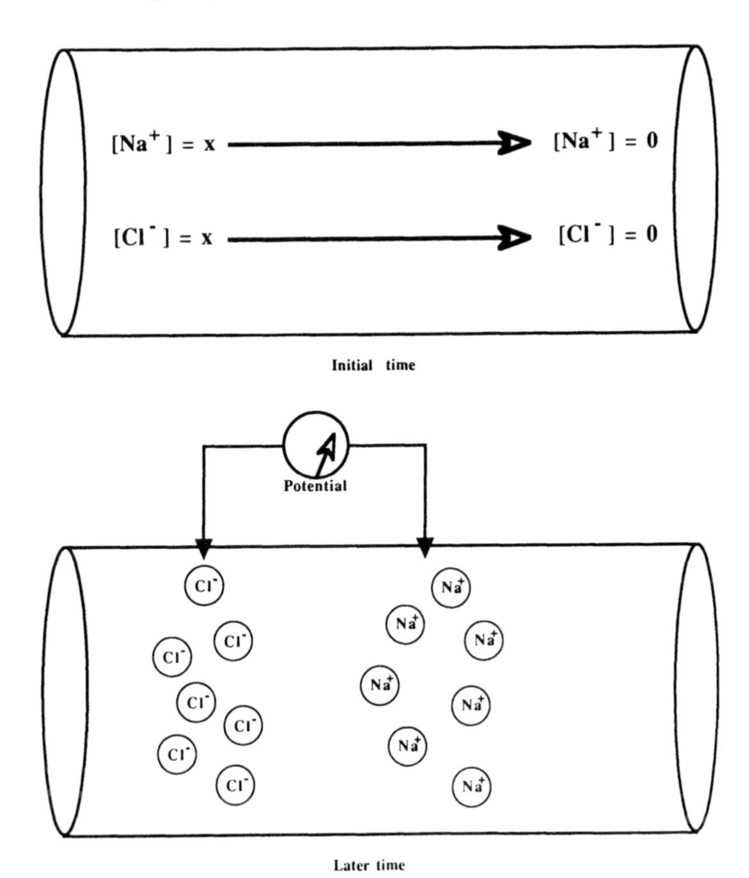

Figure 22.1-3 The diffusion potential results from the different mobilities of the ions as they move down a chemical potential gradient.

mechanism is not the same as described for the diffusion potential, this is a similar phenomenon and can be treated thermodynamically in much the same fashion. The Goldman equation is based on the principles of **electrodiffusion** and will be described. It is important to emphasize that this treatment is not without its problems and that other treatments, notably ones based on rate theory, are also currently under study. The rate theory approach will not be described here but may be found discussed in sources listed in Appendix I.

Equation (22.1-14), which describes electrodiffusion and is applicable for any freely permeable ion which distributes across the membrane according to the Gibbs–Donnan equilibrium, is the starting point for this analysis. The equation has been generalized for mammalian cells, which, in general, are much less permeable to cations than to anions. The principal membrane potential contributions come from the Na^+ and K^+ gradients, which are maintained by active pumps, while Cl^- and HCO_3^- anions have much larger membrane permeabilities and distribute according to the Gibbs–Donnan equilibrium. Generally, this is written

$$\frac{[Cl^-_{in}]}{[Cl^-_{out}]} = \frac{[OH^-_{in}]}{[OH^-_{out}]} = \frac{[HCO_3^-{}_{in}]}{[HCO_3^-{}_{out}]}$$

Neither H^+ nor Ca^{2+} ions contribute measurably to the plasma membrane potentials of such cells, even though membrane transport systems and pumps driven in part by these cations exist. The generalized equation, known as the **Goldman–Hodgkin–Katz constant field equation**, expresses the transmembrane potential at equilibrium in terms of the specific membrane permeabilities for each ion, P_i, and their intra- and extracellular concentrations, $c_{in}^{M^+}$ and $c_{out}^{M^+}$, respectively, for cations and $c_{in}^{N^-}$ and $c_{out}^{N^-}$ respectively for anions, and the change z_i on the macromolecular species of concentration, c_p. This treatment incorporates several assumptions including:

1) the partial permeability of the membrane to charged species,
2) the uniformity of charge distribution across the membrane,
3) the net equality of charge flux across the membrane,
4) the applicability of the Donnan equilibrium for each ionic species to which the membrane is permeable enough to permit free distribution across the membrane, and
5) the absence of an electrogenic pump and its resultant driving force.

The transmembrane potential, $\Delta\Psi$, can then be written:

$$\Delta\Psi = \frac{RT}{F} \ln \frac{\Sigma\, c_{in}^{N^-}}{\Sigma\, c_{out}^{N^-}} = \frac{\Sigma\, c_{out}^{M^+} + z_i c_p}{\Sigma\, c_{in}^{M^+} + z_i c_p} \tag{22.1-15}$$

The limited membrane permeability P_i for any species i is defined as

$$P_i = \frac{RT}{Fd} \mu_i b_i \tag{22.1-16}$$

where d is the thickness of the membrane, μ_i is the mobility of the ion in question, and b_i is a measure of the ease of distribution of the ion between the outside and the inside of the cell. The assumption is that as the thickness of the membrane varies, the permeability varies in a constant fashion. The Goldman–Hodgkin–Katz constant field equation is the result:

$$\Delta\Psi = \frac{RT}{F} \ln \frac{\sum_i P_i c_{out}^{M^+} + \sum_j P_i c_{in}^{N^-}}{\sum_i P_i c_{in}^{M^+} + \sum_j P_i c_{out}^{N^-}} \tag{22.1-17}$$

While this equation holds for most mammalian cells at or near physiologic conditions, it tends to fail when such cells are suspended in very nonphysiologic buffers (e.g., external $K^+ >> 50$ m*M*). The Goldman equation has a number of significant limitations, as well as useful extensions including to ions of other valences and electrogenic pumps. These details can be found discussed in references listed in Appendix I.

22.1.5. Electrostatic Profiles of the Membrane

The cell membrane now emerges as a lipid bilayer membrane with several predominant sources of electric field operating around and through it. There is a significant transmembrane potential that is generated from the unequal partitioning of ionic charge. In addition, there is a large amount of charge that is carried on the surface of the membrane. The surface charge has a magnitude of 10^{-1} to 10^{-3} C m^{-1} and is of mixed cationic and anionic charge groups. The net charge is anionic, with approximately 10^7 to 10^8 anionic charges per cell and 10^6 to 10^7 cationic charges per cell. The development of the electrified interphase has been examined in detail in Chapter 19. A model of the interface that is sometimes considered instead of the metal–electrolyte systems described in Chapter 19 is one that contains only ions as the charge carriers at the interface. This model is called the **interface of two immiscible electrolyte** solutions (ITIES). In this model, the electrolytes partition across the interface depending on a relative difference in their hydrophilicity. In such a model, the arrangement of the ions in the interphases on both sides of the interface is given as a Gouy–Chapman type diffuse layer. Given the adsorption effects of ions, organic molecules, and aqueous dipoles, this model does not fully describe the interphase near a cell membrane. A reasonable description would be expected to have elements of the Grahame model of the interphase incorporated into it. Neither a simple nor a complete model yet exists of the electrified interphase of cellular systems. A number of possible potential profiles across membranes are illustrated in Figure 22.1-4.

22.1.6. The Electrochemical Potential

It is apparent that the movement of molecules into and out of the cell or across a cellular membrane will depend on a combination of gradients. This book has focused on the chemical and electrical potentials, which are so tightly coupled to one another that the combined free energy of their gradients is called the **electrochemical potential**. Other forces are also important in the behavior of cells such as the osmotic pressure and surface tension. In the final section, a brief examination of the transport of a simple species, water, across the membrane will be considered.

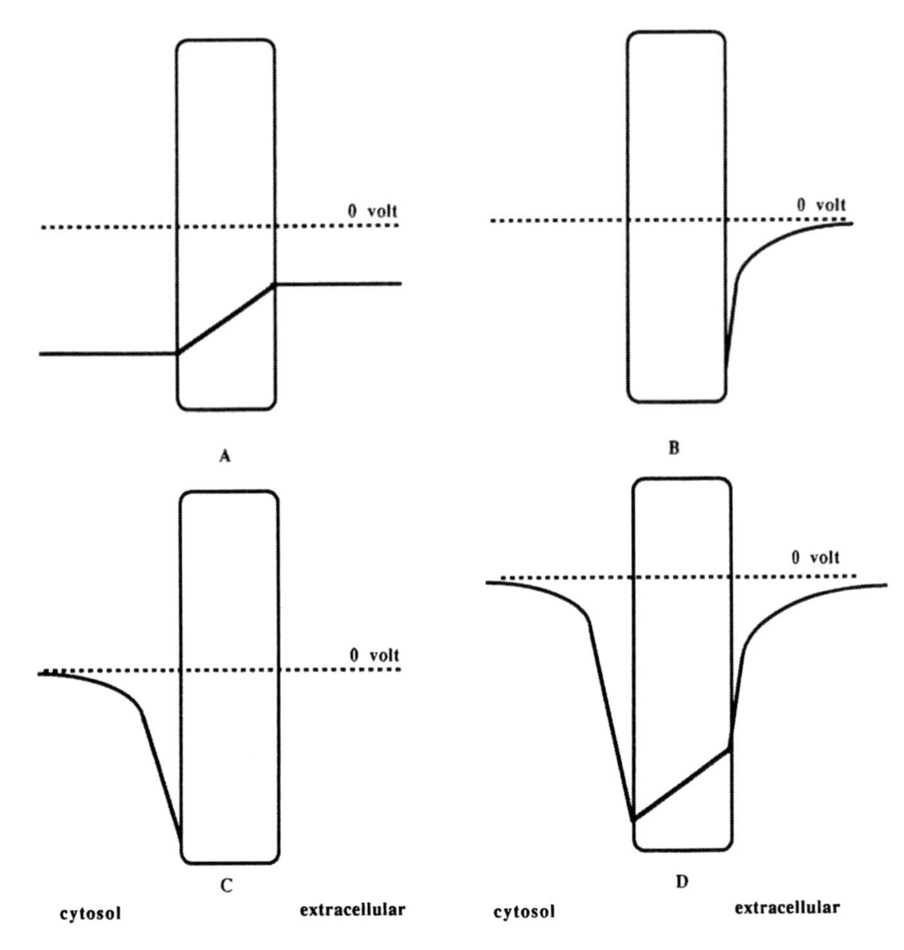

Figure 22.1-4 Model electrostatic profiles across membranes. In panel A, the electrostatic profile due strictly to the transmembrane potential is shown, and the membrane is treated as a dielectric continuum with $\epsilon = 8$; in panel B, the electrostatic profile due to the extracellular surface of the plasma membrane is drawn, based on the surface charge assumptions given in the text; in panel C, the profile due to the cytoplasmic side is shown; finally, in panel D, a total picture of the plasma membrane electrostatic picture is drawn based on the principle of superposition of each of the previous curves.

22.2. Molecules Through Membranes: Permeation of the Lipid Bilayer

The passage of any polar entity across a lipid bilayer is severely restricted and requires either the presence of pores or channels in the membrane or the existence of enough kinetic energy to drive the molecule through random (and fluctuatingly present) spaces between lipid molecules. In a mammalian membrane, transmembrane proteins can act as pores or as passive channels through which water, alone or in concert with other small molecules, can be brought across the membrane. The gradient for such passive transport will be dependent on the relative activity of H_2O in the extracellular milieu versus that in the cytoplasm. Under physiologic conditions, the driving force will clearly be inward, and, were it not for the osmotic pressure and the existence of counteracting forces attributable to cation gradients, the cell would swell and eventually burst. Therefore, in the overall free energy equation describing the movement of molecules through cell membranes, the terms involving pressure, chemical potential, surface area, and charge movement must all be considered.

Changes in cellular water content are a significant factor in the biological and clinical behavior of cells. Shrinkage of a cell due to net H_2O efflux can change the activity coefficient of the residual H_2O or change the degree of solvation of intracellular components, both cytoplasmic and membrane bound. In comparison to its permeability to other small solutes such as urea, the cell membrane as a whole is quite permeable to water. In the erythrocyte, the diffusion rate of water through the membrane is a factor of 10^2 times higher than that of urea, even though the membrane is highly permeable to urea. The exchange rate of water between the inside and outside of an erythrocyte is quite rapid, on the order of fractions of a second. In spite of the relatively high permeability of the membrane to water, a significant barrier does in fact exist. In the absence of a membrane, the relative diffusion rate of water is higher by a factor of 10^4 or 10^5.

Water can enter a cell through several mechanisms: passive diffusion directly through the lipid bilayer, passive diffusion through transmembrane protein channels or pores, or co-transport with other entities (ions or uncharged polar species). All three modes of permeation are gradient driven. Although water is co-transported across a membrane with many ions or neutral molecules via channels or carriers, it can be carried alone through any of these modes of transport.

Model systems for the passive permeation of water into intracellular spaces are readily developed. They take account of the fact that, of the four possible modes of membrane traversal by water (osmosis, pinocytosis, electro-osmosis, and vacuole contraction), osmosis is the predominant one. Implicit in all models for osmotically driven water permeation are the following assumptions:

1) The water flux is controlled at the membrane.
2) The membrane is semipermeable, passing water but no solutes.
3) The cell is at equilibrium with its surroundings.
4) The protein components of the cell interior (to which the membrane is impermeable) behave as if they were in a cell-free aqueous milieu.

The first two assumptions permit the use of synthetic liposomes as the model system for passive diffusion directly through the membrane lipid

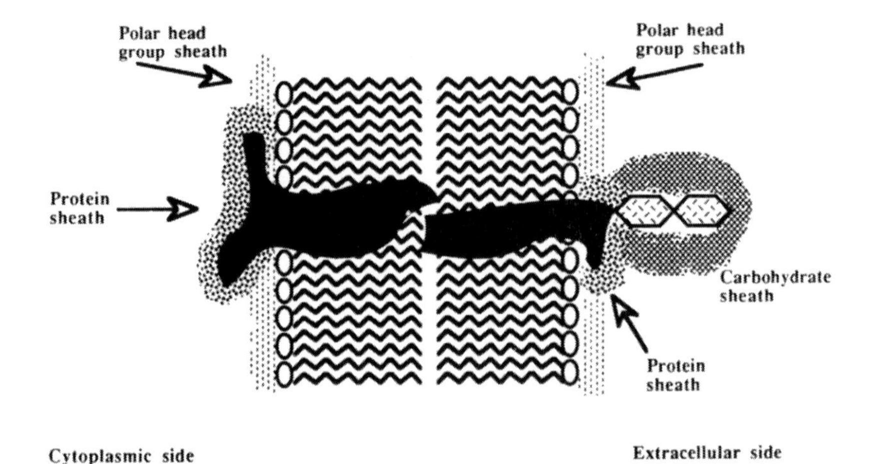

Figure 22.2-1 Regions traversed by a molecule traveling from the extracellular milieu across the cell membrane.

bilayer, whose lipids are in the crystalline or liquid crystalline state. Liposomes have a very low water permeability since the free energy requirement for forcing water through the hydrophobic portion of the bilayer is high. A molecule traveling from the bulk water in the extracellular milieu must, as indicated in Figure 22.2-1, traverses six barrier regions:

1) The irrotational layer or sheath solvating the carbohydrate ends of the membrane proteins.
2) The irrotational layer or sheath solvating the proteins themselves.
3) The irrotational layer or sheath solvating the polar head groups of the outer-layer phospholipids.
4) The nonaqueous lipid layer consisting of the fatty acid components of the outer and inner layers.
5) The irrotational layer or sheath solvating the polar head groups of the inner-layer phospholipids.
6) The irrotational layer or sheath solvating the proteins on the inner cytoplasmic face of the membrane.

The activation energy for this process through a phospholipid bilayer is approximately 37 kJ mol^{-1}. The process is thought to be facilitated by oscillations of the membrane structure, which create localized "holes" into which a water molecule can insert itself. The water permeation process is facilitated at higher temperatures where there is higher kinetic energy and a higher degree of kinking or *cis*-configuration formation. However, for a mammalian cell membrane containing embedded proteins, the activation energy has been found to be approximately 12 kJ mol^{-1}. If the cell is treated with *p*-chloromercury benzoate or another SH group blocking agent, the activation energy again becomes approximately 37 kJ mol^{-1}. This suggests that the easier water permeation through normal cell membranes is attributable to pores or channels that are —SH-reagent sensitive.

A dynamic equilibrium between a cell and its surroundings can be achieved, provided there are enough energy stores within the cell to maintain that equilibrium. Under such conditions, the cell volume remains constant, ATP is produced metabolically at a low maintenance rate, and the actual passive influx of Na^+ and efflux of K^+ ions is counteracted by the ATPase-driven pumps. As the energy stores of the cell are used up, ATP can no longer be adequately replenished. As a result, the Na^+/K^+ concentration ratio will be perturbed, the transmembrane potential will drop (become less negative), and the cell will swell as more H_2O, co-transported with passive Na^+ influx, enters. In the actual cell, the osmotic pressure and the transmembrane electrical potential act in concert with the chemical potentials of all the components to establish an equilibrium, that is, to establish conditions under which $\Delta G = 0$. However, the cell does not remain static, nor does it necessarily maintain the same values of the individual terms, even though overall equilibrium has been established.

The molecular components of the cell do not behave ideally, rendering the assumption of dilute solution thermodynamics as a model invalid in real systems. As discussed in earlier chapters, the aqueous molecules will exist in a number of phases:

1) As bulk H_2O.
2) As H_2O in the outer and in the inner Helmholtz layers solvating the phospholipid head groups, including both charged and uncharged polar species.
3) As H_2O associating with other aspects of microscopic membrane structure, including outer membrane surface proteins and glycoproteins.
4) As H_2O within the lipid bilayer (e.g., in a channel), whether free or bound to a co-transported molecule or ion.
5) As H_2O dipoles oriented in the inner and outer Helmholtz layers on the cytoplasmic side of the membrane.
6) As H_2O interacting with the various ions within the cytoplasm.
7) As H_2O solvating the cytoplasmic proteins.
8) As H_2O interacting with granule or other organelle membranes, etc.

It is clear that H_2O does not act ideally in any of these states, with the possible exception of the bulk external H_2O where its activity coefficient will be approximately 1. In comparison, the "bulk" H_2O in the cytoplasm of an erythrocyte can have an activity coefficient as low as 0.2. In the red blood cell, water constitutes less than 25% by weight of the cell, and little truly free hexagonally hydrogen-bonded bulk water is likely to exist. The complete picture of the behavior and organization of water in a cellular system is quite clearly very complex and is really not known at the present time. In general, the same can be said of each component in the cell including both inorganic ions and organic molecules.

22.2.1. The Next Step: The Need for Some New Tools

Again, considering the example of water in the cell, if the thermodynamic chemical potential gradient is very high between the "inside" and "outside" phases, why don't all cells instantly swell and burst as water moves down its gradient? Certainly a portion of the answer is that there are potential energy barriers to the movement of molecules that are part of the expression of rate theory. Considering processes in cells on the basis of rate theory is an important extension in understanding and elucidating cellular events. However, other tools are also necessary. The value of the thermodynamic approach is that systems comprised of large populations of molecules can be modeled and studied well. However, there are limits to thermodynamic formulations where the thermodynamic boat runs aground. There are also limits inherent in the assumptions on which many of the models presented in this book have been based. Generally, when molecules get too close to one another, the models become internally inconsistent, and when there are too few molecules in the system, the thermodynamic foundations are weakened. Where does the model builder go when these problems become insurmountable? One answer is to theories that are designed to take into account very small spaces. This puts biological thinking at the doorstep of quantum theory and theories such as the quasi-lattice structure of concentrated solutions. Although this book will not pursue these exciting and interesting avenues of thinking, the reader should now be ready to explore the many new and as yet undiscovered vistas of biological science.

Appendices

APPENDIX I

Further Reading List

Traditional Physical Chemistry Texts

Atkins P.W. (1985) *Physical Chemistry.* Third Edition, W.H. Freeman, San Francisco.

Castellan G.W. (1983) *Physical Chemistry.* Third Edition, Addison-Wesley, Reading, Massachusetts.

Moore W.J. (1978) *Physical Chemistry.* Fourth Edition, Prentice-Hall, Englewood Cliffs, New Jersey.

Physical Chemistry Texts with a Biological Approach

Chang R. (1981) *Physical Chemistry with Applications to Biological Systems.* Macmillan, New York.

Eisenberg D. and Crothers D. (1979) *Physical Chemistry with Applications to the Life Sciences.* Benjamin/Cummings, Menlo Park, California.

Tinoco I., Sauer K., and Wang J.C. (1978) *Physical Chemistry (Principles and Applications in the Biological Sciences).* Prentice-Hall, Englewood Cliffs, New Jersey.

Biophysical Chemistry

Cantor C.R. and Schimmel P.R. (1980) *Biophysical Chemistry, Parts I, II, III.* W.H. Freeman, San Francisco.

Edsall J.T. and Wyman J. (1958) *Physical Biochemistry.* Academic Press, New York.

van Holde K.E. (1985) *Physical Biochemistry.* Second Edition, Prentice-Hall, Englewood Cliffs, New Jersey.

Physics and Mathematics

Boas M.L. (1983) *Mathematical Methods in the Physical Sciences.* Second Edition, John Wiley and Sons, New York.

Hoppe W., Lohmann W., Markl H., and Ziegler H. (eds.) (1983) *Biophysics.* Springer-Verlag, New York.

Page L. and Adams N.I. (1969) *Principles of Electricity.* Fourth Edition, Van Nostrand and Co., Princeton, New Jersey.

Tipler, P.A. (1982) *Physics.* Second Edition, Worth Publishers, New York.

Part I

Freeman G.R. (ed.) (1987) *Kinetics of Non-Homogeneous Processes.* Wiley-Interscience, New York.

Thermodynamics

Fowler, R.H. and Guggenheim, E.A. (1965) *Statistical Thermodynamics.* Cambridge University Press, Cambridge.

Guggenheim, E.A. (1967) *Thermodynamics.* North-Holland, Amsterdam.

Gurney, R.W. (1949) *Introduction to Statistical Mechanics.* McGraw-Hill, New York.

Mahan, B.H. (1963) *Elementary Chemical Thermodynamics.* W.A. Benjamin, Menlo Park, California.

Waldram, J.R. (1985) *The Theory of Thermodynamics.* Cambridge University Press, Cambridge.

Part II

Water

Eisenberg, D. and Kautzmann (1969) *The Structure and Properties of Water.* Oxford University Press, Oxford.

Franks F. (1983) *Water.* The Royal Society of Chemistry, London.

Franks F. (ed.) (1973) *Water, A Comprehensive Treatise.* Volumes 1–5, Plenum Press, New York.

Ions and Forces in Solution

Bernal J.D. and Fowler F.H. (1933) *J. Chem. Phys.* **1**:515.

Bockris J.O'M. and Reddy A.K.N. (1970) *Modern Electrochemistry.* Volume 1, Plenum, New York.

Born M. (1920) *Z. Physik.* **1**:45.

Debye P. and Hückel E. (1923) *Z. Physik.* **24**:185.

Harned H.S. and Owen B.B. (1950) *The Physical Chemistry of Electrolytic Solutions.* Reinhold, New York.

Pauling L. (1960) *The Nature of the Chemical Bond.* Third Edition, Cornell University Press, Ithiaca, New York.

Macromolecules in Solution

Tanford C. (1961) *Physical Chemistry of Macromolecules.* John Wiley and Sons, New York.

Part III

Lipids

Gennix R.B. (1989) *Biomembranes: Molecular Structure and Function.* Springer-Verlag, New York.
Singer S.J. and Nicholson G.L. (1972) *Science* **175**:720.
Small D.M. (ed.) (1986) *The Physical Chemistry of Lipids.* Handbook of Lipid Research, Volume 4. Plenum Press, New York.

Transport

Conway B.E., Bockris J.O'M., and Linton H. (1956) *J. Chem Phys.* **24**:834.
Conway B.E. and Bockris J.O'M. (1959) *J. Chem. Phys.* **31**:1133.
Conway B.E. (1964) "Proton Solvation and Proton Transfer Processes in Solution." In: J.O'M. Bockris and B.E. Conway (eds.), *Modern Aspects of Electrochemistry,* Volume 3, Butterworths, London.
Einstein A. (1956) *Investigations on the Theory of the Brownian Movement.* Dover Publications, New York.

Interphase Structure and Electrochemistry

Bockris J.O'M. and Reddy A.K.N. (1970) *Modern Electrochemistry.* Volume 2, Plenum, New York.
Hunter R.J. (1981) *Zeta Potential in Colloid Science. Principles and Applications.* Academic Press, London.
Koryta J. and Dvorak J. (1987) *Principles of Electrochemistry.* John Wiley and Sons, Chichester.

Forces at Membranes

Heinz E. (1981) *Electrical Potentials in Biological Membrane Transport.* Springer-Verlag, New York.
Kotyk A., Janacek K., and Koryta J. (1988) *Biophysical Chemistry* of Membrane Functions. John Wiley and Sons, New York.
McLaughlin S. (1989) *Ann. Rev. Biophys. Biophys. Chem.* **18**:113.

APPENDIX II

Study Questions

1. An inventor comes to you with a proposal for a new machine that appears to be a perpetual motion machine. Before you throw him out, he begs your indulgence for a moment. Against your better judgment, you relent. He proposes to use the positive volume change of ice on freezing to power a machine. His machine works as follows. A quantity of liquid water is located in a cylinder that has a piston and a cooling unit associated with it. Initially, the cooling unit freezes the water, and as the ice expands, work is done on the piston by raising it. This mechanical work is converted to electrical power by a generator turned by the piston's movement and stored in a battery. As the ice melts, the piston moves downward, the movement again being captured by a differential gear and turning the generator. At the instant the ice becomes liquid, i.e., 273 K, the cooling unit comes on, powered by the battery. As the ice again forms, the piston moves and work is done, and so on. Should you invest in this invention? Why or why not?
2. The interior of the nucleus and the cytoplasm of a cell can be treated as a pair of phases separated by a membrane. Derive and write an expression for the free energy change associated with the transport of a strand of messenger RNA from the nucleus to the cytoplasm.
3. Derive a formula that allows the determination of the activity of a solvent from the vapor pressure of the solvent.
4. Show how measuring the voltage associated with a galvanic cell will give the equilibrium constant of the reaction.
5. Lysis of red cells can occur when distilled water is given intravenously. Assuming the internal concentration of red cells is 0.15 *M* NaCl and the membrane is permeable only to water, what is the reason for the lysis? Quantitate how much force is acting to rupture the cells.
6. Lithium chloride and urea are used in the isolation of RNA. Based on your knowledge of aqueous solutions, why have these substances been found so effective?

7. Patch-clamping techniques have shown that individual ion gates in the resting membrane open and close in a random fashion. Propose a method by which the observation of a single ion gate may be used to predict the state of the resting or steady-state membrane, that is, gates open or closed. Propose a second method to find the resting state of the membrane.
8. Why can the assumption of an activity coefficient of unity in the phase outside the dialysis bag be made in equilibrium binding experiments? What would be the effect on a Scatchard analysis if this assumption were incorrect?
9. Perfect crystals at absolute zero are defined by the third law of thermodynamics to have an entropy of 0.

   a. What does this imply about imperfect crystals at absolute zero?
   b. What are the molecular events that lead to a rise in entropy as the temperature of the perfect crystal increases?
   c. If a typical protein is crystallized and then cooled to absolute zero, do you think that it will have an entropy of zero?
10. The process of breathing depends on the movement of air into and out of the lungs. This is accomplished by the pistonlike action of the diaphragm, which leads to an expansion and contraction of the lung space. Thoracic pressure varies with lung volume, and air moves in and out due to pressure gradients.

    a. Write a differential equation that applies to the free energy changes associated with this system.
    b. What effect, if any, does respiration have on the transport of O_2 and CO_2 across the alveolar walls. Discuss in terms of the chemical potentials of the molecules in this system.
    c. Write an expression for the free energy of transport of O_2 and CO_2 from the lungs to the red blood cells.
11. From your knowledge of water structure, explain why ice floats. Would you expect ice and water to have the same structure in Boston as at the bottom of the ocean? How would these structures differ? What implications does this have for the biochemicals that might make up a deep-water fish?
12. Describe a molecular model for the organization of water molecules around a cell. Include in your analysis the orientation, structure, dimensions, and composition of the region.
13. Using the Born model, predict what the enthalpy of hydration for Na^+, K^+, Cl^-, Ca^{2+}, and OH^- will be in:

    a. Water at 368 K.
    b. CH_3CH_2OH at 298 K.
    c. Dioxane at 298 K.
14. Derive expressions for the enthalpy of ion–solvent interaction of the form of Equations (11.7-1) and (11.7-2) for positive and negative

ions when the coordination number of the solvent sheath is six water molecules.

15. What is the ionic strength of the serum component of blood? What is the osmolarity of this component?
16. Calculate the effective radius of the ionic atmosphere for ions in the serum. Use the ions listed in Question 13 and base your answer on the work derived in Question 15.
17. Name two situations in biological systems where ion pairing as described by Bjerrum may be of consequence. Justify your answer.
18. Use a table of the enthalpies of solvation for salts to derive a table of heats of solvation for single ions. Initially, use KF as the anion–cation pair that share the heat of solvation equally as the basis for the table. Then use another salt such as NaI. Try CsF. How does changing the pair of ions assumed to be equivalent affect the enthalpies of solvation for single ions? Predict what errors might be introduced in the empirical tests of the theories presented if the enthalpies of solvation of K^+ and F^- are not exactly equivalent as assumed.
19. Show that the expression

$$\langle \mu_{\text{group}} \rangle = \frac{\mu^2(1 + g\,\overline{\cos\gamma})^2}{3kT} X$$

is a general equation that includes the special case of the electric moment of a gaseous dipole.

20. Estimate the alteration in the dielectric constant in water that has formed pentagonal rather than hexagonal arrays around nonpolar molecules.
21. The enthalpies of hydration for single ions are substantially negative whether anions or cations are dissolved, yet the enthalpies of hydration of salts comprised of these ions are usually quite small and often slightly positive. Is this contradictory? How would you account for these observations?
22. Determine the equivalent conductivity at infinite dilution of a solution of NaCl by using Kohlrausch's law of independent migration. Show your work.
23. Explain in qualitative but succinct terms the following statement: "Entropy is the primary driving force in transport phenomena." What does this imply about the time scale of transport phenomena?
24. Compared to an aqueous solution at 298 K, would you expect the mean free path of a sodium ion to be longer or shorter in:

    a. Ice at 273 K.
    b. Aqueous solution at 273 K.
    c. Aqueous solution at 373 K.

25. Explain how electroneutrality is maintained in an electrical circuit that has an electrolytic conductor as a circuit element. Be sure to

address the three general aspects of the circuit's behavior, that is, electronic, ionic, and electrodic.

26. Is the Stern or Grahame model of the interphase to be preferred over the Gouy–Chapman or Helmholtz model in biological systems?
27. Assume that an isolated preparation of cytosolic organelles has been obtained and is sitting in a test tube on your bench. Describe and discuss two methods for precipitating these organelles based on their colloidal nature. Discuss when one method might be preferred over the other.

APPENDIX III

Symbols Used

A	Area
A	Debye–Hückel limiting law constant; for water at 298 K, 1.6104×10^{-2} $m^{3/2}$ $mol^{-1/2}$
A	Constant in the Debye–Hückel–Onsager formulation for conductivity, representing the electrophoretic retardation
$a\pm$	Mean ionic activity
α	Polarizability
B	Constant in the extended Debye–Hückel limiting law; for water at 298 K, 3.291×10^{9} m^{-1} $mol^{-1/2}$ $kg^{1/2}$
B	Constant in the Debye–Hückel–Onsager formulation, representing the relaxation effect
b_i	Measure of ease of distribution of an ion between the inside and outside of a cell
C	Capacitance, in farads
C	Number of components of a system
C_p	Heat capacity in a system at constant pressure
C_v	Heat capacity in a system at constant volume
c	Concentration
D	Density
D	Diffusion constant
d	Denatured state
d	Distance
dm^3	SI notation for volume equal to 1 liter

$đ$	Inexact differential used in describing path functions
E	Electrical potential, in volts
e_o	Charge on an electron, 1.6×10^{-19} C
ϵ	Dielectric constant
ϵ^o	Standard emf potential
ϵ_o	Permittivity of free space, 8.854×10^{-12} $C^2 N^{-1} m^{-2}$
F	Faraday, charge associated with one mole of electrons, 96,484.6 C mol^{-1}
F	Force
FW	Formula weight
f	Fugacity
$f(\Delta ka)$	Henry's function
G	Conductance, in the units mho, the reciprocal ohm (Ω^{-1})
G	Gibbs free energy
$\Delta G‡$	Free energy of the transition state
$\overline{G}_i^o$	Standard free energy of a mole of i
$\overline{G}_i$	Free energy per mole of component i in a mixture
Γ	Surface excess concentration
g	Acceleration of gravity, 9.8 m s^{-2}
g	Number of nearest neighbors of a dipole
g_1	Degeneracy of an energy level
γ	Activity coefficient
γ	Surface tension
$\gamma\pm$	Mean ionic activity coefficient
H	Enthalpy
h	Height
h	Planck's constant
h	Polypeptide residue in helical configuration
η	Viscosity of the solvent
I	Current, in amperes

I	Ionic strength
J_x	Flux of material in the x direction
K	Equilibrium constant
k	Boltzmann's constant, 1.3805×10^{-23} J K^{-1}
k	Hooke's law constant
κ	Conductivity, in the units Ω^{-1} m^{-1} or Ω^{-1} cm^{-1}, the reciprocal of resistivity
κ	Isothermal compressibility
κ^{-1}	Debye length or effective radius
k_B	Henry's law constant
k_b	Boiling point elevation constant; for water, 0.51
k_f	Freezing point depression constant; for water, 1.86
κ_T	Thermal conductivity coefficient
Λ	Equivalent conductivity
L_D	Debye length or effective radius
L_{ii}	Phenomenological coefficient for flux due to related gradient
L_{ij}	Phenomenological coefficient for coupled flow caused by gradient that is not directly related
l	Mean free path
Λ_m	Molar conductivity, in the units Ω^{-1} cm^2 mol^{-1} or Ω^{-1} m^2 mol^{-1}
$\Lambda°$	Equivalent conductivity at infinite dilution
λ°_+ and λ°_-	Ionic conductivities at infinite dilution
M	Molecular weight
m	Mass
$m_\pm$	Mean ionic molality
N	Number of sites available for binding
N_A	Avogadro's number, 6.02×10^{23} particles
n	Native state
n	Number of particles

n	Electrical equivalents per mole
P	Number of phases that exist in equilibrium
P	Pressure
P_A	Vapor pressure of the solvent
P_A^o	Vapor pressure of pure solvent
P_i	Specific ion membrane permeability
$\prod_i$	Mathematical operator representing the product of a series of terms
π	Osmotic pressure
π	pi
Q	Charge
Q	Ratio of activities of components
q	Heat
R	Resistance
r	Polypeptide residue in random configuration
r	Radius
ρ	Resistivity
ρ	Charge distribution or density
ρ_r	Charge excess
S	Entropy
S_{se}	Statistical expression of entropy
s	Measure of cooperativity
σ	Particle diameter
σ	Measure of initiation or the start of a conformational change
T	Temperature
T	Total number of ways that a population can be distributed
T_m	Transition or melting temperature
t	Time
U	Internal energy of a system

U	Potential energy
$\langle u \rangle$	Average distance a particle will travel in unit time
μ	Chemical potential
μ	Dipole moment, in debyes
μ	Electrophoretic mobility, in units of $m^2 s^{-1} V^{-1}$
μ	Mobility
μ^o	Chemical potential at standard conditions
$\mu^o_\pm$	Mean ionic standard chemical potential
$\mu_\pm$	Mean ionic chemical potential
V	Volume
$\overline{V}$	Molar volume of the solvent
v	Drift velocity
ν_+	Stoichiometric equivalents of cationic charge
ν_-	Stoichiometric equivalents of anionic charge
W	Total number of choices of an energy level available for a particle
w	Work
X	Electric field vector
$X_\pm$	Mean mole fraction
X_A	Mole fraction of the solvent
x	Length of a spring
$\langle x \rangle$	Average distance particle moves from the origin
$\langle x^2 \rangle$	Mean square distance
X_i	Mole fraction of a species
ψ	Electrostatic potential
$\Delta\Psi$	Transmembrane potential
Z	Partition function which sums over all the system's energy levels
z	Number of intermolecular collisions per unit time
ζ	Zeta potential
z_i	Charge number

$\oint$	Cyclical integral
∞	Infinity
∂	Partial derivative
$\sum_i$	Mathematical operation representing the sum of a series of terms
$\int_1^2$	Integral of function between 1 and 2

APPENDIX IV

Glossary

Activity. The effective concentration of a component in solution. Often expressed as the product of the **activity coefficient** and the analytic concentration of a component. The activity of a pure substance is, by definition unity.

Activity coefficient. Term that relates the analytic concentration of a component to its effective concentration. It is theoretically derived from considerations of ion–ion and ion–solvent interactions and represents deviations from ideality. The activity coefficient of a pure substance or an ideal component is unity.

Adiabatic walls. Boundaries across which no heat exchange occurs. In an adiabatically bounded system, $đq = 0$ and the **first law of thermodynamics** reduces to $dU = đw$.

Anisotropic environment. Environment where forces experienced depend both on direction and magnitude.

Anode. By convention, the positive electrode or where negative charge is withdrawn from a system. When the negative charges are electrons, oxidations occur at the anode.

Bilayer. Term used to describe the structure of a lipid membrane that forms to separate two aqueous phases.

Black membrane. Planar synthetic lipid bilayer used in modeling mammalian membrane properties and made by introducing the lipids under water at a small hole in a barrier erected between two aqueous compartments.

Boltzmann distribution. Describes the most probable distribution of particles among a set of energy levels. Usually takes the form $n_i/n_j = \exp[-(\Delta U_{i\text{-}j}/kT)]$.

Boltzmann statistics. Statistics which assume that each particle is distinguishable, and that there are no limits to the number of particles at each level. They take the general form of: $t_1 = g_1^{n_1}$.

Born model. Approximation of ion–solvent interactions based on the assumptions that ions may be represented as rigid spheres of charge, that the solvent into which the ion is to be dissolved is a structureless continuum, and that all interactions are electrostatic in nature.

Bose–Einstein statistics. Statisitcal distribution for the case when particles are considered to be indistinguishable but no limit is placed on the number of particles in a given level. Bose–Einstein statistics are not subject to the **Pauli exclusion principle**.

Boundary. Confines a **system** to a definite place in space, separating it from the rest of the universe.

Cathode. By convention, the negative electrode or where negative charge is put into a system. When the negative charges are electrons, reductions occur at the cathode.

Cerebroside. Sphingolipid whose polar head group is comprised of a neutral saccharide.

Change in state. Alteration in state from a specified initial state to a specified final state.

Chemical potential, μ. Property of matter that leads to the flow of a material from a region of high potential to a region of low potential.

Clapeyron equation. Describes the **equilibrium** between phases of a pure substance:

$$\frac{dP}{dT} = \frac{\Delta S}{\Delta V}$$

Classical or **Newtonian mechanics.** Describes the motion of large material objects, for example, planets and satellites, and the trajectory of balls and missiles.

Classical thermodynamics. See **Thermodynamics**.

Clathrate. A regular pentagonal array of water molecules that surrounds a nonpolar molecule, corresponding to a stable hydrate in which there is a given stoichiometry of H_2O molecules and nonpolar solute molecules of a specified size.

Clausius–Clapeyron equation. Quantitates the increase in the **activity** of a substance in the condensed phase with respect to the pressure in the **system**:

$$\ln \frac{P_2}{P_1} = \frac{-\Delta H_{vap}}{R}\left(\frac{1}{T_2} - \frac{1}{T_1}\right)$$

It is useful for cases of condensed phase–gas equilibria.

Closed system. See **System.**

Colloidal electrolyte. A compound that gives rise to at least one ionic macromolecular species that becomes the ionic conductor.

Concentration potential. Potential generated by the concentration difference between two solutions.

Conductance, *G*. Expresses the relationship between the current and voltage of a material: $G = I/E$. It is the reciprocal of the **resistance**.

Conductivity, κ. A measure of the ease with which current flows in a conducting medium: $\kappa = 1/RA$. It is the reciprocal of **resistivity** and has the units of $\Omega^{-1}\ m^{-1}$ or $\Omega^{-1}\ cm^{-1}$.

Configuration. Grouping within a **microstate**.

Contact potential. Voltage arising from the contact of two phases. An example is the voltage measured when two dissimilar metal wires are twisted together at one end and a voltmeter is used to complete the circuit between the two.

Convection. Fluid flow where molecules are moved by a streaming process produced by an external force, such as a pressure gradient ($\partial P/\partial x$).

Cooperativity, *s*. The interdependence of the stabilizing interactions in each state. It is a measure of the increased likelihood that once a given residue in a **polymer** has undergone a conformational change associated with denaturation, the next residue along the chain will also undergo such a change. Graphically, cooperativity is represented by the slope of the denaturation curve at its midpoint.

Copolymer. Comprised of two or more monomers. They can be random or ordered.

Coulomb's law. $F = k\ q_1q_2/r^2$, where $k = 1$, q is given in electrostatic units (esu) and r in centimeters; the force described is in dynes. In the SI system, k is equal to $1/4\pi\epsilon_o$, changing the equation to $F = q_1q_2/4\pi\epsilon_o r^2$, where ϵ_o is the **permittivity of free space** and has a value of $8.854 \times 10^{-12}\ C^2\ N^{-1}\ m^{-2}$.

Critical micelle concentration. The concentration of lipid at which a solubilized monomeric form of lipid forms a **micelle**.

Crystalline water. Ice.

Cycle. A process that returns to the original state.

Debye. Unit of the **dipole moment**, found by multiplying the separated charge by the distance of separation.

Debye–Falkenhagen effect. Term given to the effect whereby the **conductivity** of ions is higher in a high-frequency electric field than in a field of lower frequency. The explanation of the effect is that removal of the retarding force responsible for the **relaxation effect** does not occur under high-frequency conditions.

Debye–Hückel limiting law. Theoretical formulation that predicts the mean ionic activity coefficient. It links experimental values and the **Debye–Hückel model:** $\log \lambda_\pm = -A(z_+z_-)I^{1/2}$. For water, the constant A has units of $dm^{3/2}\ mol^{-1/2}$. In SI units, the value of A for water at 298 K is $1.6104 \times 10^{-2}\ m^{3/2}\ mol^{-1/2}$.

Debye–Hückel model. Describes **ion–ion interactions** by assuming that everything other than the central ion is to be treated as a nonstructured continuum of charge residing in a **dielectric** continuum. It assumes that (1) a central reference ion of a specific charge can be represented as a point charge, (2) this central ion is surrounded by a cloud of smeared-out charge contributed by the participation of all of the other ions in solution, (3) the electrostatic potential field in the solution can be described by an equation that combines and linearizes the **Poisson** and **Boltzmann equations**, (4) no **ion–ion interactions** except the electrostatic interaction given by a r^{-2} dependence on are to be considered, and (5) the solvent simply provides a dielectric medium and that the ion–solvent interactions are to be ignored, so that the bulk **permittivity** of the solvent will be used.

Debye–Hückel–Onsager theory. A model theory to explain the conductivity of ions. It is based on the fact that both the **electrophoretic** and the **relaxation effects** depend on the density of the ionic atmosphere and can be shown to increase with $\sqrt{c}$: $\Lambda = \Lambda° - (A + B\Lambda°)c^{1/2}$.

Debye length, L_D, κ^{-1}, or **effective radius.** Distance from the central ion to a point charge representing the charge atmosphere surrounding the central ion.

Deelectronation reaction. Occurs when a charge carried by an ion is transferred to the **anode** via oxidation.

Degeneracy. Number of distinguishable states produced when a set of particles is distributed among defined energy levels.

Depsipeptide. Peptidelike compound that possesses alternating peptide (amide) and ester linkages.

Diathermal walls. Boundaries where heat interactions are allowed to occur.

Dielectric. Nonconducting material which, when inserted in the space between separated charges, decreases the force per unit charge.

Dielectric constant, ϵ. Unitless term that compares the relative dielectrical constant of any material to the dielectric constant of a vacuum, which is unity.

Dielectric strength. Measure of a substance which maintains an electric potential field across it without breaking down.

Diffusion. Transport process in which molecules travel down their chemical potential gradient, $\partial c/\partial x$, until the gradient is removed or neutralized.

Diffusion potential. The difference in potential that results when two similar solutions (but of different concentration) are brought together and differences in the rates of diffusion of the ions give rise to a potential difference.

Dipole. A molecule that carries no net charge yet has a permanent charge separation due to the nature of the electronic distribution within the molecule itself. For example, oxygen is more strongly electronegative than either hydrogen or carbon, and the electrons shared by oxygen and hydrogen or carbon in a molecule will spend more time with the oxygen atom. The result of this preference of the negatively charged electrons for the oxygen atom will result in a partial negative charge on the oxygen atom, while the companion carbon or hydrogen atoms will carry a partial positive charge.

Dipole–dipole interactions. Electrostatic interactions that occur when molecules with **permanent dipoles** interact.

Dipole moment, μ. The magnitude and direction of the charge asymmetry in a **dipole.**

Dispersion. A material that is uniform on a macroscopic scale but is made up of distinct components imbedded in a matrix.

Dispersion forces. See **London forces.**

Donnan potential. See **Gibbs–Donnan potential.**

Dorn effect. See **Sedimentation potential.**

Double layer. Model of the electrified interface analogous to a **parallel plate capacitor** composed of two plates of opposite charge. By traditional usage, it also refers in general to the electrified interphase region.

Drift velocity. At the drift velocity, the forces acting to accelerate the ion, $F_{electric}$, and to retard it, $F_{viscous}$, are equal: $v = z_i e_o E/6\pi r\eta$. It is related to an index of the ease with which an ion can move, the **mobility,** μ.

Effective radius. See **Debye length.**

Einstein–Smoluchowski equation. Equation that relates the **random walk model** to the diffusion coefficient in **Fick's laws:** $\langle x^2 \rangle = 2Dt$, for the case in a single dimension.

Electrical conduction. The transport of charge in an electrical field $(\partial y/\partial x)$.

Electrical double layer or the **electrified interface.** A region where charges are separated across a **boundary** between two phases. The electric field extends not only across the phase boundary but also for a distance into each phase.

Electrical work. The movement of charge through an electrical gradient or potential.

Electricity. The study of the behavior of charges either at rest or in dynamic motion.

Electrified interface. See **Electrical double layer.**

Electrochemical studies. Describe the transport, reactions, and behavior of charge in chemical systems.

Electrodics. The study of the charge transfer reactions at electrodes, the focus of modern electrochemistry.

Electrokinetic effects. Effects associated with motion relative to the **double layer.**

Electrokinetic potential. Potential field resulting from the motion of the bulk phase with respect to the surface. The electrokinetic potential is named the **zeta potential** and given the symbol ζ. The magnitude of ζ is found by measuring the potential difference between the portion of the diffuse layer that can be removed by a shearing force and the bulk of the solution.

Electrolyte. As used in biomedicine and electrochemistry, refers to both the ionically conducting medium and to the substances that produce it when they are dissolved or liquefied.

Electromagnetics. Describes the electrodynamic phenomena of charges in motion.

Electromagnetism. Explains the behavior of electricity, magnetism, and electromagnetic radiation.

Electronation or **reduction reaction.** Charge transfer reaction occurring at the **cathode**–electrolyte interface in which the electron, which is responsible for carrying charge in the external circuit, is transferred to an ion in the **electrolyte** phase.

Electronics. The study of electron behavior that includes the conduction and behavior of the electrons in circuits.

Electro-osmosis. Electrokinetic phenomenon whereby an applied current causes a net **flux** of fluid. The relationship of the electric field to the volume of fluid flow is described by $V = \zeta\epsilon_0\epsilon I/\eta\kappa$.

Electrophoresis. Electrokinetic phenomenon in which application of an electric field causes relative motion of the surface and the diffuse layer; the solvent is stationary and the particles move. The particles must be large enough to have a **double layer** associated with them. The velocity of the particles with respect to field strength is the **electrophoretic mobility,** μ, and has units of $m^2\, s^{-1}\, V^{-1}$: $\mu = v/E$.

Electrophoretic effect. The effect in **electrophoresis** of an increased viscous force due to the interaction of ionic clouds, that is, the sum of positive charge and negative ions moving in opposite directions, as they drag past one another.

Electrophoretic mobility, μ. See **Electrophoresis.**

Electrophoretic retardation. The major force retarding the movement of the particles in **electrophoresis**; it is derived from streaming of the diffuse layer in a direction opposite to the movement of the particle.

Electrostatic system. System with stationary charges separated in space.

Electrostatics. The study of the behavior of resting charges.

Endothermic. A chemical reaction in which heat is drawn from the surroundings as the reaction proceeds toward equilibrium.

Ensemble. Group of **microstates.**

Enthalpy, *H*. State function (see **Functions of state**) derived from the **first law of thermodynamics** at constant pressure: $H = U + PV$. Enthalpy is a valuable state function because it provides a method for determining a realistically constrained biological or aqueous phase system's energy simply by measuring the heat exchanged with the **surroundings.**

Entropy, *S*. State function (see **Functions of state**) of the **second law of thermodynamics** that indicates whether a **system** is moving as predicted toward a set of unchanging state variables.

Entropy of mixing. See **Free energy of mixing.**

Equilibrium. Point at which the state values of an **isolated system** no longer change with time. Movement of a system toward equilibrium is considered the "natural" direction.

Equivalent conductivity, Λ. Used in the comparison of molar conductivities (see **Molar conductivity**) to account for the fact that different types of **electrolytes**, i.e., 1:1, 1:2, etc., will contain anions and cations of different charge: $\Lambda = \kappa / c z_i$.

Equivalent conductivity at infinite dilution, Λ°. A reference state of maximum equivalent **conductivity**, useful for comparisons between any **electrolytes.**

Extensive properties. See **Properties of a system.**

Faraday's law. Expresses the quantitative relationship for the transfer of electronic charge to ionic charge that occurs at the electrodes: $m/\mathrm{FW} = it/|z_i|F$, where m is the mass of an element of formula weight FW liberated or consumed at an electrode, i is the current passed in amperes in t in seconds, z_i is the charge on a given ion, and F is the Faraday, equal to 96,484.6 C mol^{-1}.

Fermi–Dirac statistics. Statistical distribution for the case when particles are indistinguishable but only one particle per box is considered. The **Pauli exclusion principle** is obeyed.

Fick's first law. Describes the relationship between **flux** and concentration gradient: $J_i = -D\,(dc/dx)$. Indicates that the transport of species is in a direction opposite to the concentration gradient. The concentration gradient will have units of mol m^{-4}, and the diffusion coefficient is written as $\mathrm{m}^2\ \mathrm{s}^{-1}$.

Fick's second law. Describes the change in concentration in a small volume that varies with respect to time: $(\partial c/\partial t)_x = D\,(\partial^2 c/\partial x^2)_t$. It is the basis for the treatment of many non-steady-state or time-dependent transport problems.

First law of thermodynamics. Dictates that the energy of the **system** and its **surroundings** will remain constant. For an infinitesimal change, the algebraic behavior of heat and work is known as the first law of thermodynamics: $dU = đq + đw$.

Fluid-mosaic model. Modern model of the biological membrane which describes the membrane as a phospholipid bilayer with an array of proteins floating in or spanning the bilayer structure. The membrane proteins act as the carriers, pumps, channels, and mediators of cell function and are the sites of cell–cell interactions and cell recognition.

Flux. Net movement of matter in unit time through a plane of unit area normal to the gradient of potential.

Fourier's law. Law that describes the flow of heat in a temperature gradient.

Free energy of mixing. Free energy change (see **Gibbs free energy**) resulting from the relationship between the **entropy** and **enthalpy** when one substance is mixed with another: $\Delta G_{\mathrm{mix}} = NRT\Sigma_i X_i \ln X_i$ Derived from the **entropy of mixing,** $\Delta S = -NR\ \Sigma_i X_i \ln X_i$, and the **heat of mixing,** $\Delta G_{\mathrm{mix}} = \Delta H_{\mathrm{mix}} - T\Delta S_{\mathrm{mix}}$.

Fugacity, *f*. A state function (see **Functions of state**) similar to pressure that measures the escaping tendency of a real gas or any real substance.

Functions of state. Thermodynamic properties at **equilibrium**.

Fundamental properties. See **Properties of a system**.

Galvanic cell. A system consisting of an **electrolyte** and two electrodes, with the two electrodes connected together through an external electrical circuit.

Ganglioside. Sphingolipid whose polar head group is comprised of a polysaccharide.

Gedanken experiment. Thought experiment.

Gegenions. Small ions associated with **polyelectrolytes** which ensure electroneutrality.

Gibbs adsorption isotherm, Γ_i. Quantifies the excess solute adsorbed on a surface: $\Gamma_i = (-1/RT)(\partial\gamma/\partial \ln a_i)$. It has the units of mol m^{-2}.

Gibbs–Donnan potential. The potential that is derived from the behavior of a polyelectrolyte in an electrolyte system of several phases separated by a membrane that limits the free exchange of the polyelectrolyte.

Gibbs–Duhem equation. $\Sigma_i n_i \, d\mu_i = V\,dP - S\,dT$.

Gibbs free energy, *G*. Indicates the direction and **equilibrium** position of a **system** that undergoes radical changes in energy, measured by both **enthalpy** and **entropy**: $G = H - TS$.

Gibbs–Helmholtz equation. Used in situations where **enthalpy** or **entropy** varies with temperature: $[\partial(G/T)/\partial T]_P = -(H/T^2)$.

Glycerol esters. Lipids with a three-carbon glycerol backbone. Their properties depend on length and degree of unsaturation of the fatty acids attached.

Goldman–Hodgkin–Katz constant field equation. An expression that gives the transmembrane potential at equilibrium in terms of the specific membrane permeabilities for each ion and their intra- and extracellular concentrations:

$$\Delta\Psi = \frac{RT}{F} \ln \frac{\Sigma P_i c_{out}^{M^+} + \Sigma P_i c_{in}^{N^-}}{\Sigma P_i c_{in}^{M^+} + \Sigma P_i c_{out}^{N^-}}$$

Gouy–Chapman diffuse layer. See **Gouy–Chapman model.**

Gouy–Chapman model. Model of the electrified interface in which the decay of the potential derived from the electrode versus the distance, x, into the bulk **electrolyte** depends in part on the charge, z_i, on the ion and the ionic strength of the solution. A thickness for the **Gouy–Chapman diffuse layer** exactly analogous to the reciprocal length parameter, κ^{-1}, of the **Debye–Hückel model** is described as $\kappa^{-1} = (\epsilon_0 \epsilon \kappa T/2N_A e_0^2 I)^{1/2}$. The Gouy–Chapman treatment of the double layer ignores the effect on the **dielectric constant** of the high potential fields (10^8 m^{-1}) present at the interface, the fact that ions have finite size, and the tendency of molecules to adsorb on surfaces through forces other than electrostatic interactions.

Heat. An energy transfer that results in a rise (or fall) in temperature of the **surroundings** that is not recoverable as useful or usable energy. Heat that raises the temperature of the system is given a positive sign (an **endothermic** event).

Heat capacity, C_v. The ratio $đq_v/dT$ at constant volume: $C_v = đq_v/dT = (\partial U/\partial T)_V$.

Heat conduction. The transport of heat from a hotter to colder reservoir that results from a net flow of particles with a higher kinetic energy down a temperature gradient $(\partial T/\partial x)$.

Heat of mixing. See **Free energy of mixing.**

Helmholtz–Perrin model. The earliest model of the electrified interface, developed at the turn of the century. The entire countercharge to the electrode is considered to reside in a single rigid layer of counterions. This compact and immobile layer of ions was thought to completely neutralize the charge on the electrode. The Helmholtz model is generally not adequate for describing the **electrified interphase** region.

Henry's function, $f(\kappa a)$. A correction factor used in the calculation of the electrophoretic mobility that relates the retardation of movement to the viscosity of the medium and to the size of the effective radius of the ionic cloud.

Henry's law. Expresses the vapor pressure of the solute fraction for a volatile solute, $P_B = k_B X_B$, where k_B is **Henry's law constant.**

Henry's law constant, k_B. See **Henry's law**.

Hess's law. $\Delta G = (cG_C + dG_D) - (aG_A + bG_B)$ for $aA + bB \rightarrow cC + dD$.

Heterogeneous processes. See **Homogeneous processes.**

Homogeneous processes. Processes that occur when components are randomly distributed in a single phase. Macroscopically **heterogeneous processes** occur when components from one or more phases interact at the interface between the phases. **Nonhomogeneous** behavior occurs when at least one component is not randomly distributed in either a single or a microheterogeneous phase.

Homopolymer. Polymer of a single component or monomer.

Hooke's law. Dictates that the force applied is directly proportional to the changed length of a spring: spring force $= -k(x - x_o)$ where k is a constant for a given spring. The law can be used to describe the behavior of events, such as the contraction of muscle fibers and the interactions of atoms with one another, where the displacement acts like that of a spring.

Hydrogen bond. Relatively weak bond of between -20 and -30 kJ mol^{-1} formed by a shared proton hydrogen atoms located between two electronegative atoms. The strength of the bond increases with an increase in the electronegativity or a decrease the size of the participating atoms.

Hydrophobic interaction. The interaction of nonpolar molecules with each other to exclude water. The effect is seen when increasing concentrations of nonpolar molecules are added to an aqueous system. There is a strong gain in **entropy** by allowing the nonpolar molecules to interact and allowing water to reassociate with the bulk aqueous phase.

Hydrotactoid. Large hydrate formed when water entirely surrounds a nonpolar polymerized amino acid. The small nonpolar entities exist within the ordered H_2O structures as separate entities and act as isolated molecules.

Ideal solution. A solution of components which all obey **Raoult's law** over the entire range of concentrations.

Induced dipole. Produced when a **dielectric** comprised of molecules without a **dipole moment** is placed into an electric field. The field will cause the displacement of the electron clouds (negatively charged) away from the nuclei (positively charged), thus inducing a **dipole**. Because such dipoles are generated by the field, they are aligned with the field at the instant of generation. They will exist as long as long as the field remains.

Induced dipole–induced dipole interactions. Variation in the distribution of the electrons in a neutral molecule if at an instant of time a snapshot were taken. At this instant, the "neutral" molecule would have a **dipole moment**. This instantaneous **dipole** is capable of inducing an instantaneous dipole in another neutral molecule. These forces are also called **London** or **dispersion** forces.

Inner Helmholtz plane. The plane of closest approach by unhydrated anions as they move in toward the hydration sheath of the positively charged electrode and displace a water molecule. Whether an ion will leave the **outer Helmholtz plane** and move to a position on the inner Helmholtz plane depends on the free energy change associated with the jump from one plane to the other.

Integral spin. All particles, such as electrons and photons, possess angular momentum and may be considered to spin on their axes. Quantum theory dictates that the amount of angular momentum possessed by a particle is quantized. The fundamental quantum of angular momentum is $\frac{1}{2}\hbar$ (where $\hbar$ is Planck's constant divided by 2π). Particles such as photons have an angular momentum of $\hbar$ or a spin number of 1. Such particles are considered to have integral spin. The spin numbers of particles such as electrons or muons are nonintegral; for example, an electron possesses only $+\frac{1}{2}$ or $-\frac{1}{2}$ spin. The Pauli exclusion principle prohibits two electrons with the same spin number from occupying the same orbital. This is why only electrons of opposite spin are found paired in orbitals.

Intensive properties. See **Properties of a system.**

Interface of two immiscible electrolyte solutions (ITIES). Model of the **electrified interface** in which **electrolytes** partition across the interface depending on a relative difference in their hydrophobicity.

Interphase. At the melting of two phases, interactions occur that lead to an electrical charge separation. The electric field extends across the phase boundary and into the region immediately adjacent to it. The **properties** of this interphase are different from those of either of the phases responsible for creating it.

Ion–dipole interactions. Interactive force occurring between particles when one of them is an ion and the other a **dipole**.

Ion–induced dipole interactions. Interactive force occurring between an ion and a transient **dipole** induced in a molecule which normally has no net **dipole moment** (such as carbon dioxide or carbon tetrachloride) when it is brought into an electric field. The potential of the interaction will depend on the ability of the neutral molecule to be induced into a dipole, that is, its **polarizability**, α.

Ion–ion interactions. Interactive force exerted by two charged particles.

Ion pair. Ion–ion interactions that are formed when an ion acting as a member of an ionic atmosphere gets close enough to the central ion that they become locked in a coulombic embrace; neither ion is independent of the other.

Ionic strength, *I*. Quantification of the charge in an electrolyte solution, given by $I = \frac{1}{2}\Sigma\, n_i z_i^2$.

Ionics. The study of ionic interactions with their environment, that is, charge transfer by ions in solution.

Ionophores or **true electrolytes.** Materials which conduct a current in the solid state whether or not they are in solution. By inference, such materials must be comprised of ions in their pure state. A common nomenclature also calls such electrolytes **strong electrolytes,** because they conduct strongly when dissolved in water. This is in contrast to substances called **weak electrolytes**, which do not give rise to high conductances when dissolved in water.

Irreversible processes. Real processes that inevitably lead to an increase in **entropy**.

Irrotational or **immobilized water**. Water molecules torn from their association with other water molecules, which become immobilized or trapped around an ion, forming a solvent sheath surrounding the ion. So tightly held are these water molecules that together with the ion they become a new kinetic body.

Isolated system. See **System.**

Isotropic environment. Environment where direction is inconsequential.

Joule heat. Irreversible heat which is produced due to the resistance of a metal to the flow of charge.

Kink. Term given to the bend in the hydrocarbon chain of a lipid that causes a *cis*-like configuration of the chain around the kinked carbon–carbon bond.

Kirkwood equation. Equation for finding the **dielectric constant** of a condensed medium:

$$\epsilon - 1 = 4\pi n \frac{3\epsilon}{2\epsilon + 1}\left[\alpha_{\text{deform}} + \frac{\mu^2(1 + g\ \overline{\cos\gamma})^2}{3kT}\right]$$

Latent heat of condensation. The **work** required for the return of one mole of water vapor to the liquid form. It is about 40 kJ mol^{-1}.

Lateral diffusion. Process where a lipid molecule experiences net translational movement in the plane of the membrane by executing a series of jumps from one vacancy in the liquid crystal array to another.

Lennard-Jones potential. Interactional energy resulting from the attractive and repulsive forces occurring between the electron clouds of molecules as they approach one another. The electron repulsion must be added into the formula for van der Waals interactive energy. The Lennard-Jones potential is usually written $U_{\text{interaction}} = (-A/r^6) + (B/r^{12})$.

Line integral. Function defined by an inexact integral and which depends not on the initial and final states but rather on the **path** between the states.

Linearized Boltzmann equation. $\rho_r = -\Sigma n_i^0 z_i^2 e_0^2\ \psi_r/kT$

Linearized Poisson–Boltzmann equation. The result of combining the **linearized Boltzmann equation** and the **Poisson equation**, each relating the charge density, ρ_r, in the volume element dV to the distance, r, from the reference ion:

$$\frac{d}{dr}\ r^2\left(\frac{d\psi}{dr}\right) = \frac{1}{\epsilon_0 \epsilon kT}\sum n_i^0\ z_i^2 e_0^2\ \psi_r.$$

Liposomes. Spherical entities bounded by a phospholipid bilayer with water trapped inside.

Liquid crystal. Intermediate, in degree of order, between a solid and a liquid state.

Liquid ionic conductor. A liquid salt formed by the disruption of a crystalline lattice structure upon melting.

London or **dispersion forces.** Force induced between neutral molecules by **induced dipole–induced dipole interactions.** These forces fall off very rapidly with distance but can be quite significant for molecules in close proximity.

Lyophilic colloids. "Solvent-loving" colloids which form systems called gels.

Lyophobic colloids. "Solvent-hating" colloids which form systems called sols.

Mean free path, *l*. The average distance a particle will travel between collisons. It is obtained by dividing the average distance a particle will travel in unit time if unimpeded, $<u>$, by the number of intermolecular collisions it will experience in that time, z, i.e., $l = <u>/z$.

Mean ionic activity coefficient, $\gamma_{\pm}$. Approximates the **activity coefficient** of a single ion by averaging the sum of the chemical potentials of both ions in an electrically neutral electrolyte solution. The general form for relating the measurable mean ionic acitivity coefficient to the activity coefficient for each individual ion is $\gamma_{\pm} = (\gamma^{\nu+}\gamma^{\nu-})^{1/\nu}$.

Mechanical work. Work characterized by physical movement. Two major forms of mechanical work must be considered for biological systems. The first is the work done by moving a weight in a gravitational field, and the second is work done when a force causes displacement of a structure acting like a spring.

Melting temperature. See **Transition temperature**.

Micelle. Association of amphiphilic molecules with each other into structures having the hydrophobic portions of the molecules separated from the aqueous solvent by a "self-generating" hydrophobic milieu, while the polar groups face the solvent and hydrogen bond with the water molecules.

Microstate. Each arrangement of a distribution of a group of molecules among a set of energy states in an **isolated system**. The groupings within a microstate are called **configurations** while the whole group of microstates is called an **ensemble**.

Microviscosity. Measure of the ease with which lateral motion can occur in a membrane or bilayer. It is a function of the liquid crystalline structure of the membrane.

Mobility, μ. A proportionality constant that defines the **drift velocity** with respect to a unit force: $\mu = v/F$. The mobility of an ion is related to its ability to contribute to carrying a charge in an electric field.

Molar conductivity, Λ_m. Used when comparing the conductivities (see **Conductivity**) of solutions that may not have the same concentrations: $\Lambda_m = \kappa/c$. The units of Λ_m are Ω^{-1} cm^2 mol^{-1} or Ω^{-1} m^2 mol^{-1}.

Molar free energy, μ. Relationship of the free energy to components in a system on a per mole basis: $\mu = \mu^{o} + RT \ln (P_2/P_1)$.

Native conformation. The conformation in which biopolymers are isolated.

Nernst equation. Relates the activity of the components in the system comprising the galvanic cell to the standard state emf, ϵ^{o}:

$$\epsilon = \epsilon^{o} - \frac{RT}{nF} \ln \frac{(a_C)^c\,(a_D)^d}{(a_A)^a\,(a_B)^b}$$

Newtonian mechanics. See **Classical mechanics.**

Nonequilibrium thermodynamics. See **Thermodynamics**.

Nonhomogeneous processes. See **Homogeneous processes**.

Ohm's law. Describes the relationship between current flow, potential, and **resistance** to flow of current when an electrical potential field is imposed across a conducting material: $I = E/R$, where I is the current expressed in amperes, E is the electrical potential field expressed in volts, and R is the resistance expressed in ohms (Ω).

Oncotic pressure. The osmotic pressure due to the presence of colloidal particles in solution.

Onsager reciprocity relation. The general derivation of relationships between the transport rate constants shows that there is a pairing or coupling of the processes during flux. For example, the phenomenological coefficients describing the flow of heat caused by the thermal gradient and those describing the flow of heat caused by the electrical gradient are equal: $L_{ij} = L_{ji}$. This indicates the interaction of one flow with another.

Open system. See **System**.

Osmosis. The movement of a solvent through a membrane that occurs secondary to a driving force of diffusion.

Ostwald's dilution law. $K = \Lambda^2 c/[\Lambda^\circ(\Lambda^\circ - \Lambda)]$.

Outer Helmholtz plane. Layer of molecules in a hydration sheath comprised primarily of hydrated ions of appropriate countercharge. In the case of the positive electrode, this layer is populated by anions, and in the case of the negatively charged electrode, cations. This layer of ions is the dividing line between the outer **diffuse Gouy–Chapman layer** and the inner compact region of the **interphase**.

Parallel plate capacitor. Plates separated by a **dielectric** material which has the capacity to store electrical energy as a stored charge. Capacitors do not allow current to flow through them because the material separating the plates, the dielectric, is electrically insulating.

Path. Description of the change in state of a **system** when the initial state, the sequence of the intermediate states arranged in the order traversed by the system, and the final state are defined.

Pauli exclusion principle. Principle of **quantum mechanics** that states that particles of non**integral spin**, such as electrons, cannot have the same four quantum numbers.

Peltier heat. A reversible heat flow associated with the flow of current.

Permanent dipole. See **Dipole**.

Permittivity of free space, ϵ_0. See **Coulomb's law**.

Phase. A state in which there is both chemical and physical uniformity.

Phase rule. States that the maximum number of phases that can exist at **equilibrium** in the absence of external fields is equal to the number of components plus two.

Phosphatidic acid. Glycerol ester lipid with esterified fatty acid on C-1 and C-2 and phosphorylated carbon in the C-3 position.

Phospholipid. Derivatives of **phosphatidic acid** in which the phosphate group is esterified with an amino acid, an aliphatic amine, a choline residue, a carbohydrate residue, etc.

Poiseuille's law. Describes fluid movement.

Poisson equation. Relates the electrostatic potential to the excess charge density in a volume element dV. The solution in a spherically symmetrical system is written:

$$\frac{1}{r^2}\frac{d}{dr}r^2\left(\frac{d\psi}{dr}\right) = -\frac{1}{\epsilon_o\epsilon}\rho_r$$

where ρ is the charge distribution and ψ_r is the electrostatic potential.

Polarizability, α. The ease with which a dipole moment can be induced in a molecule experiencing an external electric field.

Polyelectrolyte. Results when a portion of a **polymer** is comprised of some charged residues.

Polymer. Chain of smaller molecules covalently attached to each other.

Potential electrolytes. Electrolytes which depend on a chemical reaction to generate their ionic nature.

Pressure–volume work. Occurs with the expansion or contraction of a gas and is important in gas-exchanging organs. The expression $w = -\int P_{ext}\, dV$ will provide the work done on a system if the pressure applied and the change in volume in the system are known.

Principle of electroneutrality. At equilibrium, the charge on the ion will be exactly countered by a countercharged atmosphere that will be arranged in some charge distribution around the ion. Locally, there will be regions of excess charge density, but taken as a whole the solution will be electroneutral since each of the central ions will be surrounded by atmospheres of charge that are exactly equal in magnitude but opposite in sign. Mathematically, the principle can be expressed as $\Sigma\, z_i e_o X_i = 0$, where z_i is the number of elementary charges, e_o, carried by each mole fraction, X_i, of the species making up the entire solution.

Principle of microscopic reversibility. At **equilibrium**, any process and its reverse process are taking place on the average at the same rate.

Process. Method by which a **change in state** is effected. The description of a process consists in stating some or all of the following: (1) the **boundary**; (2) the **change in state**, the **path** followed, or the effects produced in the **system** during each stage of the process; and (3) the effects produced in the **surroundings** during each stage of the process.

Properties of a system. Measurable attributes, that is, macroscopic observables, which describe a **system**. These properties are those physical attributes that are perceived by the senses or are made perceptible by instrumentation or experimental methods of investigation. Typical examples of thermodynamic properties are temperature, pressure, concentration of chemical species, and volume. **Fundamental** properties are directly and easily measured, while **derived** properties are usually obtained from fundamental ones by some kind of mathematical relationship. **Extensive** properties are additive; their determination requires evaluating the size of the entire system. Volume and mass are examples of extensive properties. **Intensive** properties are not additive, are independent of the size of a system, and are well defined in each small region of the system; examples are density and pressure.

Pseudoforce. Certain apparent forces such as a gradient of **chemical potential** caused by an uneven distribution of the particles with respect to position will give rise to diffusion phenomena which act on particles and can be formally treated as if they were directed forces.

Quantum mechanical tunneling. See **Tunneling**.

Quantum mechanics. Theory of physics that describes the motion and behavior of submicroscopic objects on a scale where **classical mechanics** is known to be incorrect.

Raoult's law. The linear relationship between the vapor pressure of the solvent and the mole fraction of the solvent in solution: $P_A = X_A P^o_A$. If all the components of a solution obey Raoult's law for all concentrations, the solution is defined as **ideal**.

Reduction reaction. See **Electronation.**

Relativity. Describes high-speed motion and the fundamental concept that all aspects of nature obey the same set of invariant laws, particularly in cases where **classical mechanics** is known to be incorrect.

Relaxation effect. The retarding force due to the finite time it takes for ion cloud rearrangement to occur. The relaxation time for dilute solutions is on the order of 10^{-6} s.

Resistance. Impedance to flow of current. It is dependent on both the geometry and the intrinsic **resistivity** of a material to conduction of current. For a conductor of resistivity ρ, the resistance will increase as the length, l, of the current path increases and the resistance will fall as the cross-sectional area, A, of the conductor increases: $R = \rho\, l/A$.

Resistivity, ρ. Intrinsic **resistance** of a material to conduction of current. The units of ρ are Ω m or Ω cm.

Reversible process. One in which the steps of the **path** occur in a series of infinitesimal steps, the direction of which may be reversed by an infinitesimal change in the external conditions prevailing at the time of each step. The path is such that it will take an infinite time to complete, but the maximum **work** will result only from this reversible path. The maximum work that is available from a reversible path is $w = -nRT \ln (P_1/P_2)$. At best, the **entropy** of the universe remains constant by having only reversible processes at work, but in reality no reversible process can be completed in a finite time frame; hence any real process leads to an increase in entropy in the universe.

Root-mean-square velocity, $\langle u \rangle$. A measure of the average velocity of molecules for an ideal gas or solution: $\langle u \rangle = (8RT/\pi M)^{1/2}$, where M is the molecular weight.

Rotational diffusion. Describes rotational motion of a lipid molecule about its long axis.

Salting out. Precipitation of a soluble macromolecule after it has been stripped of some of the water molecules that were necessary to keep it in solution. Useful in protein purifications.

Scatchard equation. Describes the number of binding sites per molecule: $v/[i] = K(N - v)$. A **Scatchard plot** is obtained by rearranging into a $y = mx + b$ form and plotting $v/[i]$ on the y axis against v on the x axis to give a straight line. The x intercept of this plot is N, the number of binding sites per molecule, and the slope of the line is $-K$. A nonlinear plot of the equation is evidence that the binding sites on the macromolecule under study are not identical or independent.

Scatchard plot. See **Scatchard equation.**

Second law of thermodynamics. Indicates the direction of movement of a **system**. It has been described in several ways: (1) no process is possible where heat is transferred from a colder to a hotter body without causing a change in some other part of the universe, (2) no process is possible in which the sole result is the absorption of heat (from another body) with complete conversion into work, (3) a system will always move in the direction that maximizes choice, (4) systems tend toward greater disorder, and (5) the macroscopic properties of an isolated system eventually assume constant values.

Sedimentation potential or Dorn effect. Electrokinetic phenomenon where a potential is generated from the shearing forces resulting from the movement of particles in a stationary solvent and is derived from the **diffuse layer** surrounding the particle.

Seebeck emf. The electromotive force which exists between two junctions at different temperatures.

σ. A measure of the ease with which a residue exhibits a conformation different from that of its predecessor, that is, of the ease of initiating a change in conformation.

Specific adsorption. A term that refers to the adsorption of a chemical species at the surface of an electrode through chemical forces, that is, forces other than couloumbic interactions.

Sphingolipid. Complex lipid found in membranes of nerve and brain tissue which contains three components: a **sphingosine** or one of its derivatives, a fatty acid, and a polar head group.

Sphingosine. A long-chain amino alcohol.

State of the system. A thermodynamic description of a system at equilibrium. The specification of the state allows it to be reproduced at any time. A state is specified for a single phase when two of three variables are known and the identities and concentrations of all independent chemical components are given.

Stern–Gouy–Chapman model. A model of the **electrified interphase** that incorporates the questions of finite size and adsorption on the surface into the structure of the interphase. It predicts that beyond the inner layer the arrangement of the remainder of the charge needed to neutralize the electrode charge is like a **Gouy–Chapman diffuse layer.**

Stern layer. A molecular layer that is closest to the electrode, has altered charge excess, and behaves like a molecular capacitor. The amount of charge excess differs under various conditions of **electrolyte** valence and concentration.

Sterol. A membrane lipid of the steroid class.

Stirling approximation. $\ln N! = N \ln N - N$.

Stokes' law. Calculates the viscous force for a spherical object: $F_v = 6\pi r\eta v$ where r is the radius of the object, η is the viscosity of the solvent, and v is the **drift velocity.**

Streaming potential. Electrokinetic phenomenon where the generation of a potential results from the movement of the volume elements that contain b charge. A separation and accumulation of charge occurs, and, as a result, an electrical potential is generated. The streaming potential is found to depend on the viscosity (η) and **conductivity** (κ) of the **electrolyte,** the pressure gradient (P) in the capillary, and the **zeta potential** (ζ): $E = \zeta\epsilon_o\epsilon P/\eta\kappa$.

Strong electrolytes. See **Ionophores.**

Surface tension. A measure of the tendency to keep the surface area at a minimum.

Surroundings. The excluded part of the **system** being studied.

Susceptibility. Physical property of a material that is important in determining its **dielectric constant.**

System. Part of the physical universe whose properties are under investigation. A system is called **open** when material passes across the **boundary** and **closed** when no material passes across the boundary. If the boundary prevents any interaction of the system with the **surroundings,** an **isolated system** is present which produces no observable disturbance as viewed from the surroundings. An isolated system is one in which the boundary is **adiabatic,** and **work** interactions are limited. For an isolated system, $q = 0$, $w = 0$, and $\Delta U = 0$.

Thermodynamics. Describes the behavior of large numbers of particles and discusses these behaviors in terms of heat, temperature, and work. **Classical thermodynamics** describes a system at **equilibrium. Nonequilibrium thermodynamics** describes the events that occur during the approach to equilibrium.

Third law of thermodynamics. States that at absolute zero, the **entropy** of all pure, perfect crystals is 0: $S_{\text{pure-perfect xtal}}\ (0\ \text{K}) = 0$.

Thomson heat. A reversible heat associated with the flow of current which, together with the **Peltier heat,** is important in describing the thermoelectric effect.

Transition or **melting temperature,** T_m. Temperature at which the membrane undergoes transitions. The melting temperature of simple lipids depends on the chain length. The T_m of phospholipids is independent of the chain length but dependent on the nature of the polar head group.

Transmembrane flip-flop. Form of lipid motion requiring translocation of a lipid molecule from one half of the bilayer to the other leaflet.

Triple point. State where all three phases of a single component of a substance may coexist; for example, for water, where ice and water vapor, ice and water, and water and water vapor are in a three-phase **equilibrium**.

True electrolytes. See **Ionophores.**

Tunneling. Effect that results from **quantum mechanical** treatment of a particle in a potential energy well with finite barriers. There exists a distinct probability that the particle may be found outside the potential barrier without having to climb up and out of the well. In effect, the particle tunnels through the barrier. This leads to the physical consequence that particles may cross a potential energy barrier even if they do not have sufficient energy to go over the barrier in classical fashion.

Valinomycin. Ionophore which is virtually specific for K^+ transport across membranes. It is the only known ionophore which exhibits such complete selectivity toward one ionic species.

van der Waals forces. The sum of attractive forces that vary with respect to r^6 and include **dispersion forces,** permanent dipole–dipole forces, and permanent dipole–induced dipole forces (see **Dipole, Induced dipole**).

van't Hoff equation. An equation that relates the change in **enthalpy** and **entropy** to the equilibrium constant: $\Delta H/T = -R \ln K_{\text{app}} + \Delta S$.

van't Hoff equation for osmotic pressure. Describes osmotic pressure, π: $\pi = nRT/V$ or $\pi = cRT$.

Virial equation. This equation is a mathematical expression that literally means "power equation" and is a power series expansion that was developed to describe the behavior of real systems, specifically imperfect gases:

$$\frac{PV}{nRT} = 1 + \frac{B(T)n}{V} + \frac{C(T)n^2}{V^2} + \frac{C(T)n^3}{V^3} + \ldots$$

The accuracy of the description is increased as the number of terms used is increased. The coefficients B, C, D, etc., are functions of temperature. The variable B, called the second virial coefficient, is frequently used as a description of the deviation from ideality in a variety of systems. In all but the most dilute solutions, the second virial coefficient is explicitly taken into account. The second virial coefficient is usually considered in treatments of calculations of colligative properties, etc.

Weak electrolytes. See **Ionophores.**

Wien effect. The first Wien effect shows that for electric fields of high strength, i.e., 10^6–10^7 V m^{-1}, the **conductivity** is found to be increased. The second Wien effect is also called the field dissociation effect and leads to a higher equivalent conductivity for a **potential electrolyte** in a high field strength. The second Wien effect is thought to occur because of an increased dissociation of the potential electrolyte, which is due to the effect the field has on dissociating the formed Bjerrum **ion pair,** thus increasing the number of charge carriers.

Work. The displacement or movement of an object by a force, converting energy into some useful form. Traditionally, this is defined as the raising (or falling) of a weight in the **surroundings**. All forms of work take the general form w = force $\times$ displacement. Work done on a system is considered positive. If the system exerts a force that causes a displacement in the surroundings, the work is said to be negative, representing work done on the surroundings.

Zeta potential, ζ. See **Electrokinetic potential.**

Index

Printed by Publishers' Graphics LLC